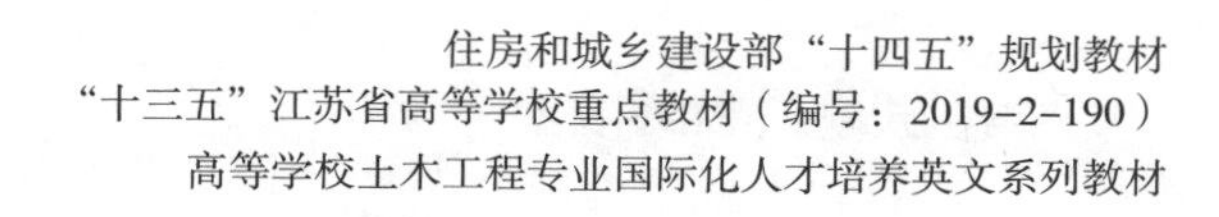
住房和城乡建设部“十四五”规划教材
“十三五”江苏省高等学校重点教材（编号：2019-2-190）
高等学校土木工程专业国际化人才培养英文系列教材

Introduction to Civil Engineering

土木工程概论

Editors in Chief

Jun Wang　Yujun Qi

王俊　齐玉军　主编

Associate Editors

Tao Lai　Huifang Wu　Chaoen Yin

赖韬　吴慧芳　尹朝恩　副主编

中国建筑工业出版社
CHINA ARCHITECTURE & BUILDING PRESS

图书在版编目（CIP）数据

土木工程概论 = Introduction to Civil Engineering：英文 / 王俊，齐玉军主编．—北京：中国建筑工业出版社，2021.9

住房和城乡建设部“十四五”规划教材 “十三五”江苏省高等学校重点教材（编号：2019-2-190） 高等学校土木工程专业国际化人才培养英文系列教材

ISBN 978-7-112-25702-7

Ⅰ.①土… Ⅱ.①王… ②齐… Ⅲ.①土木工程—高等学校—教材—英文 Ⅳ.①TU

中国版本图书馆CIP数据核字（2020）第243595号

责任编辑：郭希增　赵　莉　吉万旺
书籍设计：张悟静
责任校对：赵　菲

住房和城乡建设部“十四五”规划教材
“十三五”江苏省高等学校重点教材（编号：2019-2-190）
高等学校土木工程专业国际化人才培养英文系列教材

Introduction to Civil Engineering

土木工程概论

Editors in Chief
Jun Wang　Yujun Qi
王俊　齐玉军　主编
Associate Editors
Tao Lai　Huifang Wu　Chaoen Yin
赖韬　吴慧芳　尹朝恩　副主编

*

中国建筑工业出版社出版、发行（北京海淀三里河路9号）
各地新华书店、建筑书店经销
北京锋尚制版有限公司制版
北京市密东印刷有限公司印刷

*

开本：787毫米×1092毫米　1/16　印张：23　字数：675千字
2021年9月第一版　2021年9月第一次印刷
定价：**68.00**元（赠教师课件）
ISBN 978-7-112-25702-7
（36747）

From the perspective of status and trend of the civil engineering, this book is finished after reading and consulting a large number of references and documents home and abroad. This book is devoted to a general introduction of civil engineering discipline. Chapter 1 presents the historical evolution of civil engineering and a brief introduction to various civil engineering disciplines. Chapter 2 provides properties of civil engineering materials especially new materials as well as progress and trends. Chapter 3 shows building design and construction techniques especially for high-rise buildings. Chapter 4 focuses on the development of bridge engineering and its construction technology. Chapter 5 gives a general information to road design and associated works. Chapter 6 presents construction techniques of urban underground space structures and case studies. Chapter 7 discusses municipal engineering in terms of urban public transport water supply and drainage as well as heating and ventilation. Chapter 8 provides development of hydraulic engineering as well as construction of hydraulic structures. Chapter 9 presents developments of ocean engineering and features of offshore structures as well as protection measures. Chapter 10 discusses characteristics of earthquake, wind and fire disasters as well as mitigation methods. Chapter 11 is devoted to new concepts and techniques of intelligent construction and management.

This book has been written not only to serve as a textbook of introduction to civil engineering for college undergraduate students, but also as reference book for practicing civil engineers.

基于对土木工程的发展现状和未来的思考，在查阅大量国内外文献资料的基础上，撰写了此书。本书主要内容为土木工程专业相关学科的概述。第 1 章为土木工程发展概述及主要的土木工程学科简介；第 2 章为土木工程材料性能特点特别是新材料以及未来发展趋势；第 3 章概述了建筑工程，特别是高层建筑的建筑设计和建造，第 4 章主要介绍了桥梁工程的发展和桥梁施工技术；第 5 章概述了公路工程中道路设计方面的内容；第 6 章主要介绍了城市地下空间结构的施工技术及案例分析；第 7 章概述了市政工程中交通给水排水、暖通等方面的内容；第 8 章为水利工程的发展及水工结构的建造流程；第 9 章概述了海洋工程发展及海洋结构的特点和防护措施；第 10 章概述了地震灾害、风灾害及火灾的特点和防灾减灾措施；第 11 章主要介绍了智能建造与管理的新理念和新技术。

本书可以作为高等学校土木工程概论课程的教材和参考书，也可供广大土木工程相关科研和从业人士参考。

本书配备教学课件，我们可以向采用本书作为教材的老师提供教学课件，请有需要的任课教师按以下方式索取课件：1. 邮件：jckj@cabp.com.cn 或 jiangongkejian@163.com（邮件主题请注明《土木工程概论》英文版）；2. 电话：（010）58337285；3. 建工书院：http://edu.cabplink.com。

序 1

“一带一路”倡议的实施为我国高等教育国际化的进一步发展带来了强劲的驱动力。过去几年，“一带一路”沿线国家来华留学生人数激增，成为支撑全球来华留学人数不断增加的主体区域。“一带一路”建设需要培养优秀的来华留学生，这对我国教育国际化的发展提出了新的挑战。

喜闻由王俊教授和齐玉军教授主编的《土木工程概论》(英文版）一书即将出版，我很高兴为此书作序。全书由绪论、土木工程材料、建筑工程、桥梁工程、道路工程、城市地下空间工程、市政工程、水利工程、海洋工程、防震与减灾工程、智慧建造与管理施工等 11 章节构成，涵盖内容十分广泛，既有对传统土木工程学科的基本概念和基础理论的概述，又有对智慧建造与管理等新兴学科的介绍。未来人工智能、BIM 等先进技术与土木工程建造技术的深度融合，将深刻改变现有建造方式，促进建筑业的数字化和智能化转型升级。

作为土木工程专业的启蒙课程，《土木工程概论》是土木类专业学生接触到的第一门专业基础课，承担着培养学生学习土木工程专业的兴趣和工程意识、激发学生的学习积极性和求知欲、树立献身土木工程事业的理想和信念的重任，同时，也是学习后续专业课程的基础。

《土木工程概论》(英文版）一书深入阐述了土木工程的重要性，基本涵盖了土木类专业所涉及的所有学科内容，并对若干分支领域的发展前景进行了深入拓展。该书结构层次清晰，内容新颖充实、图文并茂、语言通顺，展示了国内外土木工程发展的历程与前景。该书可作为来华外国留学生学习土木工程专业的英文教材，帮助来华留学生系统全面地掌握土木工程的基本概念、理论、知识和方法，并为其今后进一步从事土木工程相关领域的工程应用和科学研究打下坚实基础。

东南大学

2021 年 1 月 27 日

Foreword 2

Prof. Jun Wang and her colleagues are highly qualified to undertake such a massive and impressive project on "Introduction to Civil Engineering". They should be commended for providing a glimpse of world history and current advances dealing with Civil Engineering Systems. It is impossible to get to the minute details of massive undertakings of Civil Engineering Systems in one textbook; but certainly, this book provides enough depth for readers to dive into the depths of science and technology of civil engineering if a reader is further interested in any of the topics covered in 11 chapters of this book.

Efficient functioning of civil infrastructure is the key to any successful economy. Inadequacies in the movement of people and goods slow the economic pace of any country. Since mid-nineteen hundreds, societies have recognized the need for efficient dwelling and transportation systems and have been investing heavily in this arena. Furthermore, durability of these systems is occupying the center stage due to skyrocketing construction and maintenance costs. This book summarizes the past glory of civil engineering and the future challenges facing the humankind to have better quality life by suggesting a few technical leads. The book deals with challenges of efficient land use and challenges in the transportation and other sectors with the use of modern technologies including advanced materials, computer aided design and intelligent construction techniques for durability, flexibility and cost effectiveness.

The authors' information is readable and reliable and encompassing the facts without any bias to fact and figures and devoid of fiction. It includes an overview of the historical perspective. Frankly, it is the collective understanding of all these authors about the subject and its relevance to contemporary reader that makes this book truly marvelous.

Hota GangaRao, PhD, P.E., F. ASCE

Maurice A. and Jo Ann Wadsworth Distinguished Professor of CEE, CEMR
Director, Constructed Facilities Center
Director, Center for Integration of Composites into Infrastructure
West Virginia University, USA

Preface
前言

Civil engineering is considered as one of the oldest and broadest engineering professions, and it deals with the planning, analysis, designing, construction, and maintenance of the infrastructure. The works include roads, bridges, buildings, dams, canals, water supply and numerous other facilities that affect the life of human beings. With the development of science and technology, new materials, intelligent construction and management technology, and disaster prevention and mitigation method are introduced and applied in civil engineering works. Actually, civil engineering is a multiple science encompassing numerous sub-disciplines that are closely linked with each other. The various sub-disciplines of civil engineering are covered. This book is intended to provide the engineering students with a general and extensive introduction to civil engineering.

This book consists of 11 chapters and Professor Jun Wang contributed Chapter 1, Chapter 4 and Chapter 6; Professor Yujun Qi contributed Chapter 2, Chapter 3 and Chapter 11; Assistant Professor Tao Lai contributed Chapter 8~10; Professor Huifang Wu contributed Chapter 7 and Lecturer Chaoen Yin contributed Chapter 5. Professor Jun Wang is responsible for the whole book. Sincerely thank Zhilin Chen, graduate of Nanjing Tech University, for his help in gathering materials for this book. We thank all the authors for their contributions.

土木工程是最古老和最广泛的工程专业之一，涉及基础设施的规划、分析、设计、建造和维护。基础设施囊括道路工程、桥梁工程、建筑工程、大坝工程、运河工程、供水系统工程以及许多其他影响人类生活的工程结构。随着科学技术的发展，新材料、智慧建造和管理以及防灾减灾等新技术被引入和应用在土木工程结构中。土木工程是一门多学科交叉的学科，各学科之间有着密切的联系，涵盖了土木工程的各个子学科。本书旨在为工程专业的学生介绍土木工程概论。

本书共 11 章，王俊教授负责第 1 章、第 4 章和第 6 章，齐玉军教授负责第 2 章、第 3 章和第 11 章，赖韬助理教授负责第 8 ~ 10 章，吴慧芳教授负责第 7 章，尹朝恩讲师负责第 5 章。全书由王俊教授统稿。南京工业大学研究生陈志林帮忙收集资料，谨致以衷心感谢。我们非常感谢所有作者的贡献。

Contents

Chapter 1
Introduction

1.1 Overview

Civil engineering is one of the oldest engineering disciplines, affecting many of our daily activities: the buildings we live in and work in, the transportation facilities we use, the water we drink, and the drainage and sewage systems that are necessary to our health and well-being. Civil engineers of one form or another have been around ever since humans started constructing major public works such as roads, bridges, tunnels and large public buildings. It is also an incredibly broad discipline, spanning treatment of environmental issues, transportation, power generation, and major structures.

Civil engineering is much more than erecting skyscrapers or bridges. Civil engineers are trained in the interactions among structures, the earth and water, with applications ranging from highways to dams and water reservoirs. Deeply involved with specifying appropriate construction materials, many civil engineers and others are also employed by the manufacturers of those materials. Since constructing a large building or public-work project can involve elaborate planning, civil engineers can be outstanding project managers. They sometimes oversee thousands of workers and develop advanced computerization and planning policies.

Most significantly, many civil engineers are involved with preserving, protecting, or restoring the environment. Most water treatment and water purification projects are designed and constructed by civil engineers (in these two areas, many of them are known as environmental engineers). A growing number of civil engineers are involved in billion-dollar projects to clean up toxic industrial or municipal wastes at abandoned dump sites. Civil engineers engage in such diverse projects as preserving wetlands or beaches, maintaining national forest parks, and restoring the land around mines, oil wells or factories.

Civil engineering has a significant role in the life of every human being, though one may not truly sense its importance in our daily routine. The function of civil engineering commences with the start of the day when we wash and brush, since the water is delivered through a water supply system including a well-designed network of pipes, water treatment plant and other numerous associated services. The network of roads on which we drive while proceeding to school or work, the huge structural bridges we come across and the tall buildings where we work, all have been designed and constructed by civil engineers. Even the benefits of electricity we use are available to us through the contribution of civil engineers who constructed the towers for the transmission lines. In fact, no sphere of life may be identified that does not include the contribution of civil engineering. Thus, the importance of civil engineering may be determined according to its usefulness in our daily life.

1.2 Development of Civil Engineering

1.2.1 Civil Engineering in Ancient Times

The history of civil engineering is traced back to the creation of mankind. From prehistory, shelters were built using stones and tree trunks to protect human beings from the environment. As the population grew, proper hydraulic and transportation system are needed. After the invention of wheel in 3500

B.C., rapid development in transportation sector took place. People grouped themselves in tribes and states, thus needing defense system to protect their terrain from rival.

Initially, the engineers were in fact military engineers with expertise in military and civil works. During the era of battles or operations, the engineers were engaged to assist the soldiers fighting in the battlefield by making catapults, towers and other instruments used for fighting the enemy. However, during peace time, they were concerned mainly with the civil activities such as building fortifications for defense, constructing bridges, canals, etc.

The construction of pyramids in Egypt (about 2700 B.C.~2500 B.C., Fig. 1-1) were some of the first instances of large structure constructions. Around 2250 B.C., Imhotep, the first documented engineer built a famous stepped pyramid for King Djoser located at Saqqara Necropolis. With simple tools and mathematics, he created a monument that stands to this day. His greatest contribution to engineering was his discovery of the art of building with shaped stones. Those who followed him carried engineering to remarkable height using skills and imagination. Famous ancient historic civil engineering constructions include the qanat water management system (Fig. 1-2) (the oldest is older than 3000 years and longer than 71km), the Parthenon (Fig. 1-3) by Ictinus in Ancient Greece (447 B.C.~438 B.C.), the Appian Way (Fig. 1-4) by Roman engineers (about 312 B.C.), the Great Wall of China (Fig. 1-5) by General Meng Tian under orders from Emperor Qin Shi Huang (about 220 B.C.), the stupas constructed in ancient Sri Lanka like the Jetavanaramaya (Fig. 1-6) and the extensive irrigation works in Anuradhapura.

Other remarkable historical structures are Sennacherib's Aqueduct at Jerwan built in 691 B.C., Li Bing's irrigation projects in China (around 220 B.C.), Julius Caesar's Bridge over the Rhine River built in 55 B.C., etc. Machu Picchu (Fig. 1-7), Peru, built at around 1450,

Fig. 1-1 Pyramids in Egypt
Source: http://roll.sohu.com/20111215/n329161821.shtml.

Fig. 1-2 A qanat tunnel near Isfahan
Source: http://dict.eudic.net/mdicts/wiki/Qanat.html.

Fig. 1-3 Parthenon in Ancient Greece
Source: https://10wallpaper.com/cn/view/Athens_Greece-Parthenon_wallpaper.html.

Fig. 1-4 Appian Way, Rome to Brindisi
Source: http://blog.sina.com.cn/s/blog_c39045200102v2zx.html.

at the height of the Inca Empire is considered an engineering marvel. It was built in the Andes Mountains assisted by some of history's most ingenious water resource engineers. The people of Machu Picchu built a city on the top of a mountain with advanced running water, draining systems and stone structures which have endured for over 500 years.

A treatise on architecture called *De Architectura* was published at 1 A.D. in Rome and survived to give us a look at engineering education in ancient times. It was probably written around 15 B.C. by the Roman architect Vitruvius and dedicated to his patron, the emperor Caesar Augustus, as a guide for building projects. Throughout ancient and medieval history, most architectural design and construction was carried out by an artisan, such as stonemason and carpenter, rising to the role of master builder.

Fig. 1-5 The Great Wall of China
Source: https://www.sohu.com/a/234282169_493318.

Fig. 1-6 Jetavanaramaya in ancient Sri Lanka
Source: http://m.xianzang.com/spots/show_18.html.

Fig. 1-7 Machu Picchu in Peru
Source: https://baike.baidu.com.

1.2.2 Civil Engineering in the 18th~20th Century

In the 18th century, civil engineering is considered to be a separate field from military engineering. A separate workforce for these non-military and civil works needs to be created. The first self-proclaimed civil engineer was John Smeaton (Fig. 1-8), who constructed the Eddystone Lighthouse (Fig. 1-9). In 1771, Smeaton and some of his colleagues formed the Smeatonian Society of Civil Engineers, a group of leaders of the profession who met informally over dinner. Though there was evidence of some technical meetings, it was little more than a social society.

In 1818, the Institution of Civil Engineers (Fig. 1-10) was founded in London, and in 1820, the eminent engineer Thomas Telford (Fig. 1-11) became its first president. The institution received a Royal Charter in 1828, formally recognizing civil engineering as a profession. Civil engineering is defined in the charter as: *the art of directing the great sources of power in nature for the use and convenience of man, as the means of production and of traffic in states, both for external and internal trade, as applied in the construction of roads, bridges,*

Fig. 1-8 John Smeaton, the "father of civil engineering"
Source: https://m.sohu.com/a/198121440_426424.

Fig. 1-9 Eddystone Lighthouse
Source: https://www.sohu.com/a/385173349_276366.

Fig. 1-10 Headquarters of Institution of Civil Engineers
Source: http://blog.onlylady.com/blog-1086258-10505451.html.

Fig. 1-11 Thomas Telford[12]

aqueducts, canals, river navigation and docks for internal intercourse and exchange, and in the construction of ports, harbours, moles, breakwaters and lighthouses, and in the art of navigation by artificial power for the purposes of commerce, and in the construction and application of machinery, and in the drainage of cities and towns.

The first private college in the United States that included civil engineering as a separate discipline was Norwich University established in the year 1819. Civil engineering societies were formed in the United States and European countries during the 19th century, and similar institutions were established in other countries during the 20th century. The American Society of Civil Engineers is the first national engineering society in the United States. It was founded in 1852 with members related to the civil engineering profession located globally. The number of universities in the world that include civil engineering as a discipline have increased tremendously during the 19th and the 20th centuries, indicating the importance of this technology.

After the World War II, driven by economic development, urban construction, science and technology, modern civil engineering has been rapidly developed and gradually matured in the aspects of design and construction theory, materials, construction machinery, etc., thus civil engineering has entered a new era. During the 300 years from the middle of the 17th century to the middle of the 20th century, civil engineering has developed rapidly, gradually expanding to engineering construction and facilities including houses, roads, bridges, railways, tunnels, ports, municipal engineering, etc. The construction can be built not only on the ground, but also under the ground or in the water.

The world-famous structures such as the Eiffel Tower, the Empire State Building and the Golden Gate Bridge were built in this period, shown in Fig. 1-12 to Fig. 1-14.

1.2.3 Modern Concepts in Civil Engineering

After entering the 21st century, the development of computer technology made the design and calculation methods more accurate and the degree of automation of design methods improved continuously. Lightweight construction materials with high strength have been applied extensively and the construction efficiency is significantly improved.

Fig. 1-12 Eiffel Tower
Source: https://baike.baidu.com.

Fig. 1-13 Empire State Building
Source: https://baike.baidu.com.

Fig. 1-14 Golden Gate Bridge
Source: https://baike.baidu.com.

Building designing and drawing software were developed in the 21st century based on the concept of Building Information Modeling. The development of estimation and scheduling software improves the quality of construction. Researchers have been made to work out strength of materials and innovation in construction materials. However, the most prominent contributor in this field is considered to be computer-aided design (CAD) and computer-aided manufacture (CAM). Civil engineers use these technologies to achieve an efficient system of construction, including manufacture, fabrication and erection. Three-dimensional design software is an essential tool for the civil engineer that facilitates him in the efficient designing of bridges, tall buildings and other huge complicated structures.

With economic development and population growth, urban land is increasingly tight and traffic is increasingly crowded, making it necessary for human beings to make full use of space. Building construction and road traffic began to develop to the upper and underground, and urban construction showed a three-dimensional trend.

In recent decades, high-rise buildings have also developed rapidly in Asia. In 1998, the Jin Mao Tower (Fig. 1-15) was completed in Shanghai, reaching a height of 420m. And the World Financial Center (Fig. 1-16) completed in 2008 reaching a height of 492m.

Meanwhile, many roads and railways have been elevated and developed deeper under the ground. The underground railroad has been further developed in recent decades. The subway has been electrified and connected to the basement of the building to form the underground commercial street. Underground parking lot, underground warehouse, underground factory and so on also developed in succession. Under the roads of the city were densely packed cables, water supply, drainage, heat and gas pipelines, forming the veins of the city.

Since the 2000s, due to powerful technology strength and electronic information technology revolution, the social economy has been rapidly developed and the urbanization process has been accelerated. The strong infrastructure requirements continuously promoted the rapid development of civil engineering, and showed strong vitality and creativity, challenging new limits of high, big, heavy, special engineering.

Fig. 1-15 Jin Mao Tower
Source: https://www.sohu.com/a/117308179_467467.

Fig. 1-16 World Financial Center in Shanghai
Source: http://hn.cnr.cn/zt/2012/1/8/201403/t20140326_515159664.html.

Fig. 1-17 Burj Khalifa Tower
Source: https://www.photophoto.cn/pic/24071401.html.

Fig. 1-18 Sky Tree Tower
Source: https://baike.baidu.com.

In terms of high-rise structures, the Burj Khalifa Tower in Dubai (Fig. 1-17), the world's tallest high-rise building built in 2010, is 828m high, almost twice the height of the 417m high Twin Towers of the World Trade Center built in 1972. The Sky Tree Tower in Tokyo, Japan, which was completed in 2011, is the world's tallest free-standing television tower at 634m, as shown in Fig. 1-18. At present, the highest Guangzhou TV Tower in China (completed in 2009) reaches 616m, ranking the second in the world.

In addition, as the world's tallest dam, the Grande Dixence Dam in Switzerland reaches a height of 284m, as shown in Fig. 1-19. With a total length of 2309m and a maximum height of 181m, the Three Gorges Dam is the world's largest hydropower project in the world.

Civil engineering utilizes technical information obtained from numerous other sciences, and with the advancement in all types of technologies, the civil engineering has also benefited tremendously. The future of civil engineering is expected to be revolutionized by the new technologies including design software, GPS, GIS systems and other latest technical expertise in varied fields. Technology will continue to make important changes in the application of civil engineering, including the rapid progress in the use of 3-D and 4-D design tools.

Fig. 1-19 Grande Dixence Dam
Source: https://m.sohu.com/a/153806153_670549/?pvid=000115_3w_a.

1.3 Civil Engineering Education

In response to societal drivers and new requirements for accreditation, civil engineering education continues to evolve, especially in the content of practice-oriented courses. Looking back two decades, civil engineers were facing challenges due to technological advances, new learning paradigms, and the need to prepare for rapidly changing job requirements. The term civil engineers refers to all those counted in employment statistics, whether they are licensed or not. By the year 2000, the dot-com economy was expanding rapidly, the smartphone was about to emerge, and social media was in its infancy. While change has been ongoing for many years, it seems that the pace of technological advancement is faster now.

In 1990s, Accreditation Board for Engineering and Technology (ABET) was incorporating new requirements and the Civil Engineering Body of Knowledge was being developed. Now, new approaches to learning are emerging, driven by educational technologies based on information technology, distance education and social media. Trends in business practices also affect the civil engineering workforce, which is experiencing changes in the types of jobs and requirements for skills.

There are a number of sub-disciplines within the broad field of civil engineering. General civil engineers work closely with surveyors and specialized civil engineers to design drainage, pavement, water supply, sewer service, dams, electric and communications supply. General civil engineering is also referred to as site engineering, a branch of civil engineering that primarily focuses on converting a tract of land from one usage to another. Site engineers spend time visiting project sites, meeting with stakeholders and preparing construction plans. Civil engineers apply the principles of building engineering, bridge engineering, road engineering, underground space engineering, municipal engineering, hydraulic engineering, ocean engineering and construction engineering to residential, commercial, industrial and public work projects of all sizes and levels of construction.

1.3.1 Civil Engineering Materials

To a civil engineer, the behavior of materials in structures and their ability to resist various stresses are of prime importance. The performances of materials are directly related to the applicability, artistry and durability of structures and also to the cost of projects. A large number of high-quality industrial and civil buildings need to be built for the development of society. Meanwhile, a great deal of water conservancy projects, traffic engineering and port projects need to be built to adapt to the rapid development of the national economy. It requires lots of high-quality materials which accords with the application environment of projects.

There is a wide variety of civil engineering materials. According to the chemical components of civil engineering materials, they can be classified into inorganic materials, organic materials and composite materials, as follows (Table 1-1):

Modern materials, such as polymers and composites are making headway into the construction industry. Significant research on these materials has led to better understanding of these materials and improved their strength and durability performance. The traditional materials used today are far superior to those of the past, and new materials are being specially developed to satisfy the needs of civil engineering applications.

Classification of materials used in civil engineering **Table 1-1**

<table>
<tr><td rowspan="6">Civil engineering materials</td><td rowspan="3">Inorganic materials</td><td>Metal: steel, iron, aluminum, copper, various types of alloys</td></tr>
<tr><td>Metalloid: natural stone, cement, concrete, glass, burned soil products, etc.</td></tr>
<tr><td>Metal-metalloid composition: reinforced concrete, etc.</td></tr>
<tr><td>Organic materials</td><td>Wood, plastics, synthetic rubber, petroleum asphalt, etc.</td></tr>
<tr><td rowspan="2">Composite materials</td><td>Inorganic metal-organic composition: polymer concrete, fiber reinforced plastics, etc.</td></tr>
<tr><td>Metal-organic composition: light metal sandwich panels, etc.</td></tr>
</table>

1.3.2 Building Engineering

Building engineering (Fig. 1-20) is applied to the design of new buildings, the repair and maintenance of the increasing stock of existing buildings, and the development of new building materials and technologies. The issue of sustainability, pollution, energy use and health are becoming increasingly important.

Design, construction, and maintenance of yet-to-be-built and existing buildings and structures represents a significant portion of the gross domestic product for countries of the developed world. In Canada, new construction accounts for almost 14% of the total GDP, and buildings account for almost 65% of that. Operating costs (energy, maintenance, cleaning and repairs) are also a multi-billion-dollar expenditure. Building scientists and engineers seek to improve the manner in which buildings are constructed and maintained to ensure long building life, to improve building performance throughout the expected service life, and to allow demolition, reuse and recycling.

1.3.3 Bridge Engineering

Bridge (Fig. 1-21) is a structure that spans horizontally between supports, whose function is to carry vertical loads. The prototypical bridge is quite simple—two supports holding up a beam—yet the engineering problems that must be overcome even in this simple form are inherent in every bridge: the supports must be strong enough to hold the structure up, and the span between supports must be strong enough to carry the loads. Spans are generally made as short as possible; long spans are justified where

(a) (b)

Fig. 1-20 Building engineering
Sources: (a) https://www.quanjing.com/imgbuy/QJ6965976326.html; (b) http://www.xjwxgg.cn/news/427.html.

(a) (b)

Fig. 1-21 Bridge engineering
Sources: (a) https://dp.pconline.com.cn/dphoto/list_3341192.html; (b) https://www.lngd.net/tc36x7kjjg/2018/0721/214576.html .

good foundations are limited—for example, over estuaries with deep water.

Bridge engineers ensure that bridges do what they are meant to do: carry the weight of people and/or cars without breaking, buckling, or falling down. Bridge engineers may assist with building bridges from scratch, or they may be called in to inspect or help rehabilitate an older bridge.

Some bridge engineers spend their days inspecting existing bridges and tunnels to ensure that they are still safe. These engineers may write lengthy reports about their findings, and those reports must be understandable to people who are not engineers themselves.

1.3.4 Road Engineering

Road engineering (Fig. 1-22) is a branch of civil engineering that includes planning, design, construction, operation and maintenance of roads, bridges and related infrastructure to ensure effective movement of people and goods.

Road planning involves the estimation of current and future traffic volumes on the road network. For purposes of design, traffic volumes are needed for a representative period of traffic flow. The capacity is the maximum theoretical traffic flow rate that a road section is capable of accommodating under a given set of environmental, road and traffic

(a) (b)

Fig. 1-22 Road engineering
Sources: (a) https://www.linfen365.com/forum.php?mod=viewthread&tid=521516&from=album; (b) http://www.xinhuanet.com//ttgg/2018-08/18/c_1123288534.htm?object_type=webpage&pos=1&url_type=39.

conditions. The capacity of a road depends on factors such as the number of lanes, lane width, effectiveness of traffic control systems, frequency and duration of traffic incidents, and efficiency of collection and dissemination of road traffic information. Traffic conditions arising from the interplay of volume and capacity are perceived by road users in a way that is quantitatively termed level of service.

Through road design, the most appropriate location, alignment and shape of the road are selected. Road design involves the consideration of three major factors (human, vehicle and roadway) and how these factors interact to provide a safe road. Human factors include reaction time for braking and steering, visual acuity for traffic signs and signals, and car-following behavior. Vehicle considerations include vehicle size and dynamics that are essential for determining lane width and maximum slopes, and for the selection of design vehicles. Engineers design road geometry to ensure stability of vehicles when negotiating curves and grades and to provide adequate sight distances for undertaking passing maneuvers along curves on two-lane, two-way roads.

1.3.5 Underground Space Engineering

The joy of traveling through underground metros, rail and road tunnels, especially the half-tunnels in mountains and visiting caves, cannot be described. Underground space is available almost everywhere, which may provide the site for activities or infrastructure that are difficult or impossible to install aboveground or whose presence aboveground is unacceptable or undesirable. Another fundamental characteristic of underground space lies in the natural protection it offers to whatever is placed underground. This protection is simultaneously mechanical, thermal, acoustic and hydraulic (i.e., watertight). It is effective not only in relation to the surface, but also within the underground space itself. Underground infrastructure thus offers great safety against all natural disasters and nuclear wars, ultraviolet rays from holes in the ozone layer, global warming, electromagnetic pollution, and massive solar storms.

City planners, designers, and engineers have a responsibility to foster a better environment for living, working and leisure activity at the ground surface and are therefore turning increasingly to the creation of space underground to accommodate new transportation, communication and utility networks, and complexes for handling, processing, and storing many kinds of goods and materials. So to stay on top, go underground.

In different countries, various facilities have been built underground, as shown in Fig. 1-23. These facilities include: underground parking space, rail and road tunnels, sewage treatment plants, garbage incineration plants, underground mass rapid transport systems (popularly known as "underground metro"), underground oil storage and supply systems

Fig. 1-23 Underground space engineering
Source: https://www.sohu.com/a/243850493_735537.

(through pipelines in tunnels), underground cold storage, and hydroelectric projects with extensive use of underground caverns and tunnels.

1.3.6 Municipal Engineering

A municipality is a local organizational body that is administered by a municipal council, responsible for the management of the municipal affairs of a distinct geographical region, normally a city, town, or another minor group. Municipal engineering includes the environmental issues that are the responsibility of the municipalities, including tasks that contribute in the quality of life of the society. A large number of municipal tasks are associated with the municipal engineering including the management and disposal of waste, the maintenance of public parks, water supply systems, treatment of water, and numerous similar responsibilities.

Management and disposal of waste materials is an important part of municipal engineering that includes the collection, transportation and elimination of waste materials. Management of waste has become an important issue since the industrial revolution due to the enormous exodus of population from the rural areas to the urban areas. Unless the disposal of waste is carried out efficiently, it will create a serious threat to the health of the inhabitants and the environment. Due to these reasons the humans are seriously concerned with this important subject of waste disposal.

The supply of water in an efficient manner to the residents is another important responsibility of municipal engineering. Local governments are responsible for the operation and maintenance of the water supply networks located in the municipality, but these projects are also an element in the master planning of the region. An efficient water supply network is designed by the utilization of the pipe network analysis, which is a study concerned with the flow of fluids through a network of hydraulics, to ascertain the rates of flow and drops in pressure in the different sectors of the network. The water supply network includes the watershed for collection of water, water treatment facilities, a water reservoir for storage of water (Fig. 1-24), and methods for transportation of water to the consumer.

Water treatment is essentially required before the water is delivered to the consumer, so that it is medically fit for human consumption. Treatment of water is normally carried out near the ultimate area where it is consumed so as to decrease the costs of pumping and the possibility of the water being contaminated after treatment. The process of water treatment includes the separation of impurities, such as dirt and other organic substances from water, addition of chemicals to destabilize the particles, and disinfection to destroy the bacteria. The water system networks are generally grouped into sectors for ease of maintenance, repair and operation.

Public parks are part of municipal engineering, as shown in Fig. 1-25, including athletic facilities, playgrounds, pavilions, shelter facilities, tennis courts, basketball courts, sand volleyball courts, ponds, etc. In order to meet maintenance standards and requirements, the park maintenance division may specialize in a variety of activities such as irrigation, turf management, pest control, general landscaping

Fig. 1-24 A water reservoir
Source: http://lnsb.qlwb.com.cn/lnpaper/content/20160519/Articel03002ZG.htm.

Fig. 1-25 A public park
Source: https://baike.baidu.com.

and tree trimming. The mission is to provide safe, clean and useful public parks for everyone to enjoy.

1.3.7 Hydraulic Engineering

Hydraulic engineering is a specialized field within environmental and civil engineering. Hydraulic systems are operated or fueled by the pressure of a fluid (e.g., water, oil, etc.). Hydraulic engineering deals with the technical challenges involved with water infrastructure and sewerage design. This discipline is really all about fluid flow and how it behaves in large quantities.

One main area of focus for hydraulic engineers is the design of water storage and transport facilities. Some examples include dams, channels, canals and lakes are all used to store and control water, as shown in Fig. 1-26. Machinery which uses hydraulic power is also designed by engineers in this discipline. Daily activities include designing structural elements that can withstand intense pressures.

Hydraulic engineers use fluid dynamics theory to predict how flowing water interacts with its surroundings. Students learn how to use computational fluid dynamics software packages which allow for complex simulations of fluid flow. Typically, hydraulic engineers are required by utility companies for storm water and sewerage maintenance.

1.3.8 Ocean Engineering

Ocean engineering provides an important link between the other oceanographic disciplines such as marine biology, chemical and physical oceanography, and marine geology and geophysics. Just as the interests of oceanographers have driven the demand for the design skills and technical expertise of ocean engineers, the innovations in instrumentation and equipment design made by ocean engineers have revolutionized the field of oceanography.

Ocean engineering is actually a combination of several types of engineering: a mix of mechanical, electrical, civil, acoustical and chemical engineering techniques and

(a)

(b)

Fig. 1-26 Hydraulic engineering
Sources: (a) https://zhuanlan.zhihu.com/p/46215858; (b) http://yb.newssc.org/system/20180814/002485386.html.

skills, coupled with a basic understanding of how the oceans work. Not only do ocean engineers design and build instruments that must stand up to the wear and tear of frequent use, they also must design instruments that will survive the harsh conditions of the ocean environment. Salt water is highly corrosive to many materials, and high winds, waves, currents, severe storms and marine life fouling (such as barnacles) must also be factored into design plans. It has even been said that the marine environment is more hostile than outer space!

Coastal engineering (Fig. 1-27) has become an increasingly important part of ocean engineering. With more and more people living or working at or near the world's coasts, problems associated with coastal development, such as pollution and waste disposal, will require the expertise and innovation of coastal engineers. For example, increasing the capacity of a coastal community to handle the sewage and garbage generated by a growing population requires careful attention to the effect disposal methods will have on the adjacent water bodies. What may work for an oceanside community may not work for a lakefront or riverfront community. Also, waves, rising sea level and storms have a significant impact on coastlines, often causing erosion and loss of coastal property. In efforts to protect coastal structures, coastal engineers are tasked with designing and creating ways to lessen the impact of storms and other natural shoreline processes.

Fig. 1-27 Coastal engineering
Source: https://baike.baidu.com.

Fig. 1-28 Oil industry
Source: https://baike.baidu.com.

The oil industry (Fig. 1-28), military and marine navigation fields also require ocean engineering skills. Each of these sectors directly impacts our lifestyle in some way, be it a source of energy, transportation, or our nation's defense. The work of ocean engineers plays an important role in each of these employment sectors. Because technology is central to the field of ocean engineering, future career prospects seem promising. And, as the role of the ocean continues to gain the interest of business, government and the private sector, the demand for ocean engineers should grow.

1.3.9 Disaster Prevention and Mitigation

Disasters have been classified into two major categories based on the natures of the related hazards, namely, natural disasters and man-made disasters (Fig. 1-29). In many natural (Fig. 1-30) and man-made (Fig. 1-31) disasters, civil engineering works are not only the hazard-bearing bodies, but their failure also constitutes further hazards. Especially in earthquake disasters, building collapse

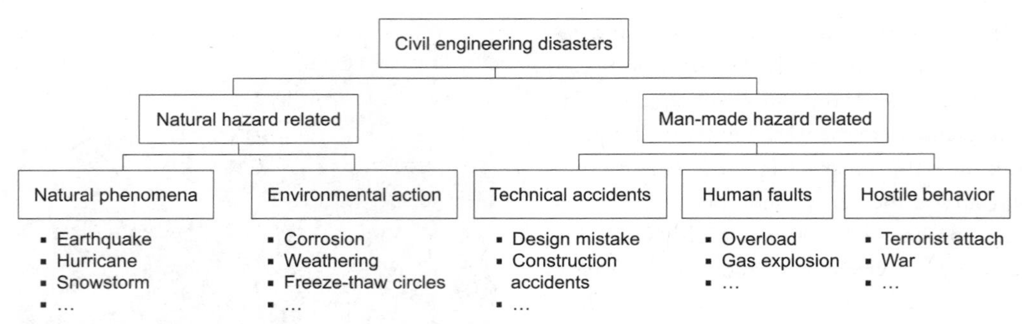

Fig. 1-29 Classification of civil engineering disasters [15]

Fig. 1-30 Beichuan County after debris flows in September 2009 [15]

Fig. 1-31 Collapsed Jiujiang Bridge under ship impact [15]

is the most significant cause of casualties and financial loss. This mechanism is yet to receive adequate emphasis in the investigation of disasters. Indeed, losses that have been attributed to many so-called natural disasters, such as earthquakes or winds, were actually caused by civil engineering factors rather than the actual natural phenomena. This is the key to developing proper methods for disaster prevention and mitigation.

Among the different types of disasters, civil engineering disasters are the most closely related to human beings and have constituted an important stimulus for civil engineering development. Many scientific and technological topics are relevant to the understanding and mitigation of civil engineering disasters, the concept of which emphasizes the transformation of civil engineering works from hazard-bearing bodies into hazards when they fail. Unlike disasters caused by natural hazards, which often cannot be predicted or controlled, civil engineering disasters can be effectively mitigated based on a thorough understanding of the associated failure mechanisms and by enhancement of the resistance capacity of engineering works.

The two goals of studies on civil engineering disasters are to understand the evolution mechanisms of the failure of the civil engineering works and to mitigate the disasters to protect human communities. The former goal can only be accomplished by accurate reproduction of the failure of the engineering works, while the latter requires comprehensive enhancement of the engineering works.

1.3.10 Construction Engineering

Civil engineering is a wide profession that comprises of several specializations including construction, structural, transportation, and environmental engineering. Expertise of each discipline is usually utilized in the accomplishment of projects related to the other disciplines of civil engineering. For example, construction engineers would be intimately involved with transport engineers while an airport is being constructed. Construction engineering concerns the planning and management of the construction of structures such as highways, bridges, airports, rail roads, buildings, dams and reservoirs. Construction of such projects requires knowledge of engineering and management principles, business procedures, economics and human behavior. Construction engineers engage in the design of temporary structures, cost estimating, planning and scheduling, materials procurement, selection of equipment, and cost control. Numerous other fields of science, management, business and other disciplines of civil engineering are also involved in the successful accomplishment of construction works. Since construction engineering is a profession that involves the contribution of a large number of human beings, skills to handle human behavior is also extremely important.

1.4 Conclusions

(1) Civil Engineering is a branch of engineering that encompasses the conception, design, construction and management of residential and commercial buildings and structures, water supply facilities, and transportation systems for goods and people, as well as control of the environment for the maintenance and improvement of the quality of life. Civil engineering includes planning and design professionals in both the public and private sectors, contractors, builders, educators and researchers. Some of the subsets that civil engineers can specialize in include photogrammetry, surveying, mapping, community and urban planning, and waste management and risk assessment. Various engineering areas that civil engineers can specialize in include geotechnical, construction, structural, environmental, water resources and transportation engineering.

(2) The development of civil engineering through ancient and modern times, has experienced three historical periods: the ancient, modern and contemporary period. Each period handed down buildings, which are not only a reflection of social development level, but also the embodiment of the human diligence and wisdom.

(3) Like other fields of work, civil engineering continues to be impacted by changes in technology and business practices. The changes reflect continuing trends that were evident in the previous study, which served as a baseline for this analysis. In addition to changes affecting other occupations, civil engineering is also influenced by the public nature of infrastructure and practices that are unique to the design and construction field. Inevitably, changes in technology and business practices will continue to impact the core work of civil engineers in planning, design and construction. Civil engineering educators will face continuing challenges, as they seek to develop the best strategies for workforce development, as delivery platforms for education and training change and new providers of alternative forms of education emerge.

Exercises

1-1 What is civil engineering and what do civil engineers do?

1-2 Describe the development of civil engineering.

1-3 List some sub-disciplines of civil engineering and briefly describe them.

References

[1] DIWAN V, PRASAD P, DIWAN A. A comparative study of ancient civil engineering Indian culture with modern civil era [C]. 2nd International Seminar on Utilization of Non-Conventional Energy Sources for Sustainable Development of Rural Areas. Parthivi College of Engineering & Management, C.S.V.T. University, Bhilai, Chhattisgarh, India, 2016.

[2] GRIGG N S. Civil engineering workforce and education: twenty years of change [J]. Journal of Professional Issues in Engineering Education and Practice, 2018, 144(4): 04018010.1-04018010.7.

[3] CACHIM P. An overview of education in the area of civil engineering in Portugal [J]. Procedia Engineering, 2015, 117: 431-438.

[4] SACK R, BRAS R, DANIEL D, et al. Reinventing civil engineering education [C]. In Vol. 3 of Proc., 29th Annual Conf. on Frontiers in Education, 1999 FIE '99. San Juan, Puerto Rico, 1999.

[5] GRIGG N S. Demographics and industry employment of civil engineering workforce [J]. Journal of Professional Issues in Engineering Education and Practice, 2000, 126(3): 116-124.

[6] SANAL I. A review on student-centred higher education in civil engineering: evaluation of student perceptions [J]. International Journal of Continuing Engineering Education and Life-Long Learning, 2018, 28(2): 205-217.

[7] FOULKE L R. ABET (Accreditation Board for Engineering and Technology) accreditation for engineering technology [C]. Transactions of the American Nuclear Society, USA, 1989.

[8] STRAVB H. A History of Civil Engineering: An Outline from Ancient to Modern Times [M]. Cambridge: MIT Press, 1964.

[9] LABI S. Introduction to civil engineering systems [M]. Hoboken: Wiley, 2014.

[10] WOOD D M. Civil engineering: A very short introduction [M]. Oxford: Oxford University Press, 2012.

[11] PARIKH P, BOUILLARD P. Civil engineering introduction [J]. Proceeding of the Institution of Civil Engineers–Civil Engineering. 2018, 171(6): 2.

[12] YAN J, TANG L. Thomas Telford: from architect to engineer [J]. The Architect, 2006, 2: 47-51.

[13] SHAMMA J E, PURASINGHE R. Introduction to sub-branches of civil engineering fields through a creative freshmen civil engineering design course [C]. 122nd ASEE Annual Conference & Exposition, Seattle, 2015: 1-12.

[14] ZHANG HM. Building Materials in Civil Engineering [M]. Cambridge: Woodhead Publishing, 2011.

[15] XIE L, QU Z. On civil engineering disasters and their mitigation [J]. Earthquake Engineering and Engineering Vibration, 2018, 17(1): 1-10.

Chapter 2
Civil Engineering Materials and Basic Members

2.1 Overview

Materials are the material factor of buildings and structures, and their cost accounts for more than 60% of the project investment. Understanding and mastering the main types, basic performance, use method and cost of materials is very necessary for ensuring project quality, strengthening financial management, practicing economy, realizing optimal allocation of resources and achieving the best benefit of investment. At the same time, it is also the necessary premise of realizing industrialization and green building.

Civil engineering materials usually refer to the materials that can meet the requirements of one aspect or some aspects of the building, reflect the specific form and performance, such as concrete, thermal insulation cotton, etc. However, it should be pointed out that in a broader sense, rammed earth, water and other materials also belong to the category of civil engineering materials. The former is usually used as the foundation of important buildings in ancient China, while the latter is an important component material of concrete.

According to the nature of use, civil engineering materials can be divided into structural materials and functional materials.

(1) Structural materials

It is usually used to form the stress skeleton used to form the building. Generally, it should have good mechanical properties such as sufficient strength, stiffness and deformation performance, it should also have better durability and cheaper price, and can be easily obtained. At the same time, the types of materials meeting the above conditions are very limited. From ancient times to modern times, there are only a few kinds of building structural materials widely used in the world, such as masonry, wood, concrete and steel.

(2) Functional materials

It mainly meets the functional requirements of a certain aspect of the building structure, such as thermal insulation, waterproof, fire prevention, etc. According to the physical composition, it can be divided into inorganic materials and organic materials. Inorganic materials generally have good fire resistance, while organic materials generally have poor fire resistance.

On the other hand, the smallest functional unit that makes up a building or structure is called a member. Building members refer to the elements of the building. If the building is regarded as a product, then the building members are the parts of the product. The members of the building mainly include: floor, roof, wall, column, foundation, etc. The concept of structural members is different from that of building members. Structural members are the elements of structural stress skeleton, which can be divided into beams, slabs, walls, columns, foundations, etc. According to the force characteristics of members, they can be divided into bending members, compression members, tension members, torsion members, compression bending members, etc.

This chapter introduces the commonly used civil engineering materials and the potential materials in the future from three parts: structural materials, functional materials and new materials, and then mainly introduces the common structural components. For the introduction of building components, please refer to Chapter 3.

2.2 Cement

2.2.1 Cementitious Materials

Cementitious material is a kind of material that can be changed from slurry into hard solid and can bond granular material (sand and stone) or block or sheet material (such as brick and stone) into a whole, also called binding material. Cementitious materials are usually divided into inorganic binding materials and organic binding materials.

Inorganic binding materials usually contain a lot of mineral components, so they are also called mineral binding materials. According to the differences of the hardening conditions, inorganic binding materials can be divided into air-hardening binding materials and hydraulic binding materials. Binding materials that can only harden in air and maintain or develop their strength only in air are called air-hardening binding materials, such as gypsum (Fig. 2-1), lime, etc. If the binding material can not only set and harden in air, but also harden better in water, it is called hydraulic binding material. This kind of material is mainly cement (Fig. 2-2), and its strength is mainly produced by the action of water.

Organic binding materials are substances of organic origin (asphalt, bitumens, pitch, polyvinyl acetate and furan resins) that have the capacity to make the transition from a plastic state to the hardness or slightly plastic states under the influence of physical or chemical processes.

Asphalt is a relatively complex polymer colloid or sol system, so people can choose the appropriate polymer materials through the appropriate production technology and synthesis methods to prepare light colored asphalt like materials that meet the performance requirements, and continuously optimize to obtain light colored asphalt binder with similar performance as ordinary asphalt.

Asphalt is a kind of viscoelastic material with complex performance. In a certain temperature range, the elastic deformation of asphalt material occurs at low temperature, and plastic flow occurs at high temperature. Asphalt binder is the most important raw material for producing emulsified asphalt, and it is also the final form of actual paving of emulsified asphalt. The quality of asphalt is directly related to the performance of emulsified asphalt.

Lime is commonly used as cementing material in Chinese ancient architecture. In

Fig. 2-1 Natural gypsum minerals
Source: https://www.1688.com.

Fig. 2-2 Cement powder
Source: https://www.hxqmj.cn/n12.html.

Zhou Dynasty, there were mausoleums built with lime. From Zhou Dynasty to South and North Dynasty, people used the mixture of lime, loess and fine sand as the surface of rammed earth wall or adobe wall, or made the floor of living room and tomb passage. According to historical records, the mixture of glutinous rice juice and lime was used as cementing material in the construction of Hezhou City in the sixth year of Qiandao in Southern Song Dynasty (1170). Lime and glutinous rice juice were also used as grouting materials in the brick walls of Nanjing City in Ming Dynasty.

In the west, the ancient Egyptians used Nile mud as cementitious material to build unburned adobe bricks. Sand and grass were added to the mud to increase strength and reduce shrinkage. This kind of building can be preserved for many years in dry areas. From about 3000 B.C. to 2000 B.C., the ancient Egyptians began to use calcined gypsum as building cementitious material, and calcined gypsum was used in the construction of pyramids. Different from the Egyptians, the ancient Greeks used calcining limestone in the construction of cementitious materials. The Roman Empire inherited the Greek tradition of producing and using lime. They first mixed lime and sand into mortar, and then used this mortar to build buildings. Later, the ancient Romans also improved the use of lime. In lime, not only sand, but also ground pozzolan was added. The strength and water resistance of the three-component mortar are much better than that of the two-component mortar of "lime+sand". The buildings built with the three-component mortar have good durability both on land and in water. Some people call this three-component mortar "Roman mortar". The mixture of Roman mortar and stone forms pozzolanic concrete, from which many buildings of the Roman Empire were built. Later, with the destruction of the Roman Empire, the Roman mortar was lost. After that, for a long period of time, the masonry materials were changed into two-component materials of lime and sand.

2.2.2 Modern Cement

In 1756, British engineer J. Smiton studied the hardening characteristics of some limes in water. He found that in order to obtain hydraulic lime, limestone containing clay must be used for firing; the most ideal composition of masonry mortar for underwater buildings is composed of hydraulic lime and volcanic ash. This important discovery laid a theoretical foundation for the research and development of modern cement.

In 1796, the Englishman J. Parker made a kind of cement from marl, which was brown in appearance, much like the mixture of lime and pozzolan in ancient Rome, and named it Roman cement. It is also known as natural cement because it is made of natural marl without ingredients. It has good hydraulic properties and quick setting characteristics, especially suitable for projects in contact with water.

In 1813, the French civil engineer Biga discovered that the cement made by mixing lime and clay by three to one has the best performance.

In 1824, Joseph Aspdin, a British construction worker, invented cement and patented Portland cement. He used limestone and clay as raw materials, mixed in a certain proportion, calcined into clinker in a shaft kiln similar to lime burning, and then ground into cement. It is named Portland cement because its hardened color is similar to the stone used for building in Portland, England. It has excellent building performance and epoch-making significance in the history of cement.

In 1871, Japan began to build cement plants.

In 1877, England's Clapton invented the rotary furnace, and in 1885 by ransom was reformed into a better rotary furnace.

In 1889, near Kaiping Coal Mine in Tangshan, Hebei Province, China, the

Tangshan "fine clay" plant, which was produced by shaft kiln, was established. In 1906, Qixin Cement Company was established on the basis of the plant, with an annual output of 40,000 tons of cement.

In 1893, Hideo Sato and Nakai Sankhen invented Portland cement which was not afraid of sea water.

In 1907, the bauxite was used to replace clay, mixed with limestone to make cement. This cement is called "bauxite cement" because it contains a lot of alumina.

In the 20th century, with the continuous improvement of Portland cement performance, a number of cement suitable for special construction projects, such as high alumina cement, special cement, etc., were developed successfully. There are more than 100 kinds of cement in the world, and the annual output of cement is about 2 billion tons in 2007. In 1952, China formulated the first national unified standard, which determined that cement production should be based on the principle of multi varieties and multi grades. Portland cement was renamed silicate cement according to its main mineral composition, and later renamed Portland cement until now.

Cement is one of the most important building materials in modern times, and also the most widely used artificial cementitious material. Cement can be mixed with sand and water in a certain proportion to make mortar, and can also be mixed with sand, stone, water and other admixtures in a certain proportion to make concrete. At present, China is the largest country of cement production in the world, and the total output of cement exceeds 50% of the total output.

2.2.3 Naming Principles and Classification of Cement

The naming of cement should be based on the main hydraulic minerals, mixed materials, uses and main characteristics of cement according to different categories, and strive to be concise and accurate. If the name is too long, abbreviation is allowed.

(1) The general cement is named after the main hydraulic mineral of cement and the name of mixed material or other appropriate name.

(2) Special cement is named for its special purpose and can be named with different types.

(3) The characteristic cement is named after the main hydraulic minerals of cement and the main characteristics of cement, and can be named with different types or mixed materials.

(4) The cement with pozzolanic or potential hydraulic materials and other active materials as the main component is named by the name of the main component with the name of the active material, and can also be named with the characteristic name, such as gypsum slag cement, lime pozzolanic cement, etc.

Based on the above naming rules, cement can be classified according to different standards.

According to its use and performance, cement can be divided into general cement and special cement. General cement refers to the cement commonly used in civil engineering, including Portland cement, ordinary Portland cement, slag Portland cement, pozzolanic Portland cement, fly ash Portland cement and composite Portland cement. Special cement is a kind of cement with special properties or uses, such as grade G oil well cement, fast hardening Portland cement, road Portland cement, aluminate cement, sulphoaluminate cement, etc.

According to the names of their main hydraulic substances, cement can be divided into Portland cement, aluminate cement, sulphoaluminate cement, ferroaluminate cement, fluoroaluminate cement and phosphate cement.

According to the main technical characteristics, cement can be divided into different types. For example, it can be divided into fast hardening cement and super fast hardening cement; according to hydration heat, it can be divided into medium heat and low heat cement; according to sulfate resistance, it can be divided into medium and high sulfate corrosion resistant cement; and according to expansion, it can be divided into expansion and self stressing cement.

2.2.4 Portland Cement

Portland cement is a kind of hydraulic cementitious material made of Portland cement clinker mainly composed of calcium silicate, limestone or granulated blast furnace slag with less than 5% and appropriate amount of gypsum. Portland cement was developed from natural cements made in Britain beginning in the middle of the 18th century. Its name is derived from its similarity to Portland stone, a type of building stone quarried on the Isle of Portland in Dorset, England.

The main mineral composition of Portland cement is: tricalcium silicate, dicalcium silicate, tricalcium aluminate and tetracalcium ferrialuminate. Tricalcium silicate determines the strength of Portland cement within four weeks; dicalcium silicate takes effect after four weeks, reaching the strength of tricalcium silicate in about one year; tricalcium aluminate plays a role in the strength of Portland cement in 1~3 days or a little longer time; the strength of tetracalcium ferrite also plays a faster role, however, the strength is low, which contributes little to the strength of Portland cement.

Different standards are used for classification of Portland cement. The major standards are the $2CaO \cdot SiO_2$ ASTM C150 used primarily in the USA, European EN 197 and GB 175—2007/XG1—2009 in China.

The five types of Portland cements exist, with variations of the first three according to ASTM C150 as following:

Type Ⅰ Portland cement is known as common or general-purpose cement. A limitation on the composition is that $3CaO \cdot SiO_2$ (C_3A) shall not exceed 15%.

Type Ⅱ gives off less heat during hydration. A limitation on the composition is that C_3A shall not exceed 8%, which reduces its vulnerability to sulfates.

Type Ⅲ has relatively high early strength. Its typical compound composition is: 57% $3CaO \cdot SiO_2$ (C_3S), 19% $2CaO \cdot SiO_2$ (C_2S), 10%C_3A, 7% $4CaO \cdot Al_2O_3 \cdot Fe_2O_3$ (C_4AF), 3.0% MgO, 3.1% SO_3, 0.9% ignition loss, and 1.3% free CaO. This cement is similar to type Ⅰ, but ground finer.

Type Ⅳ is generally known for its low heat of hydration. A limitation on this type is that the maximum percentage of C3A is 7%, and the maximum percentage of C3S is 35%.

Type Ⅴ is used where sulfate resistance is important. This cement has a very low C_3A composition which accounts for its high sulfate resistance. The maximum content of C_3A allowed is 5% for type Ⅴ Portland cement. Another limitation is that the C_4AF+2C_3A composition cannot exceed 20%. This type is used in concrete to be exposed to alkali soil and ground water sulfates which react with C_3A causing disruptive expansion. It is unavailable in many places, although its use is common in the western United States and Canada.

EN 197-1 defines five classes of common cement that comprise Portland cement as a main constituent. These classes differ from the ASTM classes.

Ⅰ—Portland cement. It is composed of Portland cement and up to 5% of minor additional constituents;

Ⅱ—Portland-composite cement. It is composed of Portland cement and up to 35% of other single constituents;

Ⅲ—Blastfurnace cement. It is composed of Portland cement and higher percentages of blastfurnace slag;

Ⅳ—Pozzolanic cement. It is composed of Portland cement and up to 55% of pozzolanic constituents;

Ⅴ—Composite cement. It is composed of

Portland cement, blastfurnace slag or fly ash and pozzolana.

Constituents that are permitted in Portland-composite cements are artificial pozzolans (blastfurnace slag, silica fume, and fly ashes) or natural pozzolans (siliceous or siliceous aluminous materials such as volcanic ash glasses, calcined clays and shale).

In China, there are two types of Portland cement, i.e. type Ⅰ Portland cement (Code: P · Ⅰ) without mixing materials; type Ⅱ Portland cement (Code: P · Ⅱ) added with limestone or granulated blast furnace slag (no more than 5% of cement mass), according to GB 175—2007/XG1—2009.

2.2.5 Aluminate Cement

Aluminate cement is a kind of hydraulic cementitious material made from bauxite and limestone, calcined and calcined with calcium aluminate as the main component and alumina content of about 50%. Aluminate brine mud is usually yellow or brown, and some are gray. The main minerals of aluminate cement are calcium aluminate (CA) and other aluminates, and a small amount of dicalcium silicate ($2CaO \cdot SiO_2$). The color is mostly gray and white. Aluminate cement includes calcium aluminate, barium aluminate and barium zirconium aluminate. Barium aluminate cement has the characteristics of rapid hardening, high strength and high fire resistance. The patent of aluminate cement was published in France in 1908, and it was first industrialized in 1908. After decades of development, aluminate cement series, including expansive cement, self stressing cement and refractory cement, has been formed. The characteristic of this series of cement is that its clinker mineral composition is mainly Ca, which endows cement with special properties such as early strength and fire resistance.

The setting and hardening speed of aluminate cement is fast. The one day strength can reach more than 80% of the maximum strength, which is mainly used for projects with urgent construction period, such as national defense, roads and special emergency repair projects. The hydration heat of aluminate cement is high and the heat release is concentrated. The hydration heat released within one day is 70%~80% of the total amount, which makes the internal temperature of concrete rise higher. Even if the construction is carried out at –10℃, the aluminate cement can quickly set and harden, which can be used for winter construction projects. Under ordinary hardening conditions, aluminate cement has strong sulfate corrosion resistance because there is no tricalcium aluminate and calcium hydroxide in cement paste, and the compactness is large. Aluminate cement has high heat resistance. If using refractory coarse and fine aggregate (such as chromite) and aluminate cement to make concrete, we can get the heat-resistant concrete with service temperature of 1300~1400℃.

However, the long-term strength and other properties of aluminate cement tend to decrease, and the long-term strength decreases by about 40%~50%. Therefore, aluminate cement is not suitable for long-term load-bearing structures and projects in high temperature and high humidity environment. It is only suitable for emergency military engineering (road and bridge construction), rush repair engineering (leakage stoppage, etc.), temporary engineering and preparation of heat-resistant concrete. In addition, the mixture of aluminate cement and Portland cement or lime produces not only flash setting, but also cracks and even damages the concrete due to the formation of high alkaline hydrated calcium aluminate. Therefore, during construction, it is not allowed to mix with lime or Portland cement, nor can it be used in contact with non hardened Portland cement.

2.2.6 Sulphoaluminate Cement and Ferroaluminate Cement

Sulphoaluminate cement is a new type of cement mainly composed of anhydrous calcium sulphoaluminate and dicalcium silicate. In the 1970s, sulphoaluminate cement was invented in China. In the 1980s, it initiated the industrial production of iron aluminate cement. If we call Portland cement series products as the first series cement and aluminate cement series products as the second series cement, then we can call sulphoaluminate cement, ferroaluminate cement and other cement varieties derived from them as the third series cement. The mineral composition of this series of cement is characterized by a large amount of $4CaO \cdot 3Al_2O_3(C_4A_3)$ minerals. This is different from other series of cement. The third series cement is characterized by early strength, high strength, high impermeability, high frost resistance, corrosion resistance, low alkali content and low energy consumption. The third series cement has been widely used in China. In 2000, the output of sulphoaluminate cement in China was only 672,500 tons. By 2005, the output of sulphoaluminate cement in China has reached 1.253 million tons. There are 30 enterprises producing sulphoaluminate cement in China, and the output of sulphoaluminate cement is basically stable at about 1.25 million tons.

The third series of cement can obtain several kinds of cement with different properties by adjusting the amount of clinker, gypsum and admixture. There are two types of cement produced: sulphoaluminate cement and ferroaluminate cement. The former includes fast hardening sulphoaluminate cement, high-strength sulphoaluminate cement, expansive sulphoaluminate cement, self stressing sulphoaluminate cement and low alkalinity sulphoaluminate cement; the latter includes fast setting ferroaluminate cement, high-strength ferroaluminate cement and self stressing ferroaluminate cement.

Sulphoaluminate cement and ferroaluminate cement have the following characteristics:

(1) Early strength and high strength. The two kinds of fast setting cement not only have higher early strength, but also have increasing later strength. At the same time, it has the setting time meeting the requirements of use. The compressive strength is 35~50MPa in 12h~1d, and the flexural strength is 6.5~7.5MPa. The 3D compressive strength can reach 50~70MPa and the flexural strength can reach 7.5~8.5MPa.

(2) High frost resistance. The two kinds of fast setting cement show excellent frost resistance. Their early strength is 5~8 times higher than that of Portland cement when it is used at 0~10℃. When the concrete is used under the negative temperature of −20~0℃ and a small amount of antifreeze is added, the concrete casting temperature is maintained above 5℃. Under the condition of alternating positive and negative temperature, the construction has little influence on the later strength growth. The strength loss of concrete is not obvious after 200 freeze-thaw cycles in laboratory.

(3) Corrosion resistance. The two kinds of cement have excellent corrosion resistance to seawater, chloride ($NaCl$, $MgCl_2$), sulfate [Na_2SO_4, $MgSO_4$, $(NH_4)_2SO_4$], especially their compound salts ($MgSO_4$ + $NaCl$). The corrosion resistance of fast setting ferroaluminate cement is better than that of fast setting sulphoaluminate cement. According to the results of 2-year corrosion resistance test in laboratory, the corrosion resistance coefficient is greater than 1, which is obviously better than high sulfur resistant Portland cement and high aluminum cement.

(4) High impermeability. The cement paste structure of the two kinds of cement is relatively dense, so the impermeability of concrete is 2~3 times of that of Portland cement concrete of the same grade.

(5) Corrosion of reinforcement. The durability of the two kinds of fast setting

cement is good. However, due to the different alkalinity of hydration liquid phase of the two kinds of cement, the corrosion of reinforcement is not completely the same. Due to the low alkalinity (pH<12) of fast hardening sulphoaluminate cement, there is no passivation film on the surface of steel bar, so it is unfavorable to protect the steel bar. In the early mixed concrete, due to more air and moisture, the concrete reinforcement has a slight corrosion at the early stage. With the growth of age, air and moisture gradually decreased and disappeared. Due to the dense structure of concrete, there is no obvious development of corrosion in the later stage. For fast setting ferroaluminate cement, due to the high alkalinity of cement hydration liquid phase (pH>12), a passivation film similar to Portland cement concrete is formed on the surface of steel bar. The fast determination of polarization curve and routine laboratory test show that the fast hardening ferroaluminate cement has no corrosion on the reinforcement.

The third series cement has the characteristics of early strength and high strength, which is suitable for rush repair and construction projects, and can save the energy cost of steam curing when applied in prefabricated components; it has good frost resistance and can be widely used in winter low-temperature construction projects; its impermeability and seawater corrosion resistance are much better than Portland cement, which is suitable for marine construction projects; it has special performance for consolidation of hazardous wastes, which is suitable for solidification treatment of hazardous and toxic wastes; the pH value of low alkali aluminate cement hydration slurry is less than 10.5, especially suitable for glass fiber reinforced concrete (GFRC). This series of cement has been successfully used in various construction projects (especially winter construction projects), seaport projects, underground projects, various cement products and prefabricated components, such as cement pressure pipes, high strength piles, large beams and columns, and various GRC products.

2.2.7 Fluoroaluminate Fement and Phosphate Cement

Fluoroaluminate cement is a kind of fast setting and fast hardening cement mainly composed of calcium fluoroaluminate. With limestone, bauxite and fluorite as raw materials, the clinker is forged at 130~1100℃ and grinded together with proper amount of gypsum and certain amount of slag. At normal temperature, the setting time of the cement is only a few minutes, and the strength of cement paste can reach 15~25MPa in one hour. Setting time can be adjusted by adding retarder. The hardening speed of this cement is fast, which is suitable for rush repair and leakage stoppage.Phosphate cement is a kind of chemically bonded cement, which is formed by chemical reaction between metal and acid solution or basic components.

Phosphate cement can be used to produce a variety of heat-resistant and thermal stable materials, such as electrical insulation coatings and high-performance adhesives. Some properties are similar to ceramic materials.

It is not necessary to calcine phosphate cemented materials at high temperature. These materials are stable in many corrosive media. Phosphate cement mainly includes aluminum phosphate cement, magnesium phosphate cement, ammonium phosphate cement, etc. One of the maturer is magnesium phosphate cement. Magnesium phosphate cement is made of magnesium oxide, phosphate, retarder and mineral admixture in a certain proportion. MgO is made of magnesite calcined at about 1700℃. Phosphate mainly provides acid environment and acid ion for hydration reaction. At present, ammonium dihydrogen phosphate is widely used in the preparation of magnesium phosphate cement. In order to make phosphate cement have sufficient operation time, borate is often used as retarder of magnesium phosphate cement. In addition, fly ash is cheap and widely used as mineral admixture of magnesium phosphate cement, which can not only reduce the cost, but also adjust the color of magnesium phosphate cement and improve the performance of magnesium phosphate cement.

2.3 Concrete

Concrete is a composite material composed of coarse aggregate bonded together with a fluid cement that hardens over time (Fig. 2-3). Concrete has high hardness, high compressive strength, strong durability, wide sources of raw materials, simple production method, low cost, strong plasticity, suitable for all kinds of natural environment. It is widely used in houses, bridges, highways, runways, retaining walls, embankments, culverts, dams, water tanks, water towers, oil tanks and channels, water ditches, wharfs, breakwaters, military engineerings, nuclear power plants and other structures.

2.3.1 Classification of Concrete

Concrete can be classified according to the apparent density, strength grade, cementitious materials, use function and other factors.

(1) According to the Apparent Density

According to the apparent density, concrete can be divided into: heavy concrete, ordinary concrete, lightweight concrete. The difference between these three kinds of concrete is the difference of aggregate. The apparent density of ordinary concrete is between 1950kg/m^3 and 2500kg/m^3. Heavy concrete is made of special dense and heavy aggregate with apparent density greater than 2500kg/m^3, such as barite concrete, steel chip concrete and so on. They have the properties of impermeable to X-ray and γ-ray. The concrete with apparent density less than 1950kg/m^3 is called lightweight concrete. It can be divided into three categories:

1) Lightweight aggregate concrete.

(a)

(b)

Fig. 2-3 Concrete
(a) Concrete mixture; (b) Hardened concrete
Source: (a) https://baike.baidu.com; (b) https://baike.baidu.com.

The apparent density of the concrete ranges from 800 to 1950kg/m^3, and the lightweight aggregate includes pumice, volcanic slag, ceramsite, expanded perlite, expanded slag, slag, etc.

2) Porous concrete (foam concrete, aerated concrete). The apparent density is 300~1000 kg/m^3. Foam concrete is made of cement slurry or cement mortar and stable foam. Aerated concrete is made of cement, water and aerogenic agent.

3) Macroporous concrete (ordinary macroporous concrete, lightweight aggregate macroporous concrete). There is no fine aggregate in its composition. The apparent density of ordinary macroporous concrete ranges from 1500 to 1900kg/m^3, which is made up of crushed stone, soft rock and heavy mineral slag. The apparent density of lightweight aggregate macroporous concrete is 500~1500kg/m^3, which is made of ceramsite, pumice, broken brick and slag.

(2) According to the Compressive Strength

According to the compressive strength, concrete can be divided into ordinary strength concrete, high strength concrete and super high strength concrete.

The compressive strength of concrete is closely related to many factors, such as the shape and size of concrete specimens, test methods and concrete age. Different countries and regions have developed their own test specifications to ensure internal consistency. According to Chinese codes, concrete with compressive strength below C60 is ordinary strength concrete, C60 to C100 is high strength concrete, and above C100 is super high strength concrete. High strength cement, ultra-fine mineral admixtures and superplasticizers are the main technical methods to prepare high strength concrete and super high strength concrete.

As a new structural material, high strength concrete and super high strength concrete are widely used in high-rise building structures, long-span bridge structures and some special structures because of their high compressive strength, strong deformation resistance, high density and low porosity. High strength concrete material provides favorable conditions for prestressed technology. High strength steel and artificial stress control can be used, which greatly improves the bending stiffness and crack resistance of flexural members. Therefore, more and more prestressed high-strength concrete structures are used in large-span buildings and bridges in the world. In addition, high-strength concrete can be used to build buildings (structures) bearing impact and explosion load, such as the foundation of atomic energy reactor. According to the characteristics of high strength concrete with strong impermeability and corrosion resistance, the industrial water tank with high impermeability and high corrosion resistance is constructed.

(3) According to the Cementitious Materials

According to the cementitious materials, concrete can be divided into cement concrete, sodium silicate concrete, bituminous concrete, polymer concrete, etc. It's easy to know the main type of cementitious material by the name of concrete.

Sodium silicate concrete is composed of sodium silicate, hardener, acid resistant powder, coarse and fine aggregate and admixture. The chemical formula of sodium silicate is $R_2O \cdot mSiO_2$, which is a glass like fusion of alkali metal silicate. It is yellow or green. The concrete made of sodium silicate has good stability to concentrated acid. After soaking in acid solution, its strength does not decrease, but increases with the extension of soaking time. It can be used not only as structural material, but also as anti-corrosive material. It is widely used in metallurgy, chemical industry, petroleum, light industry and other industries.

Bituminous concrete is a mixture of mineral aggregate with certain gradation composition, crushed stone or crushed gravel, stone chips or sand, mineral powder, etc., mixed with certain proportion of asphalt materials under strict control conditions. It is mainly used for the construction of high-grade road.

Polymer concrete refers to the concrete composed of organic polymer, inorganic cementitious material and aggregate. According to its composition and production process, it can be divided into: polymer impregnated concrete; polymer modified concrete (PMC); polymer concrete (PC), also known as resin concrete (RC). Due to the introduction of polymer, polymer cement concrete improves the tensile strength, wear resistance, corrosion resistance, impermeability and impact resistance of ordinary concrete, and improves the workability of concrete. It can be applied to cast-in-situ structures, pavement and bridge deck repair, corrosion-resistant surface of concrete storage tank, bonding of new and old concrete and other prefabricated products for special purposes.

(4) Other Classification Methods

In addition to the above common classification methods, concrete can also be classified according to the application field, including structural concrete, thermal insulation concrete, decorative concrete, waterproof concrete, fire-resistant concrete, hydraulic concrete, marine concrete, road concrete, radiation proof concrete, etc. According to the different construction technology, concrete can be divided into centrifugal concrete, vacuum concrete, grouting concrete, shotcrete, roller compacted concrete, extruded concrete, pumping concrete, etc. According to the difference of reinforcement and reinforcement types, it can be divided into plain (i.e. no reinforcement) concrete, reinforced concrete, fiber concrete, prestressed concrete, etc.

2.3.2 Normal Concrete

Normal concrete is a kind of artificial stone made of cement, coarse aggregate (gravel or pebble), fine aggregate (sand), admixture and water. Sand and stone play a skeleton role in concrete and inhibit the shrinkage of cement; cement and water form cement slurry, which is wrapped on the surface of coarse and fine aggregate and fills the gap between aggregates. The cement paste plays a role of lubrication before hardening, so that the concrete mixture has good working performance. After hardening, the aggregate is cemented together to form a strong whole structural material.

Concrete is mainly divided into two stages and states: plastic state before setting and hardening, namely fresh concrete or concrete mixture; and hard state after hardening, namely hardened concrete or concrete.

(1) Workability of Concrete

Workability refers to the performance of fresh concrete that is easy to operate in each process (mixing, transportation, pouring, tamping, etc.) and can obtain uniform quality and dense concrete. The workability of concrete is a comprehensive technical property, which is closely related to the construction technology. It usually includes fluidity, cohesiveness and water retention.

1) Fluidity refers to the performance that fresh concrete can flow and fill the formwork uniformly and compactly under the action of self weight or mechanical vibration. The fluidity reflects the viscosity of the mixture. If the concrete mixture is too dry and thick, the fluidity is poor and it is difficult to vibrate and compact; if the concrete mixture is too thin, the fluidity is good, but it is easy to separate layers. The main influencing factor is the water consumption in the concrete.

2) Cohesiveness refers to the performance that there is a certain cohesion between the constituent materials of fresh concrete, which will not cause delamination and segregation in the construction process. The cohesiveness reflects the uniformity of concrete mixture. If the cohesiveness of concrete mixture is not good, the aggregate and cement slurry in concrete are

easy to separate, resulting in uneven concrete, and honeycomb and cavity will appear after vibration. The main influencing factor is the cement-sand ratio.

3) Water retention refers to the ability of fresh concrete to keep water, which will not cause serious bleeding during construction. Water retention reflects the stability of concrete mixture. The poor water retention of concrete is easy to form a permeable channel, which affects the compactness of concrete and reduces the strength and durability of concrete. The main influencing factors are cement variety, dosage and fineness.

The workability of fresh concrete is the comprehensive embodiment of fluidity, cohesiveness and water retention. The fluidity, cohesiveness and water retention of fresh concrete are interrelated and often contradictory. Therefore, under the certain condition of construction technology, the workability of fresh concrete is the contradiction and unity of the above three aspects.

(2) Strength of Concrete

Strength is the main mechanical property of concrete after hardening, which reflects the quantitative ability of concrete to resist load. Concrete strength includes compressive strength, tensile strength, shear strength, flexural strength and bond strength. Among them, the compressive strength is the largest and the tensile strength is the smallest.

(3) Deformation of Concrete

The deformation of concrete includes the deformation under non-load and the deformation under load. The deformation under non-load includes chemical shrinkage and temperature deformation. If the amount of cement is too much, it is easy to produce chemical shrinkage and cause micro cracks in the concrete. Cracks are harmful to concrete and should be avoided as far as possible in the use process.

(4) Durability of concrete

The durability of concrete refers to the ability of concrete to resist various damage factors and maintain the strength and appearance integrity for a long time under the actual use conditions. It includes the frost resistance, impermeability, corrosion resistance and carbonation resistance of concrete.

2.3.3 Lightweight Aggregate Concrete

Lightweight aggregate concrete refers to the concrete with lightweight aggregate, whose apparent density is no more than 1950kg/m^3. The so-called lightweight aggregate is used to reduce the quality of concrete and improve the thermal effect. Its apparent density is lower than that of ordinary aggregate. Artificial lightweight aggregate is also called ceramsite (Fig. 2-4).

Lightweight aggregate concrete has the characteristics of light weight, high strength, thermal insulation, fire resistance, good seismic resistance and good deformation performance. However, the elastic modulus is low, and the shrinkage and creep are also large in general.

The application of lightweight aggregate concrete in industrial and civil buildings and other projects can reduce the dead weight of structures, save the amount of materials, improve the efficiency of component transportation and hoisting, reduce the foundation load and improve the function of buildings. It is suitable for high-rise and large-span buildings.

Fig. 2-4 Ceramsite
Source: https://m.jc001.cn/goods/3596395.html.

Fig. 2-5 Lightweight aggregate concrete insulation board
Source: https://aiqicha.baidu.com/yuqing?yuqingId=41dd7c85045ee9b109d5f6ab5a3ce1f0.

According to the different fine aggregates, lightweight aggregate concrete can be divided into full lightweight concrete and sand lightweight concrete. Using light sand as fine aggregate is called full lightweight concrete; using ordinary sand or part of light sand as fine aggregate is called sand light concrete.

Light aggregate concrete can be divided into thermal insulation lightweight aggregate concrete, structural thermal insulation lightweight aggregate concrete and structural lightweight aggregate concrete according to their different uses in construction engineering (Fig. 2-5). In addition, lightweight aggregate concrete can also be used as heat-resistant concrete instead of furnace lining.

2.3.4 Reactive Powder Concrete

Reactive powder concrete (RPC), is a new material developed in the 1990s with ultra-high strength, high toughness, high durability and good volume stability. RPC is a new type of concrete composed of defined system containing ultra fine particles and fiber reinforced materials.The application of RPC in engineering structure can solve the problems of low tensile strength, high brittleness and poor volume stability of high-strength and high-performance concrete, as well as the problems of high investment, poor fire resistance and easy corrosion of steel structure.

The outstanding technical properties of RPC materials are mainly manifested in high strength, high toughness and high durability of hardened body, good construction performance of mixture and environmental protection performance of raw materials.

(1) Mechanical Property

The strength of RPC materials can be divided into 200MPa, 500MPa and 800MPa according to the compressive strength. The 200MPa grade RPC material has been applied in the engineering, the 500MPa grade is still in the laboratory research stage, and the 800MPa grade RPC material is in the laboratory trial mixing stage. The raw materials and production process of RPC with different strength grades are quite different.

Compared with high performance concrete (HPC), RPC is characterized by high strength, high toughness, especially high tensile strength. The compressive strength of 200MPa RPC is 170~230MPa, which is 2~4 times of that of high-strength concrete, and has high deformation capacity. The flexural strength is 20~40MPa, which is 4~6 times of that of high-strength concrete, and the tensile-compression ratio can reach about 1/6 after adding fiber. The fracture toughness of RPC material with this strength grade is 100 times higher than that of ordinary concrete and high-strength concrete, and comparable with aluminum.

(2) Durability

Because of its compact structure and few defects, RPC has high durability.

The durability test of 200MPa grade RPC material used in the world's first RPC structural bridge (pedestrian/bicycle bridge) in Sherbrook, Quebec Province, Canada, shows that after 300 rapid freeze-thaw cycles (the highest temperature is 4℃, the lowest temperature is −18℃, and the temperature change rate is 6℃/h), the specimen is not damaged, and the durability factor is as high as 100%. The results of 50 freeze-thaw tests with deicing salt show that the weight loss rate of RPC plate is less than $8g/m^2$, while the allowable standard of Quebec Province is $600g/m^2$, so it can be ignored. The measured chloride ion permeability fluctuates between 6~9 Coulomb, while the chloride ion permeability of 30MPa ordinary concrete is 5000 Coulomb, and that of 80MPa high performance concrete is about 500 Coulomb. It can be seen that its anti-permeability ability is very good and can effectively prevent the invasion of harmful media. There is no loss of dynamic elastic modulus and mass of RPC specimens in 600 quick freeze-thaw tests.

(3) Workability

RPC mixture not only has good fluidity, but also has good cohesiveness. It does not segregate during transportation, pouring and tamping. In the narrow formwork and the reinforcement gap, the passing performance is good, and there is no need to vibrate after pouring.

(4) Environmental Friendliness

RPC materials have good environmental performance.

According to a test, I-beams with the same bending capacity are made of RPC, steel and reinforced concrete respectively. The results show that the cross-section dimensions of the members made of RPC are almost the same as those of steel members.

Moreover, the cement consumption of RPC material is almost 1/2 of that of ordinary concrete and HPC, so the CO_2 emission of the same amount of cement production process is only about half. The amount of non renewable natural resource aggregate in the production process accounts for only 1/3 and 1/4 of HPC and 30MPa concrete, respectively.

2.3.5 Fiber Reinforced Concrete

The cement-based composite material composed of cement slurry, mortar or concrete as base material and fiber as reinforcement material is called fiber concrete. Fiber can control the further development of cracks in matrix concrete, so as to improve the crack resistance of concrete. Due to the high tensile strength and elongation of fibers, the tensile strength, flexural strength, impact strength, elongation and toughness of fiber reinforced concrete are obviously better than those of ordinary concrete.

(1) Classification of Fiber Reinforced Concrete

The fibers used for fiber reinforced concrete can be divided into three types as follows.

1) Metal fiber, such as steel fiber (Fig. 2-6), stainless steel fiber (for heat-resistant concrete).

2) Inorganic fiber. There are mainly natural mineral fibers (chrysotile, crocidolite, iron wool, etc.) and man-made mineral fibers (alkali resistant glass fiber and alkali resistant mineral wool and other carbon fibers, shown in Fig. 2-7).

3) Organic fiber. Synthetic fibers (such as polyethylene, polypropylene, polyvinyl alcohol, nylon, aromatic polyimide, etc.) and plant fibers (sisal, agave, etc.) are mainly used. Synthetic fiber concrete should not be used in the thermal environment above 60℃.

The main varieties of fiber reinforced

Fig.2-6 Steel fiber
Source: https://baike.baidu.com.

Fig. 2-7 Glass fiber
Source: https://b2b.hc360.com/supplyself/312880244.html.

concrete are steel fiber concrete, glass fiber concrete, polypropylene fiber concrete, carbon fiber concrete, plant fiber concrete and high elastic modulus synthetic fiber concrete. Compared with ordinary concrete, fiber reinforced concrete has many advantages, but it can not replace reinforced concrete. People began to add fiber into reinforced concrete to make it become rebar-fiber composite concrete, which has developed a new way for the application of fiber reinforced concrete.

(2) Steel Fiber Reinforced Concrete

1) Materials

The preparation of ordinary steel fiber reinforced concrete mainly uses low carbon steel fiber. In order to prepare refractory concrete, stainless steel fiber must be used. The diameter of long straight steel fiber in circular section is generally 0.25~0.75mm, while the thickness of flat steel fiber is 0.15~0.4mm, the width is 0.25~0.9mm, and the length is 20~60mm. In order to improve the interface bonding, steel fibers with hooks at the end are used.

Steel fiber reinforced concrete generally uses ordinary Portland cement, and high-strength steel fiber concrete can use high-strength Portland cement or alunite cement. The maximum size of coarse aggregate used should not exceed 15mm. In order to improve the workability of the mixture, water reducer or superplasticizer must be used. Generally, the sand ratio of concrete should not be less than 50%, and the cement content should be about 10% higher than that without fiber.

2) Proportion

In order to ensure that the fiber can be evenly distributed in the concrete, the length-diameter ratio should not be greater than 100, generally 30~80. There is a limit value of the maximum content of each type of fiber, which is generally 0.5%~2% (volume ratio).

3) Stirring

Steel fiber reinforced concrete is mixed by forced mixer. In order to make the fiber evenly dispersed in the concrete, the material should be fed through shaker or disperser. The feeding sequence of mixing is different from that of ordinary concrete. One method is to add coarse and fine aggregates, cement and water into the mixer, After mixing evenly, add the fiber into the stirring. The other method is divided into three steps, the first step is to mix the coarse and fine aggregates evenly, the second step is to add fiber, and finally add cement and water to mix again.

4) Tamping

Different tamping methods have great influence on fiber orientation. The fiber is pumped to the warehouse without inserting, and the fiber is in three-dimensional disorder. If the plug-in vibration device is used for tamping, most of the fibers are in three-dimensional disorder, and a few are in two-dimensional direction. When the plane vibrator is used, most of them are in two-dimensional random direction, while a few are in three-dimensional direction. The fibers show two-dimensional random direction on the jet surface when the spray method is adopted. The fiber orientation is between one-dimensional orientation and two-dimensional random orientation by "centrifugal method" or "extrusion method". If the fiber is vibrated in magnetic field, the fiber is distributed along the direction of magnetic line of force.

5) Mechanical Properties

The mechanical properties of concrete

are improved by adding steel fiber. When the content is within the allowable range, the tensile strength can be increased by 30%~50%, the bending strength by 50%~100%, the toughness by 10~50 times, the impact strength by 2~9 times, and the compressive strength by 15%~25%. Steel fiber reinforced concrete can also reduce the dry shrinkage by 10%~30%.

At present, the cost of steel fiber reinforced concrete is still relatively high and the construction is difficult, so it must be used in the most appropriate projects, such as the important tunnel, subway, airport, elevated roadbed, spillway and explosion-proof seismic engineering.

(3) Glass Fiber Reinforced Concrete (GFRC)

The fiber used in GFRC must be alkali resistant glass fiber to resist the corrosion of $Ca(OH)_2$ in concrete. Alkali resistant glass fiber can only slow down the erosion in ordinary Portland cement, so sulphoaluminate cement should be used in order to greatly increase the service life.

Glass fiber reinforced concrete has no special requirements for coarse and fine aggregate and mix proportion, which are basically the same as steel fiber reinforced concrete. The mechanical properties of glass fiber reinforced concrete are certainly lower than that of steel fiber reinforced cement, and the compressive strength of glass fiber reinforced concrete is slightly lower than that without fiber. But its toughness is very high, which can be increased by 30~120 times, and has good fire resistance. It is mainly used for non load bearing and secondary load-bearing components.

(4) Polypropylene Fiber Concrete

Polypropylene membrane split fiber is a kind of bundle-like synthetic fiber. It can also be cut into short cuts with a length of 19~64mm. In order to prevent fiber aging, it should be packed in black packaging container before use.

The construction technology is divided into mixing method and spraying method. The content of fiber varies with the process. The cutting length by stirring method is 40~70mm and the volume mixing rate is 0.4%~1%, while the cutting length by spraying method is 20~60mm and the volume mixing rate is 2%~6%. As the fiber strength is not high, once the concrete cracks, the fiber concrete will crack and the compressive strength will not be significantly improved. Only the impact strength is higher, which can be increased by 2~10 times. The shrinkage can be reduced by 75%. It can be used for non load bearing slab, parking lot, etc.

2.3.6 Bituminous Concrete

Bituminous concrete is a mixture of mineral aggregate with certain gradation composition and certain proportion of road bituminous material.

The strength of asphalt mixture is mainly manifested in two aspects. One is the cohesive force of binder formed by asphalt and mineral powder. The other is the internal friction and locking force between aggregate particles. The large surface area of fine mineral powder particles (mostly the diameter of particles is less than 0.075mm) makes asphalt materials form a film, which improves the bonding strength and temperature stability of asphalt materials, while the locking force mainly occurs between coarse aggregate particles. When selecting the mineral aggregate gradation of asphalt concrete, both should be taken into account, so that the mixture can form dense, stable, suitable roughness and durable pavement after adding proper amount of asphalt. There are many ways to mix mineral aggregate, which can be calculated by formula or determined by experience.

The appropriate amount of asphalt in asphalt mixture should be determined by the laboratory test results and the practical situation of the construction site. Generally, the reference range of asphalt dosage is listed in the relevant specifications as the guidance of trial mixing. When the mineral aggregate variety, gradation range, asphalt consistency and type, mixing facilities, regional climate and traffic characteristics are relatively fixed, empirical formula can also be used to estimate.

The hot mix asphalt mixture should be mechanically mixed at the centralized place. In general, the fixed hot mixing plant is selected, and the mobile hot mixer should be selected when the line is long. The cold mix asphalt mixture can be mixed in a centralized way or mixed on the spot. The main equipment of asphalt mixing plant includes: asphalt heating pot, sand and gravel storage place, ore powder bin, heating drum, mixer and weighing equipment, steam boiler, asphalt pump and pipeline, dust removal facilities, etc., and some also have rescreening and storage equipment for hot aggregate (see asphalt mixture mixing base). The mixer can be divided into continuous type and batch type. In the preparation process, in the past, the sand and stone materials were dried and heated, and then mixed with hot asphalt and cold mineral powder. A method of mixing wet aggregate with hot asphalt and then heating and mixing is developed to eliminate fly ash during heating and drying. When using the latter process, it is better to prevent the residual moisture in the mixture from affecting the service life of asphalt concrete. It is better to use asphalt anti stripping agent at the same time to enhance the water resistance.

2.3.7 Reinforced Concrete

Reinforced concrete is a kind of composite material composed by adding steel mesh, steel plate or fiber into the concrete to work together to improve the mechanical properties of concrete. It is the most common form of reinforced concrete.

After the invention of modern cement, it is found that the concrete prepared with cement has a very high bearing capacity, while its tensile strength and crack resistance are very poor. In 1849, Joseph Monier, a French florist, began to use cement to cover steel wire mesh to make pots and flowerpots. However, due to the lack of mechanical knowledge, he initially placed the steel wire mesh in the center of the interface, which did not make the steel wire in a reasonable position under tension. Later, he improved this method and showed it at the Paris exhibition in 1867 and obtained the relevant patent right. The main principle is to make the iron bar bear the tension and the concrete bear the pressure. This method and concept is the prototype of reinforced concrete which has been used all the time.

The reinforced concrete Monier et al. used is made in the form of beams, slabs and other components, and then assembled for use. In 1890, French civil engineer Francois Hennebique expounded the effective effect of stirrups on shear resistance, and invented the reinforced concrete system named after him. The reinforcement cage with stirrup and longitudinal reinforcement was bound, and the original individual slab, beam and column components were poured into a whole body. The reinforcement was arranged in the concrete according to the change of tensile position. This is the earliest reinforced concrete cast-in-place frame structure. In addition, the space between the frame columns is very small and the building height is limited.

After entering the 20th century, reinforced concrete has been widely used in many fields, such as housing construction, bridges, tunnels, marine engineering, etc., and has become the most widely used structural material.

2.3.8 Prestressed Concrete

In order to make up the premature crack of concrete, before the component is used (loaded), the steel bar is tensioned in the tensile zone of concrete in advance by means of artificial force, and the retraction force of reinforcement is used to make the tensile zone of concrete subject to pressure in advance. When the component bears the tension caused by external load, the pre-pressure in the concrete in the tension area is offset first, and then the concrete is tensioned with the increase of load, which limits the elongation of concrete and delays or prevents cracks from appearing, which is called prestressed concrete.

In 1888, Dochring, a German architect, first put forward the idea of prestressed concrete for its poor tensile and crack resistance. However, due to the lack of understanding of reinforcement relaxation and concrete creep at that time, most of the prestressing force applied in the concrete was lost with time. Therefore, this assumption of Dochring did not succeed immediately in practice. It was not until 1928 that the French civil engineer Eugene Freyssinet systematized the creep theory of concrete and applied for relevant patents, which made the successful application of prestressed concrete come true. Therefore, this year is treated as the invention time of prestressed concrete in many historical necords.

(1) Classification of Prestressed Concrete

The prestressed concrete can be divided into three types according to the control degree of the pre-stress value on the member section crack.

1) Fully Prestressed Concrete

Under the action of service load, the concrete tensile stress on the section is not allowed. It is mainly used for the components that strictly require no cracks.

2) Partially Prestressed Concrete

It is a member allowing cracks but maximum crack width not exceeding the allowable value. It is mainly used for components that allow cracks.

3) Unbonded Prestressed Reinforcement

The external surface of prestressed steel bar is coated with asphalt, grease or other lubricating antirust materials to reduce friction and prevent corrosion, and is wrapped with plastic sleeve or paper tape or plastic belt to prevent coating from being damaged during construction and to isolate it from surrounding concrete. Post tensioned prestressed steel bar can slide longitudinally during tensioning.

Prestressed concrete can be divided into pretensioning method and posttensioning method according to the reinforcement tensioning technology.

(2) Tensioning Technology of Prestressed Concrete

The pretensioning method is to stretch the prestressing tendons before pouring concrete, and the tensioned prestressed reinforcement is temporarily anchored on the pedestal or steel formwork. Then the concrete is poured. When the concrete curing reaches 75% of the design strength value of the concrete, the prestressed reinforcement is relaxed, with the help of the bond between concrete and prestressed reinforcement, the concrete can be prestressed. Pretensioning method is generally only suitable for the production of small and medium-sized components in the fixed prefabricated plant.

The posttensioning method is to make the component (pouring concrete) first, and reserve the corresponding hole in the component body according to the position of the prestressed reinforcement. After the concrete strength of the component reaches the specified strength (generally not less than 75% of the design strength standard value), the prestressed reinforcement is inserted into the reserved duct for tension. The prestressed reinforcement after tension is anchored at the end of the component by means of anchorage, and the pretension force of the prestressed reinforcement is transmitted to the concrete by the anchorage at the end of the component to generate the pre-compression stress. Finally, cement slurry is poured into the duct to form a whole of prestressed reinforcement and concrete member.

(3) Advantages of Prestressed Concrete

Compared with ordinary concrete, prestressed concrete has the following significant advantages:

1) Good crack resistance and high rigidity. Due to the application of prestress to the components, the occurrence of cracks is greatly delayed. Under the action of service load, the cracks can not appear in the components, or the cracks can be delayed, so the stiffness of the components is improved and the durability of the structure is increased.

2) It can save material and reduce self weight. Because the structure must be made of high-strength materials, it can reduce the amount of reinforcement and the cross-section size of components, save steel and concrete, and reduce the weight of the structure. It has obvious advantages for long-span and heavy load structures.

3) It can reduce the vertical shear force and main tensile stress of concrete beams. The curved reinforcement (tendon) of prestressed concrete beam can reduce the vertical shear force near the support in the beam, and the main tensile stress under the load is also reduced due to the existence of prestress on the concrete section. This is beneficial to reduce the web thickness of the beam and further reduce the self weight of the prestressed concrete beam.

4) It can improve the stability of compression members. When the slenderness ratio of the compression member is large, it is easy to be bent under a certain pressure, resulting in the loss of stability and damage. If the prestressed concrete column is prestressed to make the longitudinal stressed steel bar's tension very tight, not only the prestressed steel bar itself is not easy to be bent, but also it can help the surrounding concrete to improve the ability to resist bending.

5) It can improve the fatigue resistance of the components. Because the steel bar with strong prestressing force has a relatively small change range of stress caused by loading or unloading in the service stage, it can improve the fatigue strength, which is very beneficial to the structure bearing dynamic load.

6) Prestress can be used as a means to connect structural members and promote the development of new systems and construction methods of long-span structures.

(4) Steel Bars for Prestressed Concrete

The prestressed steel bar in prestressed concrete is different from ordinary steel bar. The main requirements include:

1) It should have high intensity. The tensile stress of prestressed steel bar will cause various stress losses in the whole process of making and using components. The sum of these losses can sometimes reach more than $200N/mm^2$. If the strength of the steel bar used is not high, the stress established during tension will even be lost.

2) It should have good adhesion with concrete. Especially in pretensioning method, there must be a high bond and self anchored strength between prestressed reinforcement and concrete. For some high strength smooth steel wire, it is necessary to go through "notch", "pressure wave" or "kink" to form notched steel wire, corrugated steel wire and kinked steel wire, so as to increase the bonding force.

3) It should have enough plasticity and good processability. The higher the strength is, the lower the plasticity will be. When the plasticity of steel bar is too low, especially under low temperatures and impact loads, brittle fracture may occur. Good processing performance refers to good welding performance, and when using upsetting anchor plate, the upsetting of reinforcement head does not affect the original mechanical properties.

2.4 Steel

Steel is an important modern industrial structural material. Steel is a kind of material with various shapes, sizes and properties required by pressure processing of ingots, billets or steels.

The classification of steel products is usually classified according to the production process (e.g. cold/hot rolled products), shape (e.g. round/square bars), dimension (e.g. hot rolled strips/plates) and surface (e.g. hot rolled bars/bright bars). Specific classification can refer to *Steel Products—Vocabulary* ISO 6929: 2013, and *Steel Products Classification* GB/T 15574–2016 is usually used in China. In the world, steel is usually divided into four categories: long, flat, pipe and other steels. Long products include railway steel, steel sheet pile, various sections, cold-formed steel, bar, steel bar and wire rod; flat products mainly include thick steel plate, thin steel plate, steel strip and coated steel plate; steel pipe includes seamless steel pipe and welded steel pipe.

Because the same steel grade (such as the China national standard hot-rolled carbon manganese steel Q235B coil) can be used for automobile manufacturing and building structure, it is generally not recommended to classify steel products according to the end use and production process of steel products, but sometimes they are used as a reference for classification.

From the perspective of application requirements, as structural materials, construction steel requires certain strength, deformation ability, weldability and weather resistance. Construction steel can be divided into steel for steel structures and steel for reinforced concrete structures.

(1) Steel for Steel Structures

Steel for steel structure mainly includes ordinary carbon structural steel and low alloy structural steel. According to the carbon content, ordinary carbon structural steel can be divided into high carbon steel, medium carbon steel and low carbon steel. The higher the carbon content is, the higher the strength of the steel will be, but the lower the deformability and weldability will be. Therefore, in order to ensure good deformation ability and weldability of construction steel, low carbon steel is generally used. The higher the alloy content in the alloy steel is, the better the steel performance will be, but the more expensive the price will be. Therefore, from the economic point of view, the alloy steel used in construction is mostly low alloy steel.

The beam and column of steel structures can be made of large, medium and small section steel, cold-formed steel or steel pipe; it can also be formed by welding thick steel plate and thin steel plate (Fig. 2-8). The common section forms of steel members include H-section, box section, etc.

(2) Steel Bars for Reinforced Concrete Structures

Steel bar is a kind of steel strip, which is a kind of building materials (Fig. 2-9). It is usually used with concrete to form reinforced concrete structure. The tensile strength of the steel bar is high, but it can't bear the pressure, while concrete has high compressive strength but low tensile strength. The combination of the two has good mechanical strength. The reinforcement is protected by the concrete

Fig. 2-8 I-section steel for steel structures
Source: https://0546.79abc.com.

Fig. 2-9 The steel bar for concrete structures
Source: http://www.haogu114.com.

and will not rust. Moreover, the steel bar and the concrete have similar thermal expansion coefficient, which will not produce large internal force due to temperature changes.

According to the processing methods, steel bars for reinforced concrete structures can be divided into hot-rolled steel bars, heat-treated steel bars, cold-drawn low-carbon steel wires and steel strand pipes. According to the surface shape, it can be divided into smooth reinforcement and thread. According to the chemical composition of the steel, it can be divided into low carbon steel, medium carbon steel, high carbon steel and alloy steel.

2.5 Timber

Wood is an ancient construction material. Because of its unique advantages, wood occupies a place even in the modern engineering structure with highly developed technology.

2.5.1 Tree Species

The mechanical properties of wood are closely related to tree species. At present, the species of structural timbers can be divided into conifer and broadleaf.

The leaves of coniferous trees are as long and slender as needles. Most of them are evergreen trees with straight trunks, which are easy to form larger wood. Conifer has small bulk density, small expansion and deformation, soft wood and easy processing. It mainly includes pine, cypress and Chinese fir, which are the main tree species used in civil engineering.

Broad leaved trees are dicotyledonous trees with flat and wide leaves. Most of them are deciduous trees. The straight part of the trunk of broad-leaved trees is short, so it is difficult to obtain large-size wood. It has high bulk density, large expansion and shrinkage deformation, and it is easy to warp and crack. Wood of broad leaved trees is hard and difficult to process. It is often called hardwood. It includes elm, birch and Fraxinus mandshurica. Because most of the broad leaved trees have beautiful natural texture, they are especially suitable for interior decoration or furniture making.

2.5.2 Types of Structural Timber

According to the different processing methods, structural wood can be divided into three categories: log, sawn timber and glulam.

(1) Log

The logs are felled trunks, which are cut into sections of suitable length for sawing commercial timber after pruning and timber making (Fig. 2-10). Log is often used as structural components with high requirements, such as large length of the whole component, small diameter change, good appearance and few defects. Therefore, the construction cost is high and it is not conducive to the full use of raw materials.

(2) Sawn Timber

Sawn timber is a kind of lumber with various specifications, which is made by

Fig 2-10 Log
Source: https://baike.baidu.com.

Fig. 2-11 Glulam
Source: https://www.chinatimber.org/price/57700.html.

sawing logs. According to the different section sizes, the sawn timber is divided into square wood, plate and specification timber. With the development of wood processing technology, the degree of automation in the production process of sawn timber is constantly improved. Through computer control, the optimal section is divided, so as to maximize the use of log materials, produce sawn timber of various sections, and improve the production efficiency and the utilization rate of raw materials.

(3) Glulam

Glulam is a kind of wood products which are made of wood as raw material through special adhesives and pressing (Fig. 2-11). Due to the small size of the single raw material used to manufacture glulam, it is easy to remove the defects of the wood itself, so as to obtain wood products with more uniform materials, high mechanical properties and stable quality. The glulam commonly used in structures includes plywood, laminate glulam, etc.

2.6 Masonry and Mortar

Stone, brick and building blocks, and mortar are the main materials of masonry structures.

2.6.1 Stone

Rock is a natural material, which can be used only by mining and processing. Stone is one of the construction materials used earlier in the field of civil engineering and is still in use.

(1) Types of Rocks

There are many kinds of rocks in nature, which can be divided into magmatic rocks, sedimentary rocks and metamorphic rocks according to their genesis.

1) Magmatic Rocks

Magmatic rocks are formed by magma intrusion into the earth's crust or ejecting from the earth's interior. The common magmatic rocks are granite, basalt and granite porphyry.

2) Sedimentary Rocks

Under the surface conditions, rock debris, solution precipitates or organic matter formed by weathering are transported to low-lying land or sea by flowing water, wind and glacier, and then formed by diagenesis. The common sedimentary rocks are conglomerate, sandstone and limestone.

3) Metamorphic Rock

Under the action of high temperature, high pressure and active chemical substances, the rocks that make up the crust have changed their mineral composition and structure due to crustal movement and magmatic activity, thus forming new rocks. The common metamorphic rocks are marble and quartzite.

(2) Stone in Masonry Structure

In masonry structure, the commonly used stones are granite, sandstone and limestone without obvious weathering. According to the regular degree of shape after processing, it can be divided into two kinds: rubble and dressed stones. Irregular shaped stone is called rubble stone and regular hexahedron stone is called dressed stone.

(3) Decorative Stone

There are three kinds of decorative stone materials used in architectural decoration engineering: natural granite, natural marble and artificial stone. The content of silica in granite is more than 60%, so its acid and aging resistance are good, and its service life is long. The content of calcium oxide and magnesium oxide in marble is more than 50%. Natural marble is easy to process and can be polished. It is often made into polished plate with rich color and fine material.White marble, a kind of marble, which is pure white, fine-grained and hard texture, is the first-class construction material. Artificial stone is a kind of artificial stone which is formed by molding, curing and surface treatment with inorganic or organic cementitions materials as adhesive materials, natural sand, crushed stone and stone powder as filling materials.

2.6.2 Brick and Building Blocks

Brick is a small man-made building block, and is the most traditional masonry material. The shape of brick is mainly cuboid. It has been developed from clay as the main raw material to industrial waste such as coal gangue and fly ash, from solid to porous and hollow and from sintering to non sintering. At present, the brick can be divided into fired brick, unburned brick and concrete brick.

Fig. 2-12 Standard brick made of clay
Source: https://zixun.jia.com/jxwd/683075.html.

(1) Fired Brick

Fired brick includes standard brick, porous brick and hollow brick.

Ordinary brick, also known as standard brick, is a solid brick made of clay (including shale, coal gangue and other powder) as the main raw materials, after mud treatment, molding, drying and roasting (Fig. 2-12). According to the different raw materials, it can be divided into sintered coal gangue brick, sintered shale brick, sintered fly ash brick and sintered clay brick.

Porous brick is a kind of brick with through-hole on the larger side of common brick. The hole size is small and the number is large, and the hole rate is not more than 35%. It is mainly used for load-bearing parts.

Hollow brick is a kind of brick with hole ratio not less than 40%. Due to the large opening, it can not bear large loads, so it is mainly used for infill wall and self-bearing partition wall of frame structure.

(2) Unburned Brick

Using fly ash, coal slag, coal gangue, tail slag, chemical slag or natural sand, sea mud (one or more of the above raw materials) as the main raw materials, a new type of brick is made without high-temperature calcination, which is called unburned brick.

Fly ash, coal gangue, slag and other industrial wastes contain high content of silicon oxide, alumina and iron oxide. After the raw materials are mixed and rolled, they are fully hydrated to form silicon and aluminum type glass body. This kind of glass body combines with the oxidized calcification after hydration to produce chemical reaction, which is called "pozzolanic reaction". In this chemical reaction, calcium silicate (aluminum) hydrate is a kind of colloidal glass, which is not stable, but under the action of additives, it reacts with time and solidifies gradually to form a high-strength network structure. In addition, the reasonable allocation and maintenance of raw materials can form the strength of self-curing brick.

Due to its high strength, good durability, standard size, complete appearance and uniform color, it can be used as fair faced wall or any exterior decoration.Therefore, it is a promising replacement product to replace clay brick.

(3) Concrete Brick

Concrete brick is a kind of brick made of cement, sand and stone as aggregate, and admixtures (Fig. 2-13). Concrete brick has the advantages of light weight, fire protection, sound insulation, heat preservation and impermeability, no pollution, energy saving and consumption reduction. It can be directly used in various load-bearing structures, and is an important type of new wall materials.

(4) Building Block

Building block generally refers to the block building products with large brick shape. Its raw material source is wide, and it may take the local material that the price is cheap. According to the size, it can be divided into three categories: large, medium and small. The block with a height of more than 980mm is called the large block (Fig. 2-14); a block with a height of 380~980mm is called the medium block; a block with a height of less than 380mm is the small block. According to the materials, it can be divided into concrete, cement mortar, aerated concrete, fly ash silicate, coal gangue, artificial ceramsite, slag waste and other blocks. According to the structure of the block, it is divided into dense block and hollow block. The hollow block includes several kinds: round hole, square hole, elliptical hole, single-row hole, multi-row hole and other hollow blocks. Dense or hollow blocks can be used as load-bearing walls and partitions.

2.6.3 Mortar

Mortar is a workable paste used to bind building blocks such as stones, bricks and

Fig. 2-13 Concrete brick
Source: http://baike.baidu.com.

Fig. 2-14 Large block
Source: http://www.zk71.com.

concrete masonry units, fill and seal the irregular gaps between them, and sometimes add decorative colors or patterns in masonry walls. In its broadest sense, mortar includes pitch, asphalt and soft mud or clay, such are used between mud bricks. Mortar comes from Latin mortarium, meaning crushed.

Cement mortar becomes hard when it cures, resulting in a rigid aggregate structure; however, the mortar is intended to be weaker than the building blocks and is the sacrificial element in the masonry, because the mortar is easier and less expensive to repair than the building blocks. Mortars are typically made from a mixture of sand, binder and water. The most common binder since the early 20th century is Portland cement but the ancient binder lime mortar is still used in some new construction. Lime and gypsum in the form of plaster of Paris are used particularly in the repair and repointing of buildings and structures because it is important that the repair materials are similar to the original materials. The type and ratio of the repair mortar is determined by a mortar analysis. There are several types of cement mortars and additives.

(1) Ordinary Portland Cement Mortar

Ordinary Portland cement mortar, commonly known as OPC mortar or just cement mortar, is created by mixing powdered ordinary Portland cement, fine aggregate and water.

It was invented in 1794 by Joseph Aspdin and patented on 18 December, 1824, largely as a result of efforts to develop stronger mortars. It was made popular during the late 19th century, and had by 1930 become more popular than lime mortar as construction material. The advantages of Portland cement is that it sets hard and quickly, allowing a faster pace of construction. Furthermore, fewer skilled workers are required in building a structure with Portland cement (Fig. 2-15 and Fig. 2-16).

As a general rule, however, Portland cement should not be used for the repair or repointing of older buildings built in lime mortar, which require the flexibility, softness and breathability of lime if they are to function correctly.

In the United States and other countries, five standard types of mortar (available as dry pre-mixed products) are generally used for both new construction and repair. Strengths of mortar change based on the mix ratio for each type of mortar, which are specified under the ASTM standards. These pre-mixed mortar products are designated by one of the five letters—M, S, N, O and K. Type M mortar is the strongest, and Type K is the weakest. The mix ratios are always expressed by volume of Portland cement : lime : sand. These type letters are apparently taken from the alternate letters of the words "MaSoN wOrK".

(2) Polymer Cement Mortar

Polymer cement mortars (PCM) are

Fig. 2-15 Laying bricks with Portland cement mortar
Source: https://www.zqins.com/zhishi/1774.html.

Fig. 2-16 Mixed mortar
Source: https://www.sina.com.cn.

the materials which are made by partially replacing the cement hydrate binders of conventional cement mortar with polymers. The polymeric admixtures include latexes or emulsions, redispersible polymer powders, water-soluble polymers, liquid thermoset resins and monomers. It has low permeability, and it reduces the incidence of drying shrinkage cracking, mainly designed for repairing concrete structures. One brand of PCM is MagneLine.

(3) Lime Mortar

The setting speed can be increased by using impure limestone in the kiln, to form a hydraulic lime that will set on contact with water. Such a lime must be stored as a dry powder. Alternatively, a pozzolanic material such as calcined clay or brick dust may be added to the mortar mix. Addition of a pozzolanic material will make the mortar set reasonably quickly by reaction with water.

It would be problematic to use Portland cement mortars to repair older buildings originally constructed using lime mortar. Lime mortar is softer than cement mortar, allowing brickwork a certain degree of flexibility to adapt to shifting ground or other changing conditions. Cement mortar is harder and allows little flexibility. The contrast can cause brickwork to crack where the two mortars are present in a single wall.

Lime mortar is considered breathable in that it will allow moisture to freely move through and evaporate from the surface. In old buildings with walls that shift over time, cracks can be found which allow rain water into the structure. The lime mortar allows this moisture to escape through evaporation and keeps the wall dry. Repointing or rendering an old wall with cement mortar stops the evaporation and can cause problems associated with moisture behind the cement.

(4) Pozzolanic mortar

Pozzolana is a fine and sandy volcanic ash. It was originally discovered and dug at Pozzuoli, nearby Mount Vesuvius in Italy, and was subsequently mined at other sites, too. The Romans learned that pozzolana added to lime mortar allowed the lime to set relatively quickly and even under water. Vitruvius, the Roman architect, spoke of four types of pozzolana. It is found in all the volcanic areas of Italy in various colours: black, white, grey and red. Pozzolana has since become a generic term for any siliceous and/or aluminous additive to slaked lime to create hydraulic cement. Finely ground and mixed with lime, it is a hydraulic cement, like Portland cement, and makes a strong mortar that will also set under water.

2.7 Fiber Reinforced Polymer Composites

A composite material is a combination of two or more materials (reinforcing elements such as fibers, and binders such as polymer resins), differing in form or composition. The combination of these materials can be designed to result in a material that maximizes specific performance properties. For example, fiber reinforced polymer (FRP) composites are made of thermosetting or thermoplastic resins, glass, carbon or other types (e.g., Kevlar or natural fiber flax/kenaf) of fibers (rovings), mats, and/or fabrics. The fiber network is the primary load-bearing component, while the resin helps transfer loads including shear forces through fibers and fabrics and maintains fiber orientation. The resin primarily dictates the manufacturing process and processing conditions, and partially protects the fibers/fabrics from environmental damage, such as humidity, temperature fluctuations and chemicals.

FRP composites are being promoted as

the materials of the 21st century because of their superior corrosion resistance, excellent thermo-mechanical properties, and high strength-to-weight ratio. FRP composites are also greener in terms of embodied energy (the quantity of energy required to manufacture a product) than conventional materials such as steel and aluminum, on a per-unit-of-performance basis. The use of FRP composites in civil and military infrastructure can improve innovation, increase productivity, enhance performance and provide longevity, resulting in reduced life-cycle costs and enhanced environmental protection. For over 25 years, the researchers at West Virginia University's Constructed Facilities Center (WVU-CFC) have been conducting research on fundamentals of engineering and material sciences, innovation and development including field implementation of FRP composite components and systems. The FRP implementation has touched upon a wide range of engineering applications with emphasis on enhancing performance, serviceability and durability over conventional materials.

2.7.1 Composition of FRP Material

FRP commonly used in engineering structure is composed of high performance fiber and resin matrix. Fiber is the main component of the force, the role of the matrix is to bond the fibers together, so that the fibers bear the force together. The mechanical properties of FRP are determined by the properties of fiber and resin. The commonly used fibers are glass fiber, carbon fiber and aramid fiber, and the commonly used resins are epoxy resin, unsaturated polyester resin, vinyl resin and phenolic resin. In addition, some additives, such as flame retardants, heat stabilizers, light stabilizers, colorants and fillers, should be added to FRP according to engineering requirements to improve the process properties or product properties.

(1) Fiber Material

Glass fiber is the first fiber used to make FRP, which appeared in the United States in the 1930s and 1940s. Its main chemical components are some metal and non-metallic oxides, including silica, calcium oxide, alumina, magnesium oxide, boron oxide, sodium oxide, potassium oxide, etc. Glass fibers with different properties can be obtained by changing the composition and content of oxides, including high alkali glass fiber, medium alkali glass fiber, alkali free glass fiber, high modulus glass fiber and high strength glass fiber.

Carbon fiber is a kind of fiber with the best mechanical properties and chemical stability, but its price is high. Its chemical composition is carbon, and its microstructure is graphite like bundle structure. According to the different mechanical properties, carbon fibers can be divided into general-purpose carbon fibers and high-performance carbon fibers. High-performance carbon fibers include standard type, high strength type (tensile strength greater than 4000MPa), high model (tensile modulus greater than 390GPa), ultra-high strength type and ultra-high model. At present, high-performance carbon fiber is mainly used in engineering structure.

(2) Resin Matrix

Resin is divided into thermoplastic resin and thermosetting resin. At present, thermosetting resin is mainly used in engineering structure. However, due to its fast curing speed and recyclable property, it is an ideal green resin, so it has been applied more and more in recent years.

Epoxy resin is the most common kind of resin in structural engineering, which generally refers to the polymer compound containing two or more epoxy groups. The epoxy resin has good adhesion, mechanical properties, corrosion resistance and insulation properties, and can be cured at room temperature and low pressure.

Unsaturated polyester resin is a kind of polymer containing unsaturated dibasic acid ester group. It has the advantages of low-voltage curing, good chemical resistance, good electrical insulation and large curing shrinkage.

Vinyl resin is a kind of polymer with unsaturated double bond of end group or side group. It is similar to the form of unsaturated polyester resin. It can also be regarded as an epoxy like modification of unsaturated sawing resin. Compared with unsaturated polyester resin, vinyl resin has better corrosion resistance and toughness.

2.7.2 Preparation Process of FRP

The preparation process of FRP products is the premise to ensure that the fiber and matrix work together. Moreover, the mechanical properties of FRP are highly dependent on the preparation process. FRP products produced by different processes have great differences in mechanical properties, dimensional accuracy and so on. Therefore, the preparation process must be considered in the design of FRP structure.

The preparation process of FRP includes three basic steps: mixing fiber and resin, preforming the mixture, curing and demoulding. These three steps adopt different ways to form different FRP preparation processes. At present, there are many preparation processes of FRP, and with the development of technology, new auxiliary technology and production methods are emerging. The main FRP manufacturing processes include low pressure contact molding process, resin transfer molding process, pultrusion process, winding process, molding process, etc.

2.7.3 Application of FRP in Civil Engineering

(1) Columns

Concrete-filled FRP tubes (CFFTs) are obtained by filling the FRP tube with concrete to form a hybrid composite structure. The outer FRP tube not only provides the concrete core with a stay-in-place formwork during construction, but also provides hoop confinement which results in enhancement in concrete compressive strength. Moreover, FRP tubes help protect the concrete insulate from aqueous corrosion. It has been successfully used as waterfront or bridge piles, bridge girder, buried arch bridge structures, bridge pier frames. The composite piles were first used in the construction of the Route 40 Highway bridge over the Nottoway River in Virginia. The piles consisted of 625mm diameter concrete-filled glass fiber reinforced polymer (GFRP) circular tubes with 5.3mm wall thickness. The installation process was shown in Fig. 2-17. Fig.2-18 shows an example of CFFTs used for buried arch bridge structures. The bridge consists of multiple, parallel, plumb CFFTs arches that are erected and filled in the field. The pultruded FRP tubes were fabricated with an inner layer of E-glass fiber braid and two outer layers of carbon fiber braid. The concrete core was self-consolidating and non-shrinking.

Recently, Ozbakkaloglu presented a new form design of CFFTs to enhance the effectiveness of square and rectangular FRP tubes in confining concrete. Technique of corner strengthening and provision of an internal FRP panel was developed, as shown in Fig. 2-19. The experimental results indicated that the new form offered significantly improved performance relative to conventional CFFTs with similar material and geometric properties. Manalo et al. developed a new type of FRP jacket with an innovative, easy-fit and self-locking mechanical joining system, as shown in Fig. 2-20. The full scale experimental results showed that the use of a microfiber resin provided the highest capacity joint, and the embedment of an FRP rod in the locking key resulted in an FRP jacket with a better joint performance. The steel H-piles of East Lynn Lake Bridge in West Virginia, U.S.A

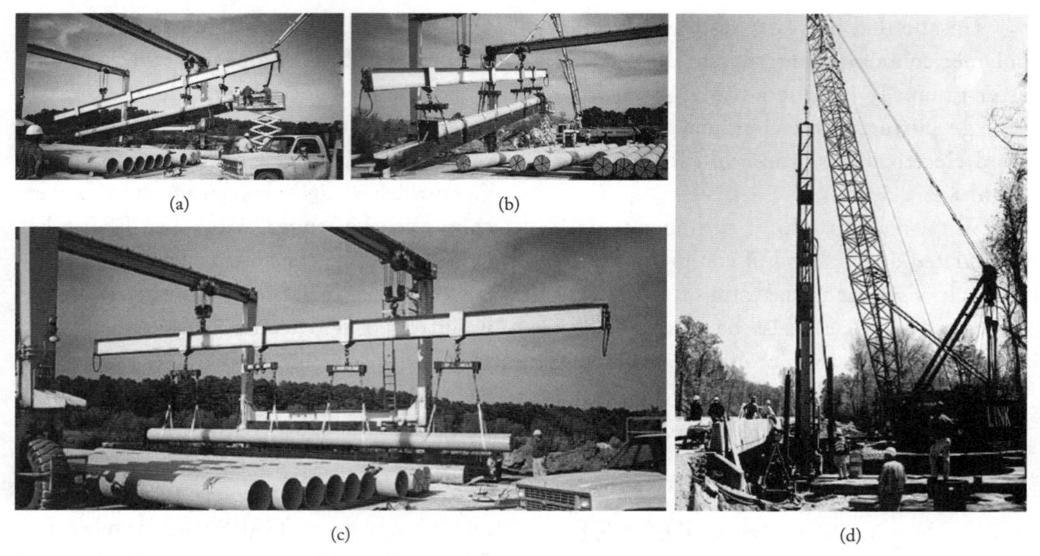

Fig. 2-17 The installation process of CFFTs piles of the Route 40 Highway bridge in Virginia
(a) Pumping concrete into the upper end; (b) Rear end showing wooden plugs; (c) Handling of the pile using eight-point supports; (d) Driving of composite test pile.
Source: FAM A, PANDO M, FILTZ G, et al . Precast composite piles for the Route 40 bridge in Virginia using concrete-filled FRP tubes. PCI Journal, 2003, 48(3):32–45.

Fig. 2-18 Buried CFFTs arch bridge in Bradley, Maine, USA
Source: DAGHER H J, BANNON D J, DAVIDS W G, et al.Bending behavior of concrete-filled tubular FRP arches for bridge structures. Construction and Building Materials, 2012, 37:432-439.

Fig. 2-19 A new form design of FRP tubes
Source: OZBAKKALOGLU T. Concrete-filled FRP tubes: manufacture and testing of new forms designed for improved performance. Journal of Composites for Construction, 2013, 17(2):280–291.

Fig. 2-20 An innovative joining system of FRP jacket
Source: MANALO A C, KARUNASENA W, SIRIMANNA C, et al. Investigation into fibre composites jacket with an innovative joining system. Construction and Building Materials, 2014, 65:270-281.

were deteriorated and the section loss was up to 60%. FRP jackets and self-consolidating concrete (SCC) were used to rehabilitated the steel piles by Constructed Facilities Center, Department of Civil and Environmental Engineering of West Virginia University, U.S.A. Circular FRP shells of 6.35mm thick×508mm diameter with tongue and groove joints were placed around the H-piles up to required heights. Additional fastening/bonding schemes were used to hold the shell in place around the piles and a two-layer circumferential GFRP wrapping over the shell. The bottom part of the FRP shell was grouted with up to 228.6mm height of epoxy grout (Fig. 2-21). The rest of the shell was filled with self-consolidating concrete and cement grout, which was pumped through the top port on the FRP shell. The design provides 3 layers of protection against further corrosion and provides pile strength over 2 times the original pile bent strength.

Concrete-filled steel tubes (CFSTs) are widely used as columns, arch ribs, piles and truss members in structural systems. Forty of these tubes were wrapped by GFRP with 3mm wall thickness, as shown in Fig. 2-22. In 2013, FRP confined hollow steel tubes were used as piles in the west and south central interchange of Taiyuan, China. The bridge crossed a chemical factory in which the groundwater was polluted seriously. It is necessary to apply an anti-corrosion jacket on the surface of these piles. Through a comprehensive comparison, the use of FRP jackets for the confinement of steel piles were adopted because it can provide a very effective means of protection of steel in corrosive environments. The outer diameter and wall thickness of steel piles were 1.7m and 10mm respectively. Finally, the

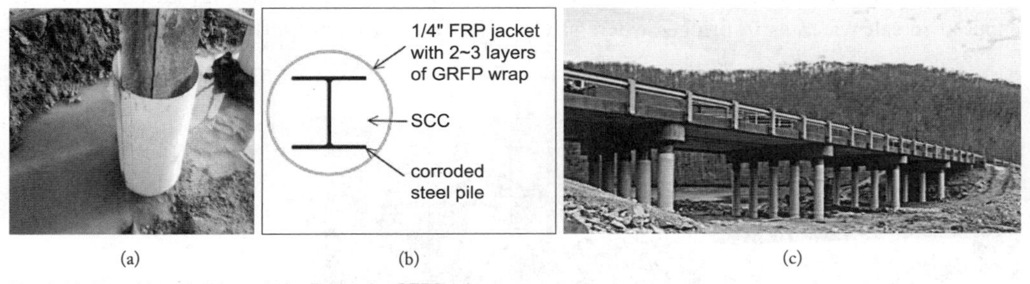

(a) (b) (c)

Fig. 2-21 Repairing East Lynn Lake Bridge by GFRP tubes
(a) Close-up of shell placed around corroded steel pile; (b) Cross-section of repaired piles; (c) The repaired bridge near completion
Source: LIANG R, SKIDMORE M, GANGARAO H V S. Rehabilitation of East Lynn Lake Bridge steel pile bents with composites. Proceeding: TRB Innovative Technologies for a Resilient Marine Transportation System, 3rd Biennial Research and Development Conference, June 24-26,2014, Washington, DC.

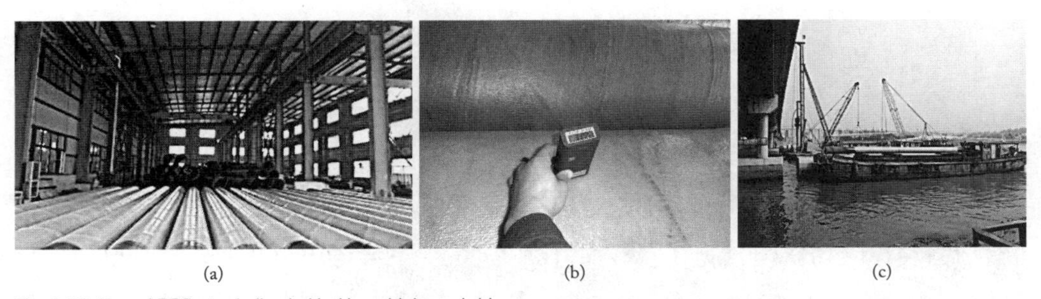

(a) (b) (c)

Fig. 2-22 Use of FRP-steel piles in Hu-Hang highway bridges
(a) Finished products; (b) Detecting the wall thickness of GFRP; (c) Driving the FRP-steel piles

outer surfaces of steel piles were wrapped with GFRP of 3mm wall thickness in the field, as shown in Fig. 2-23. There was no debonding of FRP and steel tube during the driving process.

(2) FRP rebars

Another outstanding example is the FRP rebar in lieu of the steel rebar in highly corrosive environments. Concrete is a material that is very strong in compression, but relatively weak in tension. To compensate for this imbalance, steel reinforcing bars are embedded into the concrete to carry the tensile loads. However, steel inherently corrodes (an electrochemical reaction) under salt exposure, leading to rusting. As rust takes up a greater volume than its parent material—steel rust causes severe internal pressure on the surrounding concrete, leading to cracking, spalling, and ultimately, concrete failure in tension due to rust-induced hoop stress. This is extremely serious when concrete is exposed to salt water, as in bridges where salt is applied to roadways in winter, or in marine applications.

FRP rebar appears to be the best solution to tackle this problem and offers a number of benefits to the construction of our nation's infrastructure, including bridges, highways and buildings. It is light weight (a quarter the weight of steel), strong (about twice the strength of steel), impervious to chloride ion and chemical attack, free of corrosion, transparent to magnetic fields and radio frequencies, and nonconductive for electrical and thermal loads. FRP rebars are commercially available on the market and they are mostly made from unidirectional glass fiber-reinforced thermosetting resins.

WVU-CFC researchers started FRP rebar application in 1986 for a hospital building, but it was after eight years of research when the first vehicular bridge, McKinleyville Bridge, was built in the U.S. to use FRP rebars in a concrete deck. McKinleyville Bridge is located in Brooke County, the northern panhandle

Fig. 2-23 Use of FRP-steel piles in the west and south central interchange of Taiyuan
(a) Steel piles; (b) Steel piles after polishing and sand washing; (c) Wrapped glass fiber clothes soaked with resin; (d) Driving the GFRP-steel piles

of West Virginia. It is a 180-feet long, three-span, continuous integral abutment bridge accommodating two lanes of traffic. The selection of constituent materials and the manufacturing processes for FRP rebars were given careful consideration. Screening of several types of resins and fibers under harsh environments was extensively researched at the WVU-CFC. The GFRP rebars were placed in the concrete deck as top and bottom layers in the transverse and longitudinal directions. McKinleyville Bridge deck with FRP rebars has been in service for 16 years.

With support from the sponsors and contractors, in 2007, WVU-CFC researchers completed the nation's first continuously reinforced concrete pavement (CRCP) test section with GFRP rebars, along with steel rebar CRCP test segment for comparison. These test segments are located on Route 9 in Martinsburg, in the northeastern corner of West Virginia and are being studied for their performance. Field studies show that GFRP rebar offers a low life-cycle cost option for reinforcement in concrete pavements. It is anticipated that FRP rebar reinforced pavements will offer many years of additional service life as compared to steel rebar reinforced pavements. There have been many other successful field implementations, in particular using GFRP rebars in bridge deck applications in West Virginia and many other states. WVU-CFC has been working with federal and state agencies and private industries for over 25 years in promoting and implementing FRP composite products on the US highway system, including construction or deck replacement of over 100 bridges with FRP bridge deck, FRP bridge superstructure, FRP reinforcing bar as well as FRP dowel bar for concrete pavements.

(3) Geosynthetics

The tailing dam needs to be strong but still allow for drainage without soil loss. The geosynthetic under consideration would be geocells that are made of strips and can be expanded into three-dimensional, stiff honeycombed cellular structures, resulting in a confinement system when infilled with compacted soil. The cellular confinement reduces the lateral movement of soil particles, thereby maintaining compaction, retaining the earth, and protecting the slope. Geogrids offer open, gridlike configurations and as reinforcement materials, play similar roles to those of geocells. In addition, geotextiles are flexible, porous, woven or knitted synthetic fibers/fabrics and can function as reinforcement, drainage, filtration or separation, depending on design, while geomembranes are thin, impervious sheets of polymeric material and function as containment. Geomembranes are being extensively used as linings of waste acid copper solution reservoirs in Zijin Mines.

(4) Sheet Piling

FRP composites in the form of panels, pipes and posts, in addition to pultruded standard shapes can find broad applications in every type of sea-waterfront facilities. These applications include: decking, walkways, platforms, ship-to-shore bridges, fenders, docking systems, retaining walls, crosswalks, moorings, cables, piles, piers, underwater pipes, railings, ladders, handrails and many others. For example, WVU-CFC has recently tested several composite sheet piles made by Creative Pultrusion Inc. , as an application to protect soil erosion near sea-front homes. These FRP products can survive under constant exposure to saltwater and salt air, and will not corrode, rust or create sparks.

2.8 Basic Members

A structure is the mechanical skeleton of a building, which is composed of various types of stressed units. These stressed units are called structural members. Different types of components are combined to get different structures. It can be said that members are the basic units of the structure to bear the load. There are various forms of engineering structures, but the types of basic structural members are not too many.

2.8.1 Linear Members

A linear member is a component type whose size in one direction is significantly larger than that in the other two directions. This kind of member can be simplified to linear mechanical model for mechanical calculation when we carry out force analysis of a member, so it is usually called linear member. The direction with the largest dimension is called the axial direction of the member, and the section perpendicular to the axis is called the section of the member. Typical linear members include column, beam, pole, pier and pile.

(1) Column

Column is a kind of linear member which is vertically arranged and bears vertical load directly. It is most common in all kinds of building structures, such as residential buildings, hospitals, stadiums, workshops and so on. Columns are mainly subjected to pressure and bending moment. The types of columns can be classified according to materials, such as steel columns, reinforced concrete columns, wooden columns, masonry columns, composite columns, etc. It can also be divided according to the slenderness ratio of columns, such as short columns, long columns, medium-length columns, etc. Columns with different slenderness ratios have different mechanical properties. According to the different section forms, it can be divided into rectangular section column, circular section column, I-shaped section column, etc. According to the size of the section, it can be divided into ordinary column and giant column.

(2) Beam

Beam is a kind of linear member which mainly bears the load vertically distributed along its axis. In addition to being widely used in various building structures, beams are also widely used in bridge engineering, road engineering, water conservancy engineering and port ocean engineering. The beam is mainly subjected to bending moment and shear force, and is usually not subjected to axial force. The ratio of span to section height of a beam, called span-height ratio, is an important parameter affecting the mechanical performance of the beam. When the span-height ratio of the beam is small, it is called deep beam. The deep beam needs to be analyzed according to the planar member when carrying out the stress analysis, and it is no longer a linear component.

(3) Rod

A rod is usually a linear member that only bears axial force. The member bearing only the axial compression is called the compression rod, and the member bearing only the axial tension is called the tension rod. The compression and tension rods are not subjected to bending moment and shearing force. They are often used to form a truss or grid structure (Fig. 2-24), and their characteristics are that the connection joints of rod members are hinged joints.

2.8.2 Planar Member

The members with larger plane size but smaller thickness are called planar members. The basic types include slabs and walls.

(1) slab

The slab is usually set in the horizontal

Fig. 2-24 Grid structure
Source: http://www.060ss.com/show.php?cid=20&id=96.

direction and bears the load perpendicular to the slab surface, such as floor slabs, bridge decks, various platform slabs, etc. The internal forces of the slab are mainly bending moment and shearing force. The stress of the plate is closely related to the boundary conditions of the plate. Taking the rectangular plate as an example, the common boundary conditions are four sides fixed support, four sides simply supported, opposite sides simply supported and the other opposite sides free side, one side fixed support and the other three sides free edge. In addition, according to the force transmission characteristics of the plate, it can be divided into one-way plate and two-way plate. When the load on the slab is transferred along one opposite direction of the slab, it is a one-way slab. If the load on the slab transfers the load along two opposite directions of the slab, it is a two-way slab.

(2) Wall

The wall generally is the vertical planar load-bearing member. According to the mechanical characteristics and application situation, the wall can be divided into bearing wall, shear wall, underground diaphragm wall and retaining wall. Bearing walls and shear walls are generally used in building structures. Generally, bearing walls only bear vertical loads, which are self weight loads and vertical live loads. The shear wall bears not only the vertical load, but also the horizontal load of the structure, such as the horizontal shearing force caused by wind load and earthquake. Diaphragm wall and retaining wall are usually used in foundation pit engineering and slope location, mainly bearing the action of horizontal earth pressure perpendicular to the wall (Fig. 2-25).

2.8.3 Curved Members

Curved members include arches and shells.

(1) Arch

When a beam is subjected to a load perpendicular to the member, a large bending

(a)

(b)

Fig. 2-25 Diaphragm wall and retaining wall
(a) Diaphragm wall; (b) retaining wall
Source: (a) http://biz.co188.com; (b) https://baike.baidu.com.

moment will be generated inside it. Generally, we call this kind of member like the beam bending member or flexural member. When the beam is bent, the magnitude and direction of the force on each part of the beam section are not the same, and even the tensile stress will appear on one side of the cross-section, so the material utilization efficiency of flexural members is relatively low. If the straight beam is changed into a curved member, the internal force distribution of the member can be changed and the utilization efficiency of materials can be improved. This kind of curved structural member is called arch.

When an arch is subjected to a vertical load in its own plane, there will be an outward horizontal force on the arch foot. The arch, together with its supports on arch foot or tension rod between arch feet, forms an arch structure. Due to the horizontal force in the arch foot, the arch sections are mainly compressed. Compared with the beams with the same span, the bending moment and shearing force in the arch are much smaller, so the materials can be saved, the stiffness can be improved, the span can be enlarged, and the cheap materials with good compression performance and poor tensile performance such as brick, stone, block and concrete can be effectively used. Steel arch and reinforced concrete arch can span a large space. Arch structure can be single-span or multi-span, they have been used in bridges, roofs and tunnel linings. The Zhaozhou Bridge built in Sui Dynasty in China is an example. It is the oldest, largest and most complete single-span stone arch bridge in the world.

(2) Shell

A shell is a kind of structural member with the spatial curved configuration. Similar to the arch, the internal force of the shell is mainly compression force. Compared with the plates with the same span, the bending moment and shearing force in the shell are much smaller. The shell structure has good space transmission performance. It has high bearing capacity and high rigidity with small thickness of the member. It can cover or maintain large-span space without pillars. It can also play dual roles of bearing structure and enclosing structure, so as to save structural materials.

2.8.4 Solid Members

The solid member is the similar size in three dimensions. The solid member is usually in three-dimensional stress state when it is subjected to loads, so it is difficult to analyze the mechanical behaviors. At the same time, the mechanical properties of structural materials under three-dimensional stress states are different from those under ordinary one-way stress state. This also makes the mechanical analysis of solid members more complex.

The common solid member is gravity dams. Gravity dam (Fig. 2-26) is a kind of dam. Under the action of water pressure and other external loads, it mainly depends on the self weight of dam body to maintain its stability. In addition, there is a typical solid member, which is the joint of beam and column in reinforced concrete frame structure. In the mechanical analysis of joints, the joint area should be regarded as a solid component.

2.8.5 Flexible Members

A kind of material which can only bear

Fig. 2-26 Gravity dam
Source: http://www.zbcc168.com/nd.jsp?id=12.

tension load, but not compression or shear load, like rope or membrane, is called flexible material. The members formed by such materials are called flexible members, which mainly include cable and membrane. Cables usually include steel cables, composite cables, etc., which can form various suspension structures and tension structures, such as cable-stayed bridges, suspension bridges, cable net roofs, etc. The structure composed of membrane is called membrane structure, including inflatable membrane structure and tension membrane structure.

2.9 Conclusions

This chapter mainly introduces the common civil engineering materials and basic component types. They are the basic elements of all kinds of buildings and structures.

(1) Concrete material which is a kind of cheap local materials, is the most widely used material in engineering at present with excellent comprehensive performance, good mechanical properties and excellent durability, and is irreplaceable by other materials.

(2) Steel has good strength and has almost the same mechanical properties in tension and compression. This is something that concrete materials do not have. Therefore, steel can be made into a variety of thin-walled components to use, so as to reduce the weight of the structure. It has incomparable advantages.

(3) The application range of wood and masonry materials is smaller than that of concrete and steel. This is limited by the material itself. However, under the premise of fully understanding the mechanical properties of materials, combined with structural innovation, beautiful and durable structures can also be built.

(4) The application time of composite materials in civil engineering field is relatively late. Because of its light weight, high strength and corrosion resistance, it has a good application possibility under the special engineering requirements.

(5) Structural members are the basic units for forming structure. Through the flexible combination of these basic members, rich and colorful structures can be formed.

Exercises

2-1 What is the position of materials in civil engineering? How can civil engineering materials be classified according to their properties?

2-2 Why do you think Roman mortar was lost? What are the similarities and differences between it and the modern concrete?

2-3 What are the characteristics of plain concrete? Why should reinforced bar be installed in plain concrete?

2-4 What is prestressed concrete? What are its advantages over concrete?

2-5 What are the characteristics of masonry? Do you think masonry can form large span structure?

2-6 What fields do you think composite materials can be used in civil engineering?

2-7 What are the basic components of the structure? What are the basic characteristics of mechanical properties?

References

[1] LUO F W, LIU W Q. Introduction to Civil Engineering (4th Edition)[M]. Wuhan: Wuhan University of Technology Press, 2012. (In Chinese)

[2] GAO L R, FANG E H, QIAN J R. Conceptual Design of High-rise Building Structure[M]. Beijing: China Planning Press, 2005. (In Chinese)

[3] QIAN J R. Structural Design of High Rise Buildings (3rd Edition)[M]. Beijing: China Architecture & Building Press, 2018. (In Chinese)

[4] XIONG F. Introduction to Civil Engineering (New 1st Edition)[M]. Wuhan: Wuhan University of Technology Press, 2015. (In Chinese)

[5] YI C, SHEN S Z. Introduction to Civil Engineering (3rd Edition)[M]. Beijing: China Architecture & Building Press, 2017. (In Chinese)

[6] Cement[EB/OL]. [2020-09-22]. https://baike.baidu.com/item/%E6%B0%B4%E6%B3%A5/66778?fr=aladdin.

[7] WANG H. Construction Technology of Super High-rise Steel Structure (2nd Edition) [M]. Beijing: China Architecture & Building Press, 2020. (In Chinese)

[8] FENG P, LU X Z, YE L P. Application of Fiber Reinforced Polymer in Construction: Experiment, Theory and Methodology [M]. Beijing: China Architecture & Building Press, 2011. (In Chinese)

[9] YE L P. Concrete Structures [M]. Beijing: Tsinghua University Press, 2005. (In Chinese)

[10] Fiber reinforced concrete[EB/OL]. [2020-09-21]. https://baike.baidu.com/item/%E7%BA%A4%E7%BB%B4%E6%B7%B7%E5%87%9D%E5%9C%9F/7440675?fr=aladdin.

[11] Cement[EB/OL]. [2020-09-21]. https://baike.baidu.com/item/%E7%BA%A4%E7%BB%B4%E6%B7%B7%E5%87%9D%E5%9C%9F/7440675?fr=aladdin.

[12] Concrete[EB/OL]. [2020-09-21]. https://encyclopedia.thefreedictionary.com/Concrete.

Chapter 3
Building Engineering

3.1 Overview

Building engineering, also known as architectural engineering, is the application of engineering principles and technology to building design and construction. The building is the direct object and ultimate goal of this work.

A building is a structure with a roof and walls standing more or less permanently in one place, such as a house or factory. Buildings come in a variety of sizes, shapes and functions, and have been adapted throughout history for a wide number of factors, from building materials available, to weather conditions, to land prices, ground conditions, specific uses and aesthetic reasons.

Structures other than building structures are called nonbuilding structures. A non-building structure, also referred to simply as a structure, refers to any body or system of connected parts used to support a load that was not designed for continuous human occupancy. The term is used by architects, structural engineers and mechanical engineers to distinctly identify built structures that are not buildings, such as bridges, towers, chimneys, tunnels, pools, etc.

According to the role of each component, the building can be divided into four systems: structural system, envelope system, facility and pipeline system, and interior decoration system. A good building is the organic integration of these four systems, so as to provide people with complete functions and excellent performance buildings.

3.2 Typical Building Types

3.2.1 Classification of Buildings

Buildings can be classified according to the historical period of construction, use function, construction scale, main construction materials of structural system and building height.

(1) Classification by Historical Period

According to the historical construction period of building, the buildings can be divided into ancient buildings, modern buildings and contemporary buildings. In different historical periods and regions, the buildings present different development stages and architectural styles.

For example, western ancient architecture can be divided into ancient Greek architecture (Fig. 3-1), ancient Roman architecture, Byzantine architecture, Western European medieval architecture, Baroque architecture (Fig. 3-2), classical architecture and classical Renaissance architecture.

According to the change of historical dynasties, ancient Chinese architecture can be divided into pre-Qin buildings, Qin Han buildings , Sui Tang and Song buildings (Fig. 3-3), Yuan Ming and Qing buildings (Fig. 3-4).

Fig. 3-1 Greek architecture
Source: https://www.sohu.com.

(2) Classification by Function

According to the functional classification, buildings can be divided into civil buildings, industrial buildings and agricultural buildings.

Among them, civil buildings are the general name of buildings for people to live in and carry out public activities, and civil buildings are divided into residential buildings and public buildings according to their use functions. Residential buildings are buildings for people to live in, mainly including dwelling houses and dormitory buildings. Public buildings are buildings for people to carry out various public activities, which play an important role in people's daily life, such as educational buildings, office buildings, etc.

The buildings with industrial production as the main function are industrial buildings (Fig. 3-5), such as production workshop, auxiliary workshop, power house, storage building, etc. The buildings with agricultural production as the main function are agricultural buildings, such as greenhouse (Fig. 3-6), livestock farm, grain and feed processing station, agricultural machinery repair station, etc.

(3) Classification by Project Scale

It can be divided into large quantity buildings and large scale buildings.

Large quantity buildings are mainly those buildings with large quantity and wide range and closely related to people's life, such as

Fig. 3-2 Baroque architecture
Source: http://www.ela.cn/index.php?g=portal&m=index&a=show&id=2034.

Fig. 3-3 Tang building
Source: https://baike.baidu.com.

Fig. 3-4 Qing building
Source: https://baike.baidu.com.

Fig. 3-5 Industrial building
Source: https://su.99cfw.com.

Fig. 3-6 Greenhouse
Source: www.sg560.com.

houses, schools, shops, hospitals, small and medium-sized office buildings, etc.

Large scale buildings are mainly large-scale, expensive and influential buildings. Compared with the large quantity buildings, the number of them is limited. However, these buildings are representative in a country or a region and have a great impact on the appearance of the city, such as large railway stations, air stations, large gymnasiums, museums, city halls, etc.

(4) Classification by Main Construction Materials

The structural system of the building can be constructed with different structural materials, and the corresponding buildings can be divided into concrete structure building, steel structure building, masonry structure building, wood structure building and mixed structure building.

(5) Classification by Building Height

According to the height and number of floors, buildings can be divided into single storey building, multi-storey building, high-rise building and super high-rise building.

3.2.2 Concrete Structure Buildings

Concrete structure is a kind of modern structure type with the invention of Portland cement, which has a history of more than 100 years. Concrete structure includes plain concrete structure, reinforced concrete structure and prestressed concrete structure. The so-called plain concrete structure is the concrete structure without reinforcement or without load-bearing reinforcement, and reinforced concrete structure refers to the concrete structure with ordinary load-bearing reinforcement. Prestressed concrete structure is a kind of concrete structure which is equipped with prestressed reinforcement and prestressed by tension or other methods.

Reinforced concrete structure is widely used in industrial buildings and civil buildings, with the largest amount of application and the widest range of application. It is the main form of building structure at present. Therefore, the common concrete structure usually refers to reinforced concrete structure. The main types of reinforced concrete structures include bent structure, frame structure, shear wall structure, tube structure, etc.

(1) Frame Structure

Frame structure is composed of many beams and columns to bear the full load of the building (Fig. 3-7). The horizontal member is called frame beam, the vertical member is called frame column, and the common area of frame beam and frame column is called frame joint.

The frame structure should bear not only the vertical load from the floor and roof, but also the horizontal wind force and earthquake action. Reinforced concrete frame structure is usually constructed by integral cast-in-place method, which has good integrity and resistance to deformation. Frame structure is a kind of structure widely used in housing construction.

(2) Bent Structure

The concrete bent structure can form a large building space, which is mostly used in single storey industrial workshops. Its structural system is composed of bent column, roof truss or roof beam, foundation and various supports. Among them, the bent column is

Fig. 3-7 Frame structure
Source: https://baike.baidu.com

Fig. 3-8 Bent structure
Source: http://jd.crec4.com/content-3175-2712-1.html.

generally prefabricated reinforced concrete member, and the roof truss or roof beam is usually precast prestressed concrete member. If roof slab is large, precast prestressed concrete slab is also used. The foundation is cast-in-place reinforced concrete cup foundation. The bent structure of single storey workshop under construction is shown in Fig. 3-8.

(3) Shear Wall Structure

Reinforced concrete wall not only has great vertical compression bearing capacity, but also has great horizontal shear bearing capacity. In the structure, it can effectively bear the horizontal load. This kind of wall is called reinforced concrete shear wall, which is called shear wall for short. A structure which uses shear wall to bear vertical load, horizontal wind load and horizontal earthquake action is called shear wall structure.

Shear wall has the dual functions of building and structure. It is not only a load-bearing component, but also a separation and maintenance component. Shear wall is also called seismic wall because of its strong spatial integrity and small horizontal deformation. Shear wall structure is widely used in high-rise buildings with more than ten to thirty storeys. In non-seismic design, the height that can be built is 130~150m.

(4) Tube Structure

Tube structure is a kind of spatial tubular structure. Compared with shear wall structure, its integrity is stronger and its ability to resist horizontal load is greater. It is very suitable for the construction of super high-rise buildings. There are three forms of tube: one is the solid web tube surrounded by shear wall, the second is the frame tube surrounded by dense column and deep beam, and the third is the truss tube surrounded by truss. From the structural layout plan, the tube is generally arranged in the center of the structural plane, so it is called the core tube.

According to the combination form of tube and other structural components, concrete tube structure can be divided into the following types: frame-core tube system composed of frame and solid web tube, tube-in-tube system composed of solid web tube and frame-tube, bundle tube system composed of frame-tube and/or truss-tube.

3.2.3 Steel Structure Building

Metal materials are relatively rare in ancient times. Building houses with metal materials is not only complex in technology and difficult in construction, but also very expensive in cost. Therefore, metal buildings have been unable to be popularized and applied. Located at the top of Tianzhu Peak in Wudang Mountain, Hubei Province, Wudang

Mountain Golden Hall is a representative ancient Chinese metal building. It is also the largest and highest grade bronze gilded hall in China. It was built in 1416, the 14th year of Yongle in Ming Dynasty. The building is 5.54 meters high and has an area of about 160 square meters. All the components are cast separately, installed by tenon riveting, and then gilded all over the building. The structure is rigorous, the connection is tight, and there is no mark of casting chisel. It represents the very high level of metal buildings at that time.

Steel building makes metal building move towards large-scale application. The beginning of modern structural engineering can be attributed to the fact that steel materials have become the main material of building structures for the first time, which has changed the appearance of masonry as the main structural material in the western world for a long time.

(1) Crystal Palace and Cast Iron Structure

The construction of Crystal Palace and cast iron structure is a milestone event in the world structure engineering field.

The Crystal Palace was built for the first World Expo (Fig. 3-9). It was built in 1851 and has been a great success at the London World Expo. The building is located in Hyde Park, London. It is a building with steel as the skeleton and glass as the main building material. It is one of the wonders of British architecture in the 19th century. The building is about 124m in width, 564m in length, 3 floors in height, and has a construction area of 74,000m^2. A total of 3300 iron columns, 2300 iron beams and 93,000 square meters of glass were used. It took only nine months to complete the construction. It was the largest single building in the world at that time. Its designer, Joseph Paxton, was knighted by the queen.

Fig. 3-9 Crystal Palace
Source: https://baike.sogou.com.

Subsequently, the steel truss structure also began to be used in housing construction. In the early stage, the compression member of roof truss was made of cast iron, and the tension member was made of forged iron. Later, the plane truss of forged iron appeared in the industrial plant. The span of such house can reach 30~40m.

After the mid-19th century, cast iron structure has been widely used in the United States as a new industrial country. There are many cast iron warehouses and workshops along the Mississippi River. By 1870s, Chicago School and high-rise iron frame structure were also born in the United States. The column of iron frame is made of round pipe with cast iron or square pipe with forged iron. The beams are generally made of forged iron. These members are connected by angle steel, bolts and rivets. The frame itself can bear the load, and the external wall has no bearing function. The iron frame is easy to build, and stone or concrete members can be installed outside easily. The bearing capacity of the iron frame can ensure that the number of floors of the building can reach 10 or even higher.

(2) Development of Steel Structure Joint

In the 20th century, great changes have taken place in the connection methods of steel structures. The original steel structure is mainly connected by anchor bolt. After the end of World War Ⅱ, the quality of steel welding connection has been significantly improved, and the high-strength bolt has also replaced the ordinary bolt (Fig. 3-10). Because welding connection and high-strength bolt connection have lower requirements on machining accuracy of steel components than rivet connection, and their bearing capacity is much higher than ordinary bolt connection, welding connection and high-strength bolt connection become the main connection method of steel structure, which leads to the joint of steel structure more lightweight. Great changes have taken place in the joint pattern of steel structures, and the change of joint pattern also affects the form of steel structure.

For large-span spatial structures such as grid structures and latticed shells, there are many members intersecting at the joints, and the stress state of the joint is complex, so the welded spherical joints are generally used. This kind of joint is welded and assembled at the construction site, which requires high requirements for the positioning and welding process of the member. Later, in order to avoid the trouble of welding, the high-strength bolted-spherical joint has developed for the grid structure with regular grid size (Fig. 3-10). For the super long-span reticulated shells, cast steel spherical joints and hub joints are developed. These joints are completely manufactured in the factory and can be directly installed on site to form a structure. In 1990s, with the extensive application of computer technology in civil engineering field, the accurate cutting and processing of steel pipe intersection nodes can be achieved, thus realizing the industrial manufacture of intersecting joints, and promoting the appearance and engineering application of pipe truss structures.

(3) Light Steel Structure

Modern steel structure can be divided into ordinary steel structure and light steel structure according to the section size of steel element.

(a) (b) (c) (d)

Fig. 3-10 Joints of steel structure
(a) Welding connection; (b) Bolt connection; (c) High-strength bolted-spherical joint; (d) Intersecting joints
Source: (a)、(b) https://kj.ggditu.com/Blog/Post.aspx? Post ID=255.
(c)、(d) https://www.sohu.com.

The traditional steel structure uses the common shaped steel and steel plate to make beam columns and other main structural components. The thickness of the steel plate is larger, the bearing capacity of the structure is also large, and the structure is relatively bulky. Such traditional steel structure is now also known as ordinary steel structure. With the development of steel structure design theory, steel structure materials can be made very thin in the case of small load. Small cross-section and variable cross-section members can be used to significantly reduce the self weight of the steel structure. This kind of steel structure is called light steel structure. The light steel structure not only reduces the steel consumption, but also lightens the burden of the foundation, thus the cost of the building is greatly reduced, reflecting the advantages of light steel structure.

(4) Main Advantages and Disadvantages of Steel Structure

The advantages of the steel structure are uniform material, high strength, small cross-section size, light weight, good plasticity and ductility, good weldability, simple manufacturing process, easy mechanized construction, pollution-free and renewable. Steel structure is not only suitable for single storey workshops, but also has great advantages in multi-storey and high-rise buildings, especially in super high-rise buildings with height over 100m. At the same time, the span capacity of steel structure is far greater than that of reinforced concrete structure, so it has important application value and application advantage in long-span structures.

However, the shortcomings of steel structure can not be ignored. They mainly include easy corrosion, high maintenance cost, poor weather resistance and higher cost than reinforced concrete structure. The collapse of the World Trade Center in New York caused a lot of casualties. From the perspective of engineering application, this is the tragedy caused by the poor fire resistance of steel structure. When the terrorist hijacked the plane and hit the World Trade Center in New York, because the steel structure was not fire-resistant, the steel softened rapidly when it reached a certain temperature, and two 110-storey steel tower buildings collapsed seriously. Because of the short time from the starting of fire to the collapse of the structure and the large height of the structure, most of the people in the building did not have time to escape. Eventually, 3021 people were killed and 6291 were injured. Only three survivors were rescued after the building collapsed. The losses were quite heavy.

3.2.4 Masonry Structure Building

A structure which is made of brick, stone, concrete block, and other block materials and mortar is called masonry structure.

Masonry structure is widely used in residential buildings and general civil buildings, such as office buildings, shops, hotels, restaurants, hospitals and teaching buildings, among which brick masonry is the most widely used. The construction of masonry structure mainly depends on manual operation. In recent years, there are also robots that can be used for bricklaying.

(1) Ancient Masonry Structure

Masonry structure is an early form of building structure in human history.

In China, "Qin brick and Han tile" opened the glorious history of Chinese ancient masonry structure. Kaiyuan Temple Pagoda in Zhengding County, Hebei Province, was built in 1055. The tower is octagonal in plane and 9.8m in length at the bottom. It adopts brick double-layer cylinder structure system, with a total of 11 floors and a total height of 84.2m. It is the highest masonry structure building in the world at that time. In addition, the Beamless Hall of Linggu Temple in Nanjing, which was built in the early 14th century, is a brick arch structure with a total length of 53.5m and a total width of 37.35m. It is a brick masonry system in both directions in plan, and there is no beam, so it is called Beamless Hall (Fig. 3-11).

In the west, the masonry structure is also a very mature structural form. The Pantheon and the Cathedral of Saint Sophia are the representatives. Pantheon, located in Rome, Italy, is the representative work of the early dome technology in ancient Rome (Fig. 3-12). The span of the dome is 43.5m, and the net height of the dome is also 43.5m. There is a circular daylight opening with a diameter of 9m in the center of the dome, and there is a columnar porch in the front. It is the earliest and at that time the largest large-span masonry structure in the history of architecture. Its span was not surpassed until the end of 19th century. The Pantheon was originally a temple, but changed into a Christian Church after the 7th century. Cathedral of Saint Sophia, built between 532 and 537, is located in Istanbul, Turkey, with a length of 77m in the east-

Fig. 3-11 Beamless Hall in Nanjing, China
Source: http://www.naic.org.cn/html/2017/gjjg_0814/17058.html.

Fig. 3-12 Pantheon
Source: https://baike.baidu.com.

west direction and 71.7m in the north-south direction. There is a dome with a diameter of 32.6m in the middle, all of which are made of bricks. The dome is more than 50m high. The church was originally the court church of the Orthodox Church of Byzantine Empire. After the 15th century, it was built around a light tower and changed into a mosque. In 1935, it was changed into a museum.

(2) Modern Masonry Structure

The mechanical performance of masonry has the following two main characteristics. One is that the compressive strength of masonry is far less than the original compressive strength of block due to the lower strength of mortar than that of block and the uneven paving of mortar joint. However, the more regular the block shape is, the higher the mortar strength will be; the thinner and more uniform the mortar joint is, the higher the strength of masonry will be. Another characteristic is that the composite structure material is made of blocks bonded by mortar, so it has the characteristics of good compressive strength and low tensile strength. Therefore, when it is not reinforced, it is only suitable for axial compression members and eccentrically compressed members with a small eccentricity.

Based on this experience, the masonry structure played an important role in the walls and towers of ancient buildings, especially the arch and dome structures mentioned above. The masonry structure suitable for manual construction is still used in the wall and foundation of a large number of small and medium-sized public and civil buildings because of its economic, thermal insulation and easy construction advantages.

At the same time, in order to reduce the damage to farmland caused by digging clay for bricks, the production of various concrete blocks, such as light concrete blocks, sand free macroporous concrete blocks and ordinary concrete blocks, and silicate blocks are also gradually developed.

According to the different block materials, modern masonry structure can be divided into brick masonry, block masonry, stone masonry, reinforced masonry and other masonry structures. Among them, reinforced masonry is a kind of composite masonry in which the reinforcement mesh is arranged in the horizontal mortar joint of masonry or vertical coarse reinforcement is set in the reserved groove outside the masonry and pouring the fine aggregate concrete (or cement mortar) in it. This kind of masonry can improve the strength, reduce the cross-section size of members, strengthen the integrity, increase the ductility of the structure, thus improving the seismic capacity of the structure. In addition, there is a kind of hollow wall built by solid bricks, which is called cavity wall structure. It can save materials, reduce weight and improve the heat insulation performance of the wall. However, the overall stability of the cavity wall is poor, so it is not suitable to build a house in the vibration, damp environment and in the area with larger earthquake intensity.

The main advantages of masonry structure are as follows: It is easy to obtain local materials, and save cement, steel and wood; it is with low cost, good fire resistance and durability, and good thermal insulation performance. The main disadvantages are low strength, heavy weight, heavy masonry work and poor seismic performance, which limit its utility. In the future, masonry products should be developed in the direction of high strength, porous, thin wall, large block and reinforcement.

(3) Eladio Dieste, the Great Master of Brick Structure

Dieste is a Uruguayan architect who was born in 1917 in the northern city of Artigas, near Brazil. At the age of 19, he entered the University of the Republic in Montevideo, the capital, and studied in the school of engineering. After graduation, he worked in the highway management department of the Ministry of Public Affairs of Uruguay and participated in bridge construction. In 1956, Dieste and his university alumni Montanez founded "Dieste and Montanez Office". All of Dieste's works are made of reinforced bricks. He is a well deserved master in this field. The two most important structural forms in Dieste's works are self-supporting shell tube and Gauss arch.

Dieste's self-supporting shell tube does not rely on any conventional shell bearing system (such as buttresses, flying buttresses, continuous longitudinal walls, etc.). Only a few rows of columns, sometimes even single column can ensure its stability.

The Church of Christ the Worker in Atlantida, built in 1960, is the representative work of Dieste's self-supporting shell tube (Fig. 3-13). The wall and roof of the church adopt reinforced brick structure, and the construction of the wall does not need the help of formwork. The horizontal stiffness of the wall is strengthened by undulating waves. The top of the ruled curved wall is a reinforced concrete edge beam, which helps transfer the load from the roof to the wall. The two ends of the steel connecting rod embedded in the trough of the roof are fixed on the side beam. The curved brick walls on both sides and the curved roof form a stable structure. The steel connecting rod embedded in the roof can absorb a part of the horizontal force. The shape of the transverse section of the church is consistent with the bending moment diagram of the portal steel frame under its self weight, thus giving full play to the mechanical properties of the material. This original form is purely based on the mechanical characteristics of the structure, reflecting the way of thinking of structural artists. This kind of structure designed by Dieste has a maximum horizontal span of 12.6m, a maximum longitudinal span of 32m and a maximum overhang of 16.2m. Even today, no matter from the concept of structure or architectural art, the large-span structure of these bricks is still impressive.

Gauss arch is a kind of arch with special curved surface. The section obtained by cutting the Gauss arch along its span direction is a catenary. When the section plane moves perpendicular to the span direction, the rise height of the catenary will change, showing an oblique "s" shape. Its undulating wave crest appears in the middle of the span which is prone to instability, and the wave shape at the midpoint strengthens the shell stiffness and avoids the instability of the structure. At the two ends of the span, the curved surface is completely flattened and easy to connect with the walls on both sides. At the same time, the large-span and low rise Gaussian surface will produce strong horizontal force, so it is necessary to arrange tie rods to balance the force. Because a group of shell units is connected end to end, there is a long and narrow crescent shaped space between the high and low ends of the adjacent "s" shape. It's

(a)

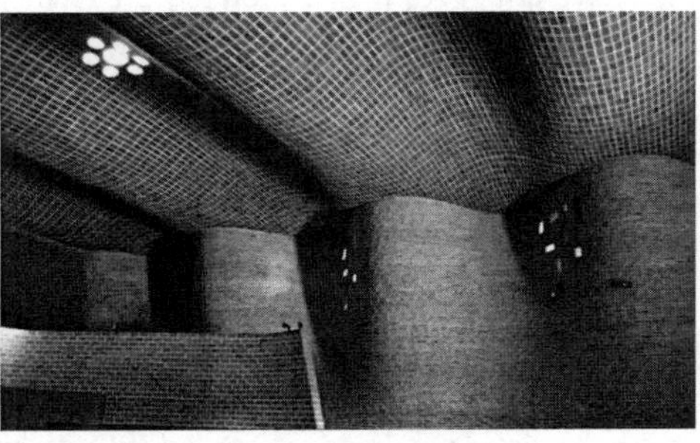

(b)

Fig. 3-13 The Church of Christ the Worker in Atlantida
Source: https://zhuanlan.zhihu.com/p/39274111.

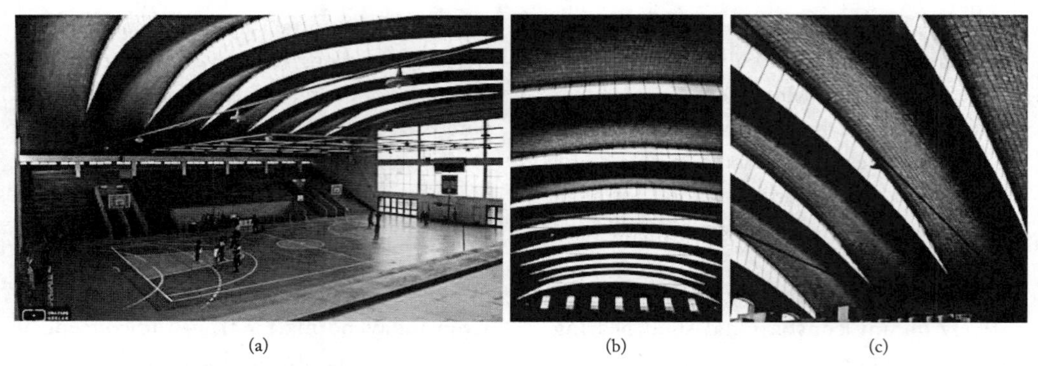

(a) (b) (c)

Fig. 3-14 Don Bosco School Gymnasium
Source: https://zhuanlan.zhihu.com/p/39274111.

also an ideal natural lighting surface, so Dieste arranged a glass skylight in this long, narrow crescent. It makes the whole roof appear transparent and light, and the natural light sprinkles on the masonry to form a natural and original halo, which adds the crowning touch to the roof (Fig. 3-14).

3.2.5 Wood Structure Building

Wood structure is a kind of structure constructed with wood as the main structural material. It is an ancient architectural structure. Especially in ancient China, wood structure architecture was widely used and achieved high artistic achievements. There are many representative ancient wooden structures in Europe. Since entering modern times, with the development of modern construction theory and technology, new forms of wood structure buildings have emerged, which radiate new vitality.

(1) Ancient Chinese Wooden Structure Architecture

In China, according to archaeological discoveries, wooden beams and columns and mortise-tenon joints have appeared in houses 6000 years ago. Because wood is easy to decay, easy to be eaten by insects, easy to catch fire, a large number of ancient wooden structure buildings have not been preserved. However, the existing ancient wooden structures are still more common throughout China. Among them, the main hall of Foguang Temple in Wutai Mountain of Shanxi Province, the wooden tower of Yingxian County in Shanxi Province (Fig. 3-15), and the architectural complex of the Forbidden City in Beijing are the representatives.

In ancient Chinese wooden structure buildings, wooden beams and columns were used as load-bearing framework, and the joints between columns and beams were mostly mortise-tenon joints. It is designed and constructed by skillful craftsmen, and made

Fig. 3-15 Wooden tower of Yingxian County
Source: https://baijiahao.baidu.com.

of bricks and stones, covered with tiles and painted with oil. It has a high value of cultural relics and ornamental value.The ancient wood structure system is generally divided into three types: Tailiang style structure, Chuandou style structure and Jinggan style structure, among which the Tailiang style structure is widely used. Tailiang style structure is to erect the beam on the column, and another beam is superimposed on the beam. This structure is often used in palaces, temples and other large buildings. Chuandou style structure is to connect rows of columns with beams to form a bent, and then connect them with beams in another direction. It is mainly used for residential buildings and small buildings. The Jinggan style structure is made of cross stacked wood. It is named because the space it surrounds is like a well. This structure is relatively primitive and simple, and is rarely used except in a few forest areas.

In the aspect of detail production of wood structure, dry wood is used to make the structure, and the key parts of the structure are exposed to the air, which can prevent moisture and decay. The foundation stone is set under the wooden column to prevent the wood column from contacting with the ground and prevent termites from climbing up and damaging the structure.

In Tang Dynasty, there was a set of strict manufacturing methods for wooden structure buildings in China. However, the "Yingzao Fashi" edited by Li Jie of the Northern Song Dynasty is the first code for the design, construction, materials and quantity quota of wood structure buildings in China and the world. Among them, according to the design of the house, the cross-sections of wood members are divided into eight types, and different sections are selected according to the span size of the house. According to the modern mechanical principle, the relationship between the cross-section and the span of the wood member conforms to the principle of equal strength, which indicates that the material can be selected according to the proportional relationship in Song Dynasty, reflecting the strength design concept of bending beam.

(2) Western Ancient Wooden Structure Architecture

The history of Western wooden architecture can be traced back to 7000 years ago. The wooden tenon mortise structure of the Neolithic Age was unearthed in Germany. Similarly, due to that the wood structure is difficult to preserve, many wooden structure buildings can not be preserved until now. The representative examples of existing wooden structures are as follows.

1) Westminster Hall

The building was built in 1097 (Fig. 3-16). It is the largest building with a clear span of 73.2m in length and 20.7m in span in Medieval England. Its ceiling is not supported by a column, and similar buildings are unique. From the 12th century to the 19th century, the Royal coronation was held here, and the House of Lords and House of Commons attended with the monarch at some major public ceremonies, such as the 25th (1977) and 50th anniversary of the accession of Elizabeth II (2002), the

Fig. 3-16 Westminster Hall in the 11th century
Source: https://baike.baidu.com.

300th anniversary of the Glorious Revolution (1988) and the 50th anniversary of the end of World War Ⅱ.

2) The Wooden Roof of Notre Dame de Paris

Cathédrale Notre Dame de Paris is a Gothic Christian Church and a symbol of ancient Paris. It stands on the Seine River, in the center of the whole city of Paris. It was first built in 1163 and completed in 1345 for more than 180 years. The front twin towers are about 69m high and the rear spire is about 90m. It is a very important representative building in the Gothic church group in France. The sculpture and painting art of altars, cloisters, doors and windows, as well as a large number of art treasures from the 13th century to the 17th century in the hall are world-famous. They are the symbol of ancient Paris. The main structure of Notre Dame de Paris is made of stone, and the church roof is made of wood. The whole wooden structure system of Notre Dame de Paris is known as "forest", because it is composed of a large number of wooden beams, and each column and beam comes from different trees. The size of the wooden beams is also amazing. The length of the main hall is more than 100m and the width is more than 13m. The cross wing of the church is 40m wide and 10m high. The earliest wooden structure system of Notre Dame de Paris was built between 1160 and 1170, but the earliest wooden structure has disappeared. From 1220 to 1240, the wooden roof system was restored using a large number of woods until modern times. In the early morning of April 15, 2019, a fire broke out on the wooden roof and it was completely destroyed (Fig. 3-17). This is a great loss in the history of human civilization.

(3) Modern Wood Structure

Modern wood structure is a new structural form developed from traditional building materials (wood) and modern advanced design, processing and construction technology. Compared with the traditional wood structure, the modern wood structure has the following differences.

1) The Processing Methods of Materials are Different

The wood used in the traditional wood structure is basically unprocessed, and the simple natural air drying method is adopted. The moisture content of wood can't be guaranteed, so it is inevitable to crack, twist, rot and moth. Modern wood structure uses engineering wood processed by modern science and technology. There are essential differences between engineering wood and log. Beams, columns and other components are processed by modern industrial means and advanced technology, which are especially suitable for modern architecture. Engineering wood can not only maintain the texture of logs, but also solve the problems of cracking and distortion of single timber, enhance the durability of the house and reduce the maintenance cost in the later stage.

2) Different Connection Modes.

The traditional wood structure is connected by mortise and tenon, while the modern wood structure is connected by metal parts.

3) High Reuse Rate

Because of the change of connection method, the disassembly and assembly of modern wood structure becomes more simple. Wood structure houses can be moved and

Fig. 3-17 The wooden roof of Notre Dame de Paris. Source: https://baike.baidu.com.

relocated, and the reuse rate of main materials is as high as 80%.

From the 1980s to now, it is a period of rapid development of wood structure in the world. Wood has the characteristics of light weight, high strength, beautiful appearance and good processing performance, so it has been favored by people since ancient times.

From solid wood, log structure to glued wood structure, and then to composite wood structure, the wood structure is no longer the traditional concept of wood structure, modern wood structure building has been comparable with steel structure building. In developed areas and countries such as Europe, America and Japan, a large number of researches and applications of wood structure also promote the benign cutting and utilization of forest resources, forming a mature forest management system.

Modern wood structure can be divided into light wood structure, heavy wood structure and various forms of mixed structure according to the different patterns of wood members. The lightweight timber structure is mainly composed of static timber frame, which is mostly used in residential buildings. Heavy wood structure mainly uses glulam or log with large section as structural timber, which is mostly used in large public buildings and commercial buildings.

In North America, wood structure housing is in the leading position in the market. In the United States, there are nearly 1.5 million new residential buildings every year, of which 90% are wood structures. In Canada, the industrialization, standardization and supporting installation technology of wood structure residence are very mature. In Japan, a large number of houses are also wood structure buildings. In Finland and Sweden of northern Europe, 90% of houses are also wood structure buildings.

The Mjstrnet Building in Norway, which was just completed in March 2019, is the tallest building in the world officially recorded as a wooden structure (Fig. 3-18). The building was designed by Voll Arkitekter in Norway for Investor AB, a Swedish investment company. It has 18 floors and is 85.4m high. It is a mixed use building with apartments, hotels, swimming pools, offices and restaurants. Precast concrete slabs are used in the upper seven floors of the building to increase the weight of the building and resist the hidden dangers caused by wind.

In addition to the high-rise buildings, modern wood structures are also widely used in large-span structures. The airport terminal is the representative building, like Fort McMurray International Airport in Canada, Mactan-Cebu International Airport in the Philippines, and the expansion projects of Wellington International Airport in New Zealand, etc.

1) Fort McMurray International Airport

Fort McMurray International Airport is a three-storey building with an area of 8040m^2 (Fig. 3-19). The building was designed an all wood structure, and the main public area is made up of heavy wood structure, using a large number of glulam and cross-laminated timbers. Due to the extremely cold winter climate in Alberta, the building is designed with air tight insulation to minimize heat loss. In this project, the most common woods in location such as SPF (short for Spruce-pine-fir) and Douglas fir were used extensively.

Fig. 3-18 The Mjstrnet Building in Norway
Source: https://www.sohu.com/a/385511485_99896221.

(a)

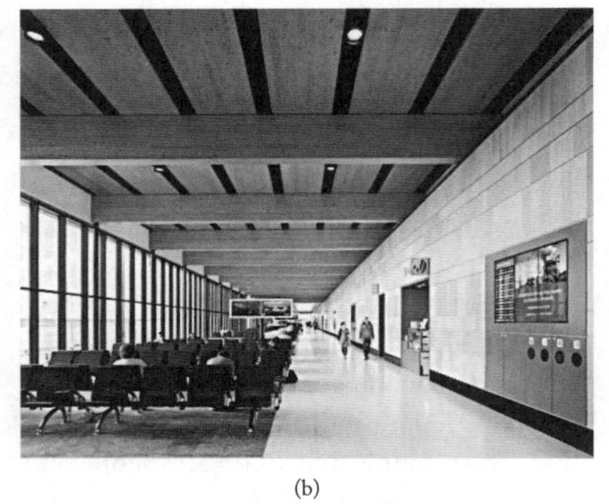

(b)

Fig. 3-19 Fort McMurray International Airport
Source: http://www.zhuxuncn.com/user page/article/detail? blog_id=212&id=73667.

The characteristics of wood carbon fixation and environmental protection reduce the impact of buildings on the environment, which not only has visual beauty, but also has cultural implications. At the time of its completion, the airport was the largest building used CLT (Short for cross-laminated timber) in North America areas.

2) Mactan-Cebu International Airport

The airport project is designed to meet the function of a transportation hub and create the feeling of a resort. It is the first large-span International Airport of all wood structure in Asia.

The roof of the terminal building is supported by glued laminated timber arches with a continuous span of 30m, which also defines its basic module composition and its overall architectural form (Fig. 3-20). The skylight of the vault introduces natural light into the room, while the valley connection between the arches is skillfully integrated the air conditioning system. The 15m high north and south main facade, under the shade of the roof overhanging eaves, enables the interior to have a continuous and unobstructed view. The wooden frame inside the airport, like a sailing ship, creates a simple and warm atmosphere, which is obviously different from the cold airport style of many other cities, symbolizing the hospitality of Filipinos.

3) Expansion Projects of Wellington International Airport

The application of glulam provides more possibilities for the design of large and complex building structures (Fig. 3-21). The expansion project of the main terminal of Wellington International Airport is a typical case. Among all the extensions, the most impressive one is the unique curved *x*-glulam column member. These glulam members are prefabricated in the factory of TECHLAM, the largest glulam wood factory in New Zealand. The advantage of prefabrication is obvious. The performance of glulam produced under the constant temperature and humidity control of the factory is more stable and the precision is very high.

(4) Shigeru Ban and his wooden structures

Shigeru Ban is a famous Japanese architect and the winner of the 2014 Pulitzer Prize.

Fig. 3-20 Mactan-Cebu International Airport
Source: www.mafengwo.cn.

(a) (b)

Fig. 3-21 The expansion projects of Wellington International Airport
Source: https://xw.qq.com/cmsid/20200826a035jd00.

(a) (b)

Fig. 3-22 Centre Pompidou-Metz
Source: http://jz.docin.com/buildingwechat/index.do?buildwechatId=9019.

Shigeru Ban designed and built a series of famous wooden structures, such as the Centre Pompidou-Metz, Swatch headquarters building and pyramid buildings in Kentucky Owl.

1) Centre Pompidou-Metz

The Centre Pompidou-Metz is located in a park in Metz, France (Fig. 3-22). Its basic plan shape is a hexagon, with a total construction area of 14,000 square meters. There is a 77-meter high spire constructed by steel structure in the center of the building plan, which is said to symbolize the Centre Pompidou-Metz's opening in Paris in 1977. The whole roof is surrounded by one-meter-thick timber beams as the outer frame of the structure, and then the 200mm thick timber beams are crisscross in three directions to form a sinuous and undulating shape in the three-dimensional space. The timber beams in each direction are made up of two layers of wood plate, and six layers of boards are closely linked to each other at each junction joint. Therefore, although the roof area of the building is 8000m^2, it only needs four wooden columns to support it. Inspired by the Chinese woven bamboo hat, the top of the building is supported by a hexagonal umbrella shaped structure made of complex steel pipes and plywood, covered with a translucent film, which plays a protective role.

2) Swatch Headquarters Building

Swatch headquarters building is located in the city of Bell, Switzerland (Fig. 3-23). The whole building is in the shape of a snake.

(a) (b)

Fig. 3-23 Swatch headquarters building
Source: https://www.sohu.com.

The facade of the building is covered with a wooden frame. The serpentine outer shell has a total length of 240m, a width of 35m and a height of 27m. The building skillfully uses the flexibility of wood, combining the curved facade with the straight trunk of the building to form a dynamic curve space. The outer shell of the building consists of 4600 individual wooden components, each of which has a unique shape and a precision of millimeter. Different components are connected by mortise-tenon at the joint, and finally form a curved reticulated shell structure. All the wooden components are made of local spruce in Switzerland, which uses 1997m^3 of wood.

(5) Protection Measures of Wood Structure Building

1) Fire Prevention

The fire prevention of wood structure and its components is mainly to determine its fire resistance limit, and according to the requirements of fire resistance rating of buildings, measures to improve the fire resistance limit of wood components are taken. The fire resistance limit of wood components is the time from the beginning to losing its original function for a certain component burning in a special furnace according to the simulated fire temperature (700~1000℃).

Wood members have relatively good fire resistance performance, especially those with large cross-section. This is because that wood is made up of hollow cells and has a low thermal conductivity. In addition, a layer of charcoal is formed on the surface of wood during combustion, and charcoal also has good heat insulation performance, which slows down the thermal decomposition of wood. Under the action of fire, the wood component is on fire in the first two minutes, and the carbonization rate in the next 8 minutes is about 0.8mm per minute. Due to the formation of charcoal layer, the carbonization rate is slowed down to 0.6mm per minute. The carbonization rate of different tree species is different. The fire resistance limit of wood members can be calculated according to the carbonization rate of different tree species besides experimental measurement.

For the wood members without the protective layer, the integral wood members with larger section size should be used as far as possible to improve the fire resistance. There are two ways to improve the fire resistance of wood structures. One is to add plastering layer or gypsum board. For example, a 30cm×30cm wooden column is covered with 2.5cm thickness of plastering layer contained steel wire mesh, its fire resistance limit can be increased to 1.5 hours. Another method is impregnating the fire retardant agent into the wood member or painting the fire proofing coatings on the surface of the wood member. For example, acrylic emulsion fire-retardant paint can decompose phosphoric acid at the temperature of 100~200°C to dehydrate and carbonize the wood, reduce the formation of combustible gas, and expand to form a honeycomb fire barrier at about 250°C, so as to prevent the expansion of the initial fire, and extinguish itself after leaving the flame.

2) Anticorrosion

Wood decay is caused by wood rot fungi. The hydrolytic enzymes in wood rot fungi can decompose the cellulose, lignin and cell contents of wood cell wall as nutrients, so that the strength of wood gradually decreases until it loses its full bearing capacity.

The wooden poles or piles buried in the soil are soaked by water in the soil and supplied with oxygen, so they are decayed. The reason why the part buried in the soil is not rotten is that it is lack of oxygen, while the reason why the higher part above the ground is not rotten is water shortage (i.e. the water content is lower than 18%). Therefore, preservatives must be used to prevent the growth of wood rot fungi for the wood structures that are often or intermittently affected with moisture, and the ends of wood beams or wood bricks that have to be enclosed in the wall.

Preservatives are prepared by toxic chemicals, including water-soluble, oil-soluble and paste, etc. For the wood components that are often affected by moisture, it is better to use mixed preservative oil, which is composed of coal creosote oil (wood preservative oil) and coal tar, which is not easy to lose when exposed to water and has long efficacy. Asphalt is black sticky in appearance, similar to anthracene oil, and is often misused as preservative. However, asphalt can only be waterproof but not antiseptic. If asphalt is applied on the wood that has not been dried, it will be counterproductive and hinder the air drying of wood.

3) Insect Prevention

Termites and beetles are the main wood borers, and termites are more harmful than beetles. Therefore, the construction measure of moisture-proof to separate wood members from water sources has some effect on reducing the harm of termites. However, the moisture-proof construction measure is only an auxiliary measure for pest control. In areas with termites or beetles, the wood structures and wood products should be treated with insecticides.

In order to ensure the durability of wood structure, all countries in the world use both anti-corrosion and anti insect agents. For example, the boron phenol mixture prepared with boric acid, borax and sodium pentachlorophenol is a water-soluble agent. Wood members can be immersed in the aqueous solution of the agent. If the wood can absorb 4.5~6kg of chemicals (dry agent weight) per cubic meter of wood, the purpose of anti-corrosion and insect prevention can be achieved. As this kind of medicament is easy to lose in water, it is only suitable for wood members which are not affected by damp.

3.2.6 Hybrid Structure Buildings

The structure mainly made of two or more materials is called hybrid structure. There are mainly several hybrid structures as follows.

(1) Brick Wood Structure

Brick wood structure usually takes brick as the wall, brick column as the vertical load-bearing member, wood beam and wood floor as the horizontal load-bearing component. Brick wood structure is often used in old multi-storey masonry buildings.

(2) Brick Concrete Structure

Brick concrete structure usually takes brick wall and brick column as the vertical load-bearing member, and reinforced concrete as the horizontal load-bearing component, including reinforced concrete beam, reinforced concrete floor, reinforced concrete roof, etc.

Brick concrete structure is widely used in modern multi-storey masonry buildings.

(3) Steel-reinforced Concrete Hybrid Structure

Steel-reinforced concrete hybrid structure is usually used in super high-rise buildings. In this kind of structure system, the structure system composed of steel frame or steel reinforced concrete frame and reinforced concrete cylinder structure bears both vertical and horizontal actions. Shanghai Jinmao Building, Taipei 101 Building, Shanghai World Financial Center and other super high-rise buildings adopt this structural system.

3.3 High-rise and Super High-rise Buildings

3.3.1 Early Development of Skyscrapers

High-rise buildings and super high-rise buildings are defined and classified according to the height and the number of floors of buildings. In August 1972, the International High-rise Building Conference was held in Bethlehem, Pennsylvania, U.S. The classification and definition of high-rise buildings are discussed and proposed in the meeting. The first type of high-rise buildings is 9~16 floors (height to 50m), the second type of high-rise buildings is 17~25 floors (height to 75m), the third type of high-rise buildings is 26~40 floors (up to 100m) and the super high-rise buildings is more than 40 floors (over 100m in height).

Now, for high-rise buildings, different countries and regions have different specific provisions. For example, in the United States, 24.6m or above is considered as a high-rise building; in Japan, 31m or 8-storey or above is considered as a high-rise building; in Britain, a building equal to or greater than 24.3m is regarded as a high-rise building; in China, the reinforced concrete structure with 10 floors and more or more than 28m in height is called high-rise building structure.

Modern high-rise buildings originated in the United States. The Chicago Family Insurance Company Building, built in 1883, is the first high-rise building in the history of modern architecture, with 10 floors above the ground and 1 floor underground. In 1891, another 16-storey brick masonry building with elevator was built in Chicago, which is still in use today.

Super high-rise building is also known as skyscraper. For super high-rise buildings, according to the new standards of the Council on Tall Buildings and Urban Habitat, buildings over 300m are super high-rise buildings. The Council on Tall Buildings and Urban Habitat (CTBUH) is the world's leading resource for professionals focused on the inception, design, construction and operation of tall buildings and future cities. As a non-profit organization, CTBUH developed internationally accepted criteria for measuring tall building heights, and is recognized as the arbiter for bestowing such designations as "The World's Tallest Building".

In the early 20th century, there was a competition to build world skyscrapers in Manhattan, New York. In 1908, the headquarters of American singer company in New York was established. The Singer Building, the first skyscraper in the world, was also the tallest building in the world at that time, with a total of 44 floors and 187m high. In 1911, the Metropolitan Life Insurance Company Tower was completed, with a total of 50 floors and a height of 213m. In 1913, Woolworth, a retail giant, built the Woolworth Building with 57 floors, which is all steel structure and a height of 241m. It won the title of the world's tallest building and created the era of super high-rise buildings. In 1931, the Empire State Building was built in New York with a height of 381m, setting a new skyline height and keeping this record for 40 years until the completion of Sears Tower in 1973. Sears Tower is located in Chicago, U.S., with 110 floors and a total height of 442m. It became the first building in the world to break through the height of 400m (Fig. 3-24).

In 1998, 25 years later, the Petronas Twin Towers were built in Kuala Lumpur, Malaysia, with a total height of 452m, becoming the tallest building at that time (Fig. 3-25). The building is composed of two buildings side by side, so it is also called Twin Towers. In 1999, Jinmao Building was built in Shanghai, China, with a total height of 420m, becoming the tallest building in China. Until entering the 21st century, the height of high-rise buildings built by human beings has not exceeded 500m.

In the 21st century, the construction of skyscrapers in the world has witnessed a rapid development in height and quantity. In terms of quantity, in the 20 years from 2000 to 2019, there are 82 skyscrapers over 300m built in the world, of which 41 are located in China. In the aspect of building height, it broke through 500m in one fell swoop and soon exceeded 800m. At present, there are 10 skyscrapers over 500m in the world, of which 6 are located in China.

The first skyscraper to break through 500m is Taipei 101 Building, which was built in 2004. The building has 101 floors with a total height of 508m, setting three world records: the world's tallest building, the world's highest use floor and the world's highest roof height. The height of the Taipei 101 Building, which has been the highest in the world for six years, was broken by the Burj Khalifa Tower in 2010. The Burj Khalifa Tower is located in Dubai, with 828 meters high and 163 floors in total. It is the tallest building in the world and the highest artificial structure on earth.

3.3.2 Empire State Building

The Empire State Building in New York is a landmark in the history of architecture (Fig. 3-26). The building is of all steel structure. Construction started in 1930 and completed in 1931. It took only 410 days to complete the project. The reason why the construction speed is so high is due to two effective measures taken in the construction process. First of all, the contractor adopted the method of "design while construction", which was very avant-garde at that time. Before the complete design drawing of the building was released, the construction work was carried out first, that is, when the structure of the lower floor began to be constructed, the architect did not even draw the construction drawings of the top floors. Secondly, steel reinforced concrete is adopted in the structure of Empire State Building, that is, the steel is connected

Fig. 3-24 Sears Tower
Source: https://baike.baidu.com.

Fig. 3-25 Petronas Twin Towers
Source: https://www.sohu.com.

Fig. 3-26 Empire State Building
Source: https://travel.sina.com.

into frame beam column by rivets, and then the steel reinforced concrete column is formed by formwork pouring concrete outside the steel column. Because the steel frame made of section steel can bear the load, the construction of the floor does not need to wait for the concrete strength of the lower floor to be fully formed. Therefore, the construction speed of the Empire State building was 4.5 storeys a week at an amazing speed.

73 elevators are installed in the building. These elevator shafts form relatively closed steel reinforced concrete tubes with large lateral stiffness. The periphery of the tubes is multi-circle frame columns, which forms a typical frame-core tube structure system in high-rise buildings. The whole building was completed and put into operation on May 1, 1931, and its cost was 10% lower than the expected $50 million. The decorative materials used include 5660 cubic meters of Indiana limestone and granite, 10 million bricks, 730 tons of aluminum and stainless steel.

The Empire State building is still in safe use and its structural form has withstood the test of time and disaster. On July 28, 1945, a US Air Force B-25 bomber, which was lost in the fog, crashed into the 78~79 floors of the building at a speed of 320km/h, causing a large hole of 5.5m in width and 6m in height. A fire broke out on the 79th floor, and the fire spread to the 86th floor. The accident resulted in 13 deaths and 26 injuries. But the rest of the building was barely affected, and the building remained intact, indicating that the structure of the Empire State building was very effective and reasonable. The concrete wrapped outside the steel structure effectively protects the internal steel structure. However, the World Trade Center built after 40 years was hit by a Boeing 757 on September 11, 2001, which caused a fire. Because the structure is all steel structure, there is no concrete protection outside the column, and the steel column lost its bearing capacity when the temperature reached 600℃. Finally, it led to the collapse of the building, caused serious disasters and caused the reflection of the structure of skyscrapers.

3.3.3 Burj Khalifa Tower

The Burj Khalifa Tower is currently the tallest building and artificial structure in the world (Fig. 3-27).The tower is 828 meters high, with 162 floors and a cost of $1.5 billion. The construction cost of the building itself is at least $1 billion, excluding the construction costs of its large shopping center, lakes and slightly lower towers. A total of 330,000 cubic meters of concrete, 62,000 tons of reinforced steel and 142,000 square meters of glass were used in Khalifa. About 4000 workers and 100 cranes were mobilized to pump concrete to a height of more than 606m, breaking the record of 492m when the Shanghai World Financial Center was constructed. There are 56 elevators with a maximum speed of 17.4m/s in the building, and also double deck sightseeing lifts, which can carry up to 42 people at a time.

Fig. 3-27 Burj Khalifa Tower
Source: https://www.sohu.com.

3.3.4 High-rise and Super High-rise Buildings in China

The development of high-rise and super high-rise buildings in China is slow. Before the founding of the People's Repulic of China, the Shanghai International Hotel was built in Shanghai in 1934. It was a 24-storey high-rise building. Its record of the highest building in China has been maintained for more than 40 years. In 1976, Guangzhou Baiyun Hotel was built, with a total of 33 floors and a height of 112.45m, exceeding 100m for the first time. In 1989, the Capital Mansion was built, with 52 floors above the ground and 4 floors underground. The total height of the building is 183.5m, which still has not exceeded 200m.

With the promotion of reform and opening up, the construction technology is constantly improved, the construction equipment is rapidly updated, and the economic strength is significantly enhanced. The construction speed of skyscrapers in China is also very amazing. In the 1990s, the height of the building rapidly reached 200m, 300m, even breaking through 400m. Shanghai Jinmao Building was started in 1994 and completed in 1999. It has 88 floors on the ground and 3 floors underground, with a total height of 420.5m. It became the tallest building in China at that time. The steel-concrete composite structure is adopted in the building. The reinforced concrete core tube and eight steel reinforced concrete columns are connected together by an overhanging truss.

Shanghai World Financial Center was started in 1997, and then stopped work due to the Asian financial crisis. It was resumed in February 2003 and completed in August 2008, which lasted 11 years. The building has 101 floors above the ground and 3 floors underground, with a building height of 492m.

It is also a steel concrete hybrid structure.

After entering the 21st century, the height of super high-rise buildings in China has exceeded 500m and even 600m. At present, there are 10 buildings over 500m in the world, and 6 in China. The tallest Shanghai Center Tower, with a height of 632m, is the second tallest building in the world (Fig. 3-28). Shanghai Center Tower is one of the most sustainably advanced tall buildings in the world. A central aspect of its design is the transparent second skin that wraps around the entire building. The ventilated atriums it encloses conserve energy by modulating the temperature within the void. The space acts as a buffer between the inside and outside, warming up the cool air outside in the winter and dissipating heat from the interior in the summer. The tower also notably employs a tri-cogeneration system, a grey water/rainwater system, and several renewable energy sources.

3.3.5 The Future of Super High-rise Buildings

The development of high-rise buildings and super high-rise buildings can save land and urban infrastructure costs, which is an important way to solve the problem of more people and less land in the process of urbanization. High-rise building is also a symbol of economic prosperity and technological progress of a country and region, so every country and region has high enthusiasm to build high-rise buildings and super high-rise buildings.

In 2013, a company in Changsha, China, proposed to build a 838m world's tallest building, which was later called off. It is not known whether the project has any chance

Fig. 3-28 Shanghai Center Tower
Source: https://www.163.com.

to restart. At the same time, people are also challenging the height of 1000m, Japan's planned aerial city is up to 1000m; Saudi Arabia is building a skyscraper with a height of more than 1000m—the Jeddah Tower. The construction of the project began in 2013, and one third of the project was completed in the whole five years of 2013~2018. After 2018, it was once in a state of stagnation.

The continuous renewal of skyscraper height reflects people's continuous pursuit of the spirit. Challenging the limit has always been the driving force for social progress and development.

In addition, the construction of skyscrapers is a complex and huge engineering project. Both the construction cost and the later maintenance and operation cost are very high. The cost of building construction and maintenance will increase exponentially with the increase of height. A city without strong economic power can not build and hold skyscrapers. Building skyscrapers is an inevitable measure to save land and ask for space from the sky, but blindly pursuing height will bring huge economic burden to the society. Therefore, under the guidance of green development goals, the rational construction of skyscrapers is the embodiment of human architectural rationality.

3.4 Spatial Structures

3.4.1 Concept and Characteristics of Spatial Structure

Human beings live in the three-dimensional space world, any building structure has the three-dimensional space attribute in essence; however, for the purpose of simplifying design and construction, people decompose them into plane structures for construction and calculation in many occasions. At the same time, there are some types of structures, which have the characteristics of three-dimensional stresses and force transmission, and can not be decomposed into simple plane elements. This kind of structure is called spatial structure.

On the one hand, the excellent working performance of spatial structure is manifested in three-dimensional force. Generally speaking, this kind of stress state is more efficient than the plane structure. On the other hand, it can effectively resist the effect of external load through reasonable curved shape. When the span increases, the more spatial structures can show their excellent technical and economic performance. In fact, when the span reaches large size, the general plane structure is often difficult to become a reasonable choice. From the engineering practice around the world, most of the long-span buildings adopt various forms of spatial structure system.

The span of spatial structure can reach tens of meters to hundreds of meters, and the span capacity is far greater than the plane structures such as bent frame, so the spatial structure is called large-span structure. Stadiums, movie theaters, exhibition centers and other buildings are the main application fields of spatial structure.

According to the spatial stress mechanism, spatial structure can be divided into folded plate structure, thin shell structure, grid structure, space truss structure, tension structure and so on.

3.4.2 Folded Plate Structure

Folded plate structure is a kind of thin-walled space system, which is composed of a number of narrow thin plates intersecting at a certain angle. It not only has the function of bearing loads, but also plays the role of enclosure. It can be used as the roof of industrial and civil buildings such as workshops, warehouses, stations, shops, schools, residences, pavilions and stands of stadiums. In addition, folded plate structure can also be used as the external wall, foundation and retaining wall.

(1) Composition and Classification of Folded Plate Structure

According to the structure system, folded plate structure can be divided into common folded plate surface, folded plate frame and spatial folded plate. According to the form of folding surface, it can be divided into V-shaped folded plate, multi-folded plate, conical folded plate, corner cone folded plate, combined folded plate, etc. According to the number of spans, there are single span plate, multi-span plate and cantilever folded plate. According to the covering plane, it is divided into rectangular, fan-shaped, circular and circular plane folded plates. According to the materials, they are divided into reinforced concrete folded plate, prestressed concrete folded plate and steel fiber reinforced concrete folded plate. If the folded plate is also a broken line or arc along the span direction, it will form a folded plate arch, which is one of the forms of large-span roof structure.

The typical folded plate structure is mainly composed of three parts: folded plate, edge beam and diaphragm. Folded plate is mainly used for bearing and maintenance. The edge beams connect the adjacent folded plates to strengthen the longitudinal stiffness of the folded plates and increase the out of plane stiffness of the folded plates at the same time. The diaphragms are generally arranged at the ends to form a geometrically invariant system of the folded plate structure and serve as the longitudinal support of the folded plate edge beam.

Folded plate structure and shell structure belong to thin shell space structure, so they have similarities. They all take advantage of the shape to improve the overall stiffness. The folded plate structure is composed of plane plates. Although the internal force analysis of folded plate structure is simpler than that of curved shell structure, the folded plate structure is not a perfect arch line, so it inevitably involves bending stress, which is the main difference from the shell structure mainly composed of membrane stress.

(2) Mulimatt Sports Training Center (Fig. 3-29)

This project is designed by Swiss architect Livio Vacchini (1933-2007). Architects want to make the structure exposed, to express the beauty of the structure, but also to be durable. In addition, because there are trains passing through the view height, the architect also hopes that the structural form of the facade and roof can form a whole. According to the requirements of the architect, the concrete folded slab frame was adopted.

The building is 80m long and 55m wide. The interior is a column free space. The structural wall and roof are all V-shaped folded plates, which are composed of 27 units. Each unit is composed of two groups of V-shaped walls and a group of V-shaped roofs. The folded slab is made of precast prestressed concrete, the thickness of roof slab is only 16cm, and prestressed steel strand is set inside. By adopting self compacting concrete and applying prestress, the waterproof and durability of concrete can be improved.

The interior of the building is divided into a number of sports venues and multi-function hall. People can't feel the complex structural form inside, and they won't be distracted by it. On the other hand, the structure of the outer side is very strong, which attracts the attention of pedestrians. The folded plate is not only a structural load-bearing component, but also an important part of architectural expression, which makes the facade as the skin of the whole building and become a sculpture. From this, we can not only see the excellent structural design level of structural engineers, but also reflect the close cooperation between architects and structural engineers.

(a)

(b)

Fig. 3-29 Mulimatt Sports Training Center
Source: http://www.360doc.com.

(3) The Passenger Transport Center of Qingdao International Cruise Home Port

The passenger transport center of Qingdao International Cruise Home Port was designed by CCDI. The overall design of the building takes the meaning of "sail". The construction process adopts the method of unit assembly. It was completed in 2015 (Fig. 3-30).

The building is 23m high, 338m long and 96m wide. The steel roof structure adopts variable cross-section truss folded plate structure, which is composed of 18 basic units arranged with a spacing of 18 meters. The indoor span is 55m and the outdoor span is 36m. The results of structural calculation are consistent with the conceptual analysis characteristics of folded plate frame. Firstly, the lateral and vertical stiffness of the structure is large due to the spatial action of the folded surface, and the *y*-direction displacement is only 25% of the allowable displacement. Secondly, the axial force is the main force of the structure, but the bending moment is large at the turning point of the structural fracture surface, which is the main load-bearing member of the structure and needs to be strengthened accordingly.

3.4.3 Thin Shell Structure

Thin shell structure is a kind of thin-

Fig. 3-30 The passenger transport center of Qingdao International Cruise Home Port
Source: http://k.sina.com.cn/article_3914163006_e94d633e02000caom.html.

walled structure with continuous curved surface, which is usually constructed by reinforced concrete or prestressed reinforced concrete. Thin shell structure has a typical space force characteristics, which can evenly distribute the pressure to all parts of the object. The shell can make full use of the material strength, and at the same time, it can integrate the bearing and enclosure functions. In practical engineering, the cutting and combination of space curved surface can also be used to form buildings with peculiar and novel shapes and can adapt to various plane structures, but it is more labor-intensive and cost-effective.

(1) Classification of shells

Shell is divided into thin shell, medium thick shell and thick shell according to the ratio of shell thickness to minimum curvature radius. Generally speaking, thin shell with ratio less than 1/20 is usually used for roof of a house; the medium thick shell and thick shell are mostly used for underground structure and protective structure.

The common thin shell structures in engineering include the following:

1) Cylindrical Shell Structure

It is composed of shell body, side edge member and diaphragm. The distance between the transverse compartments is the span of the shell, and the distance between the side members is the wavelength of the shell. When the span is larger than the wavelength, it is called the long shell; when the span is less than the wavelength, it is called the short shell.

2) Dome Shell Structure

It is composed of shell surface and support ring. The thickness of shell surface is very thin, generally 1/600 of the radius of curvature, and the span can be large. The support ring acts as a hoop on the dome shell, through which the whole thin shell is placed on the supporting member.

3) Hyperbolic Shallow Shell

A surface formed by the translation of a parabola along another orthogonal parabola is called hyperbolic shallow shell (micro bending plate). The ratio of the vector height at the apex to the length of the short side at the bottom should not exceed 1/5. The hyperbolic shallow shell is composed of a shell body and four diaphragms. The diaphragms are arches with tie rods or beams with variable heights. It is suitable for buildings covering square or rectangular plane with a span of 20~50m (the ratio of long and short sides should not exceed 2).

4) Hyperbolic Paraboloid Shell

A curved surface formed by a vertical parabola (generatrix) moving parallel to another parabola (conductor) opposite to it is called hyperbolic paraboloid shell. All kinds of torsion shell in engineering are also one of the types. This shell structure is easy to make, stable and easy to adapt to the needs of architectural function and modeling, so it is widely used.

(2) Palazzetto Dello Sport of Rome

Palazzetto Dello Sport of Rome was built for the Olympic Games held in Rome in 1960. It is also used for basketball, tennis, boxing and other competitions. It was built during 1956~1957 (Fig. 3-31).

The designers are Italian architect Annibale Vitellozzi and engineer Pier Luigi Nervi. This simple and beautiful gymnasium is one of the representative works of structural design of Nervi, which occupies an important position in the history of modern architecture.

It is an excellent work of art with ingenious combination of architectural design, structural design and construction technology. Its plane is round with a diameter of 60m. The roof is a spherical dome, which is separated from the grandstand structurally. The top of the shell is made up of 1620 rhombic trough plates prefabricated with steel mesh cement. The reinforcing bars are arranged between the plates and cast into "ribs", and a layer of concrete is poured on it to form an integral waterproof layer. The size of the prefabricated trough plate is determined by the building

Fig. 3-31 Palazzetto Dello Sport of Rome
Source: https://baike.baidu.com.

scale, structural requirements and the lifting capacity of construction equipment. The arch ribs crisscross to form a beautiful pattern, like a blooming chrysanthemum, simple and elegant. The dome is like a reverse lotus leaf, supported by 36 "Y" shaped diagonal braces evenly distributed along the circumference, transmitting the load to a circle of ground beams buried in the ground. There is a circle of white reinforced concrete "belt" in the middle of the diagonal bracing, which is the roof of the auxiliary building and also serves as the connecting beam. The lower edge of the shell top is evenly divided by the fulcrum and arched upward to avoid adverse bending distance. From the perspective of architectural effect, it can not only enrich the outline, but also prevent the subsidence caused by visual illusion.

(3) Centre National des Industries et Techoniques

The Centre National des Industries et Techoniques (CNIT) looks like an inverted shell. It is the first building built in La Défense, and it is also a building with unique shape (Fig. 3-32). The project was completed in 1959. The roof is an integrated precast thin shell structure of reinforced concrete. The building plane is triangular, with a span of 218 meters on each side, and the shell top is 48 meters above the ground. The double-layer corrugated arch shell is supported on three corner piers, which are connected with prestressed tie rods. The thickness of the shell of the upper and lower layers is only 64mm, 1.8m apart, and the middle is connected by vertical plates with a spacing of 9m. The total shell thickness is only 180mm, and the thickness span ratio is 1 : 1200, which is far less than the ratio of eggshell thickness to length (1 : 1000). Moreover, the architectural form is novel, which fully demonstrates the superiority of the concrete shell structure. The building is a masterpiece integrating structure, materials and natural force. It has bold innovation in structural calculation and construction, and is a great progress in the field of world architecture.

The thin shell structure is the largest shell in the world at that time, and it is also the largest reinforced concrete public building in the world.

(4) Sydney Opera House

Sydney Opera House, located in Sydney Harbour, Australia, started construction in March 1959 and was officially put into use on October 20, 1973. It is a landmark building in Australia (Fig. 3-33). The Sydney Opera House consists of three groups of huge shells, which stand on the base of cast-in-place reinforced concrete structure. The base is 186 meters of long from north to south and 97m of wide from east to west. The first group shells is on the west side of the site. Four pairs of shells

Fig. 3-32 Centre National des Industries et Techoniques
Source: https://m.sohu.com/a/212162583_612969.

Fig. 3-33 Sydney Opera House
Source: https://www.sohu.com.

are arranged in a series, three pairs of shell face north and one pair of shell faces south. The interior is a large concert hall. The second group is on the east side of the site, which is roughly parallel to the first group, with the same form, but with a slightly smaller scale. The interior is an opera hall. The third group, in their southwest, is the smallest, consisting of two pairs of shells, with a restaurant inside. The other rooms are cleverly arranged in the base. These "shells" are arranged in order. The first three shells cover one another and face the bay. The last one stands with its back to the bay. It looks like two groups of clams with their lids open. At the same time, the appearance of the Sydney Opera House is like a white sail that is about to sail out on the wind, which sets off an interesting contrast with the surrounding scenery.

The Sydney Opera House has a total building area of 88,000m^2. There is no direct relationship between the external shape and the internal function. The shape of the interior is determined by the steel truss on the suspended reinforced concrete shell. It has a 2700-seat concert hall, a 1550-seat opera house and a 420-seat small theater. There are 900-rooms of exhibition, recording, bar and restaurant. Sydney Opera House is one of the most distinctive buildings in the 20th century. It is the landmark building of Sydney city. It was awarded the World Cultural Heritage by UNESCO in 2007.

3.4.4 Grid Structure

Grid structure is a kind of reticulated spatial member structure, which is composed of many members arranged in two or several directions according to certain rules.

Among them, the members are mostly made of circular cross-section steel tubes, and occasionally square cross-section steel tubes are also used. A grid structure with flat shape is called flat grid structure, while a grid structure with curved shape is called reticulated shell structure, also called grid shell structure (Fig. 3-34).

The grid structure has large space stiffness, good integrity and stability, good seismic performance and good building modeling effect. It is suitable for industrial and civil buildings with various supporting conditions, plane shapes, and large and small spans. The members and joints of grid structure are relatively simple, which is easy to manufacture and install. This kind of structure is mainly made of steel, which is light in weight and has the advantages of saving steel consumption.

(1) Flat Grid Structure

There are many kinds of grid structures, which can be classified according to different standards.

1) According to the structure patterns of the grid, it can be divided into single-layer grid structure, double-layer grid structure and three-layer grid structure. Among them, single-layer grid structure and three-layer grid structure are suitable for small span (no more than 30m) and extra large span (more than 100m) respectively.

2) According to the construction materials, it can be divided into: steel grid, aluminum grid, wood grid, plastic grid, reinforced concrete grid and composite grid, among which steel grid is the most widely used.

3) According to the support conditions,

Fig. 3-34 Grid structure (in construction)
http://www.hhssgroup.com.

it can be divided into four types: peripheral support, four-point support, multi-point support, three-side support, opposite-side support and mixed support.

4) According to the different composition, it can be divided into cross truss grid, triangular cone grid, quadrangular pyramid grid and hexagonal cone grid.

There are many types of flat grid structure, so many factors should be considered when choosing a suitable type. The selection should adhere to the following principles: safe and reliable, advanced technology, economic and reasonable, beautiful and applicable. The grid height and grid size of the grid structure should be determined according to the span size, load conditions, column grid size, supporting conditions, grid form, structural requirements and building functions. The height span ratio of the grid structure can be 1/18~1/10. The grid number of short span of grid structure should not be less than 5. When determining the mesh size, the angle between adjacent members should be less than 45° and not less than 30°.

(2) Reticulated Shell Structure

Reticulated shell is a kind of spatial truss structure similar to the flat grid structure. It is a spatial frame which is based on members, forms a grid according to certain rules, and is arranged according to the shell structure. It has the properties of skeletal structures and shell structures. The reticulated shell structure includes single-layer reticulated shell structure, prestressed reticulated shell structure, plate cone reticulated shell structure, rib ring cable supported reticulated shell structure, single-layer fork tube reticulated shell structure.

With the development of reticulated shell structure and a large number of engineering applications, reticulated shell structure provides a new and reasonable structural form for buildings, which mainly has the following advantages:

1) Reticulated shell structure is a typical spatial structure. Reasonable curved surface can make the structure force flow uniform, and the structure has large stiffness, small deformation, high stability and material saving.

2) Reticulated shell structure has beautiful architectural modeling, so designers can have full creative freedom in terms of architectural plane and outer shape of the building. The shape that can not be realized by the thin shell structure or the flat grid structure can be realized almost by reticulated shell structure.

3) It can be used not only in medium or small span civil and industrial buildings, but also in large-span buildings, especially in super large-span buildings.

4) It is possible to form a large spacial structure using small components, and the types of members are just a few. These components can be prefabricated in the factory to achieve industrial production, simple and rapid installation, adapt to the use of various conditions of construction technology, with no need of large equipment, so the comprehensive economic indicators are better.

5) Because of the curved shape of the reticulated shell structure, the natural drainage function is formed, so it is not necessary to use small columns for slope making as the grid structure.

3.4.5 Tension Structure

Tension structure is the use of flexible materials through a certain way to exert tension, as part of the roof, covering a large space. The tension steel cable can be made into suspension structure, the tension membrane material can form membrane structure, and the combination of the two can form cable membrane structure.

(1) Suspension Structure

Suspension structure is a kind of structure that directly bears the roof load by flexible cables or cable nets arranged according to certain rules. These cables or cable nets are generally suspended on the edge members of the supporting structure system. Under the vertical load, the cable or cable net bears the axial tension force, which is transmitted to the foundation of the building through the edge member or supporting structure. Suspension structures are mostly used in construction and bridge engineering.

The flexible cable is used to bear the axial tension. The cable can be composed of wire bundle, wire rope, steel strand, chain, round steel and other wire rods with good tensile performance. The edge member is used to anchor the cable net and bear the tension of the cable at the support. The edge members can generally choose ring beam, arch, truss, steel frame and other rigid components. The supporting structure is used to bear the bending moment caused by the compression and horizontal force from the edge members. Generally, the supporting structure can be reinforced concrete independent column, frame, arch and other structural forms.

1) Hannover Exhibition Center

The Hannover Exhibition Center in Germany was famous for the world's only suspension structure pavilion with a span of 36 meters.The most distinctive Exhibition Center is No.26 hall (Fig. 3-35). The exhibition hall was once called "the first exhibition of Expo 2000". The appearance of the exhibition hall is the product of the combination of the most advanced construction technology at that time and the optimal environmental and sustainable energy utilization. The roof is suspended in the air by 300mm×40mm tension bars spaced 5.5m to support the wooden roof panel. The span of suspension cable is about 55m, and the minimum tensile strength of reinforcement is 520N/mm^2. The load of the tension bar is concentrated on the truss beam on the roof plane. It has a good reputation as one of the best trade exhibition halls in the world.

2) Raleigh Stadium

In 1953, the Raleigh Stadium in North Carolina, U.S., was built, which is the first modern suspension roof in the world (Fig. 3-36). It is a saddle shaped "orthogonal cable net" structure with two inclined parabolic arches as edge members, and its plane size is 92m × 97m.

The oval site is supported by two cross arches, and the saddle shaped cross cable system is supported by two cross arches. The anchor head of the cable system is anchored on the arch ring with prestressed anchor, and the arch feet are connected with tie rods to reduce the reaction force of the arch foot.

3) Yoyogi National Gymnasium

Tokyo Yoyogi National Gymnasium was built for the 18th Olympic Games (Fig. 3-37). The venue has achieved the perfect unity of materials, functions, structure, proportion, and even the concept of history. It was once called one of the most beautiful buildings in the 20th century.

The gymnasium adopts the high strength

Fig. 3-35 No.26 hall of Hannover Exhibition Center
Source: https://www.uniqueway.com.

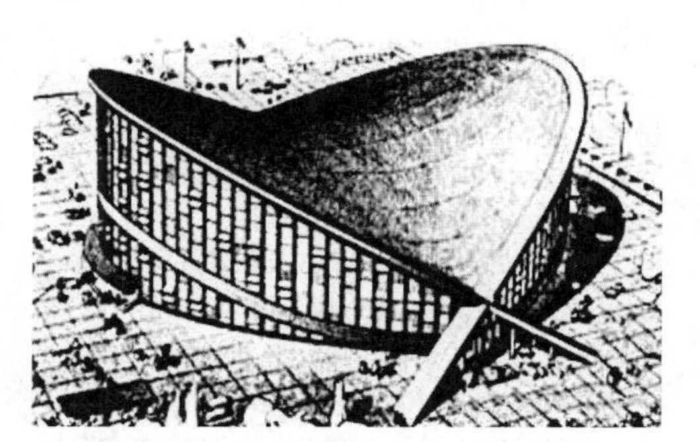

Fig. 3-36 Raleigh Stadium
Source: https://zhuanlan.zhihu.com/p/60158973.

Fig. 3-37 Yoyogi National Gymnasium
Source: https//www.uniqueway.com.

Fig. 3-38 Air supported membrane structure
Source: https://baike.baidu.com.

cable as the main suspension roof structure. The large span of indoor space is realized by the load-bearing structure formed by flexible cables and its edge components. The roof of the first gymnasium is tensioned by a pair of steel cables between two tower columns at both ends of the building to realize the main structure like suspension bridge. The two main cables are expanded in a spindle shape in the middle of the span. The second gymnasium is slightly different from the first gymnasium in the form of suspension structure. The roof adopts an independent high concrete tower column located at one side passageway behind the audience stand to realize the rotary suspension.

(2) Membrane Structure

Membrane structure is composed of flexible membrane materials, which can be divided into air supported membrane structure and tension membrane structure.

In the space covered by the membrane structure, the pressure difference between the inside and outside is used to bear the external force and form a structure together with the steel cable, which is called air supported membrane structure (Fig. 3-38).

The membrane is tensioned on the steel cable and fixed on the edge member to obtain the determined shape, which is called tension membrane structure.

Membrane structure is the most representative architectural form in the 21st century. It breaks the mode of pure linear architectural style. With its unique beautiful curved surface modeling, the perfect combination of simplicity, lightness, rigidity and softness, strength and beauty, membrane structure provides architectural designers with more imagination and creation space.

(3) Cable Dome Structure

Cable dome is a kind of tension structure formed by reasonable combination of cable, pole and membrane. Because a large number of prestressed steel cables are used in the structure, the compression bar is few and short, so it can give full play to the tensile strength of steel, and the structural efficiency is very high.

Cable dome structure comes from the idea of tensegrity put forward by American architect R.B.Fuller in 1962: to reduce the compression state in the structure as much as possible and keep the structure in continuous tension state, so as to realize the assumption that "the island of compression bar exists in the ocean of tension bar".

Geiger, an American engineer, developed Fuller's idea of tensegrity and creatively put forward the concept of cable dome, which was put into practice for the first time in gymnasium and fencing gymnasium of Seoul Olympic Games in 1988.

Since then, Levy, an American engineer, has improved the out of plane stiffness and instability of the cable net in the cable dome designed by Geiger. He changed the radial notochord into a triangular united square layout, and successfully designed the Georgia Dome, the main stadium of the Atlanta Olympic Games in 1996 (Fig. 3-39). At present, more than ten cable dome structures have been built in the world, mainly distributed in the United States, South Korea and other developed countries.

Fig. 3-39 Georgia Dome
Source: https://www.tripadvisor.com.

3.5 Conclusions

Building is an artificial structure closely related to human life. The building framework composed of beam, slab, column, wall and foundation is called building structure. According to the main materials used in the construction, the building structure can be divided into concrete structure, steel structure, masonry structure, wood structure and mixed structure. Different materials have direct and even decisive effects on the mechanical properties and service functions of the building structure.

The development of human architectural structure is also the process of human being challenging and coordinating with natural forces. Since ancient times, the architectural structure has gradually developed from short and small to tall and great, which reflects people's continuous pursuit. Now people can easily build high-rise buildings of a few hundred meters, and can also build large-span structures with a span of more than 100m.

The implementation of building structure is a comprehensive project involving science, technology, economy, humanities, aesthetics, environment, etc. In the future, the development of architecture will not only pursue bigger and stronger, but also advance towards the direction of green construction, so as to realize the sustainable development of human society.

Exercises

3-1 Who is the inventor of reinforced concrete structure? Who is the inventor of prestressed concrete and what difficulties has he overcome?

3-2 What revolutionary significance does Crystal Palace have in architectural structure?

3-3 Try to analyze the different consequences of the Empire State Building and the World Trade Center after being hit by aircrafts.

3-4 What structural scheme has been adopted by Sydney Opera House? How to understand the relationship between structural scheme and architectural scheme?

3-5 Talk about the future development direction of human architecture.

3-6 What is the difference between modern wood structure and traditional wood structure?

3-7 What are the forms of joint connection of steel structure?

3-8 What are the main advantages and disadvantages of tension structure?

References

[1] The Intricate, Undulating Brickwork at Eladio Dieste's Cristo Obrero Church in Uruguay. [2020-09-21].http://www.archdaily.com/890362/the-intucate-undulating-brickwork-at-eladio-diestes-cnsto-obrero-church-in-uruguay.

[2] XIONG F. Introduction to Civil Engineering (New 1st Edition)[M]. Wuhan: Wuhan University of Technology Press, 2018. (In Chinese)

[3] YI C, SHEN S Z. Introduction to Civil Engineering (3rd Edition)[M]. Beijing: China Architecture & Building Press, 2017. (In Chinese)

[4] QIAN J R. Structural Design of High Rise Buildings (3rd Edition)[M]. Beijing: China Architecture & Building Press, 2018. (In Chinese)

[5] Overview of global high rise timber buildings[EB/OL]. [2020-09-21]. https://baijiahao.baidu.com/s?id=1669099689516209448&wfr=spider&for=pc.

[6] Palazzetto Dello Sport of Rome[EB/OL] [2020-09-21] http://www.metrostudio.it/m/workinfo_en.aspx?id=740.

[7] A Brief History of Rome's Luminous Rotundas[EB/OL]. [2020-09-21] https://www.archdaily.com/775844/a-brief-history-of-romes-luminous-rotundas.

[8] SHIGERU BAN ARCHITECTS [EB/OL]. [2020-09-21] http://www.shigerubanarchitects.com/.

[9] Wood structure[EB/OL]. [2020-09-21]. https://baike.baidu.com/item/%E6%9C%A8%E7%BB%93%E6%9E%84/343356?fr=aladdin.

[10] 100 Tallest Completed Buildings in the World by Height to Architectural Top[EB/OL]. [2020-09-21]. http://www.skyscrapercenter.com/buildings.

[11] Empire State Building[EB/OL]. [2020-09-21]. http://www.skyscrapercenter.com/building/empire-state-building/261.

[12] Mulimatt Sports Training Center [EB/OL]. [2020-09-21].https://www.futureview360.com/2019/06/05/mulimatt-sport-education-center/.

Chapter 4
Bridge Engineering

4.1 Overview

Bridges are the structures built for carrying mostly road and railway traffic or any other moving loads over a depression, a gap or an obstruction such as a river, channel, canyon and valley. The required passage may be for pedestrians, a road, a railway, a canal, a pipeline, etc. The obstacle can be rivers, valleys, sea channels and other constructions (i.e. bridges, buildings, railways, roads, et al.). The covered bridge at Cambridge in Fig. 4-1 and a flyover bridge at Osaka in Fig. 4-2 are typical bridges according to above definition systems, and generally serving as "lifelines" in the social infrastructure systems.

Bridges have been gradually developed with the progress of society. As the advent of new inland vehicles, and the improvement of people's requirements on convenience and comfort of transport, the performance requirement of bridge is improved in load-bearing, span capacity and artistic design. Simultaneously, a huge source power for the progress of bridge is provided by the development of high-performance materials, new devices and technology, along with the great improvement of computer and computational capacity.

Nowadays, bridge is not only a key part of transport line, but also a special art with distinct characteristics of the times. Its huge span, strong body expressive force and extraordinary scale remarkably affect the city appearance and the landscape.

Fig. 4-1 The Bridge of Sighs, Cambridge, United Kingdom [1]

Fig. 4-2 A flyover bridge, Osaka, Japan [1]

4.2 Development of Bridge Engineering

4.2.1 Ancient Bridges

In the beginning, people used tree trunks, vines or rock arches to cross the waterway or valley. Before the 17th century, bridges were generally constructed with wood, ropes and stone. These bridges usually have a simple geometry and very limited uses.

The Zhaozhou Bridge, located in the southern part of Hebei Province of China, is the world's oldest open-spandrel segmental arch bridge of stone construction, as shown in Fig. 4-3. There are smaller arches at each end of the bridge that transmit the load of the deck down to the main arch. The open spandrels allow some water to flow over the main arch when the river floods. Credited to the design of a craftsman named Li Chun, the bridge was constructed in the years 595~605 during Sui Dynasty (581~618). It has a span length of 37.02m and a total length of 50.82m. The rise is only 7.23m, providing a low profile. It is comprised of 28 thin, curved limestone slabs, joined with iron dovetails, so the arch could gently yield and adjust to the rise and fall of the abutments as they responded to the weight of traffic.

The Precious Belt Bridge or Baodai Bridge is a stone arch bridge near Suzhou, Jiangsu Province, China, as shown in Fig. 4-4. The bridge has a length of 317m, a width of 4.1m and a total of 53 arches. The three central arches are enlarged to allow for the passage of larger river vessels without masts. The average span of each arch is 4.6m.

Anping Bridge is a stone girder bridge in Fujian Province constructed in Southern Song Dynasty (1127~1279), as shown in Fig. 4-5. The length of the bridge is 2070m. The bridge is also known as the Wuli Bridge because its length is about 5 li (1 li is about 500m). It is a nationally protected historic site registered with the National Cultural Heritage Administration of China. Anping Bridge consists of 331 spans of granite beams resting on top of stone piers, among which the largest beam weighing 25t. The width of the bridge varies from 3m to 3.8m. It originally had five pavilions. Now, only one pavilion still exists.

Five Pavilion Bridge at Shouxi Lake, part of Slender West Lake in Yangzhou, is one of the most iconic sights in the city, as shown in Fig. 4-6. It was built in 1757 by a wealthy salt merchant. The Wuting (Five Pavillions) was formally called Lotus Flower Bridge because of its resemblance to the open petals of the lotus flower.

Fig. 4-3 Zhaozhou Bridge
Source: http://club.kdnet.net/dispbbs.asp?id=13380406&boardid=1.

Fig. 4-4 Suzhou Precious Belt Bridge
Source: https://baijiahao.baidu.com.

Fig. 4-5 Anping Bridge
Source: http://qz.fjsen.com/2016-03/31/content_17582488.htm.

In European countries, there are also many famous old bridges.

The Chapel Bridge is a 204m long bridge crossing the Reuss River in the city of

Fig. 4-6 Five Pavilions Bridge
Source: http://www.naic.org.cn/html/2017/jdtj_0815/17672.html.

Fig. 4-8 Guardian Aqueduct in France
Source: https://www.sohu.com/a/108472191_164590.

Fig. 4-7 Chapel Bridge
Source: https://www.sohu.com/a/328242269_822860.

Lucerne in Switzerland (Fig. 4-7). It is the oldest wooden covered bridge in Europe, and one of Switzerland's main tourist attractions. The covered bridge, constructed in 1333, was designed to help protect the city of Lucerne from attacks. Inside the bridge are a series of paintings from the 17th century, depicting events from Lucerne's history. Much of the bridge, and the majority of these paintings, were destroyed in a fire in 1993, though it was quickly rebuilt.

Guardian Aqueduct (Fig. 4-8) with the maximum span of 24m is an ancient Roman aqueduct bridge built in the 1st century to carry water over 50km to the Roman colony of Nemausus. It crosses the Gardon River near the town of Vers-Pont-du-Gard in southern France. The Guardian Aqueduct is the highest of all Roman aqueduct bridges, and one of the best preserved. The stones are cut with incredible precision and no mortar was used to render the aqueduct stable or watertight. The larger blocks weigh a staggering 5.5 tons.

During the 18th century, there were many innovations in the design. A major breakthrough in bridge technology came with the erection of the Iron Bridge in Coalbrookdale in England during 1779, using cast iron for the first time as arches to cross the River Severn (Fig. 4-9). It is a single arch iron

Fig. 4-9 The first iron bridge
Source: http://www.360doc.com.

Fig. 4-10 Brooklyn Bridge
Source: https://dy.163.com/article/DQCO0I1B0524SS65.html.

bridge with a main span of 30.63m. It became a UNESCO World Heritage Site in 1986 and remains an iconic feature of Britain's industrial past.

4.2.2 Modern Bridges

With the Industrial Revolution, steel, which has a high tensile strength, replaced wrought iron for the construction of larger bridges to support large loads, and later welded structural bridges of various designs were constructed.

Brooklyn Bridge, completed in 1883, was the longest suspension bridge in the world at the time of its opening, with a main span of 486.3m and a deck located 38.7m above mean high water (Fig. 4-10). It connects Manhattan and Brooklyn by spanning the East River, and has been designated a National Historic Landmark, a New York City landmark, and a National Historic Civil Engineering Landmark. The bridge has a wide pedestrian walkway open to walkers and cyclists. This walkway takes on a special importance in times of difficulty when usual means of crossing the East River have become unavailable during several blackouts and most famously after the attacks of September 11, 2001.

Tower Bridge is a combined bascule and suspension bridge in London, over the River Thames (Fig. 4-11). It is close to the Tower of London, which gives it its name and has become an iconic symbol of London. Construction started in 1886 and took 8 years to build. The bridge consists of two towers which are tied together at the upper level by means of two horizontal walkways which are designed to withstand the forces of the suspended sections of the bridge.

Sydney Harbour Bridge is one of Australia's most well-known landmarks that carries rail, vehicular, bicycle and pedestrian traffic between the Sydney central business district (CBD) and the North Shore (Fig. 4-12). It is the world's largest (but not the longest) steel arch bridge with the top of the bridge standing 134m above Sydney Harbour. It was also the world's widest long-span bridge, at 48.8m wide, until construction of the New Port Mann Bridge in Vancouver was completed in 2012. Because the steel expands or contracts depending on whether it is hot or

Fig. 4-11 Tower Bridge
Source: https://dy.163.com/article/DQCO0I1B0524SS65.html.

Fig. 4-12 Sydney Harbour Bridge
Source: https://dy.163.com/article/DQCO0I1B0524SS65.html.

Fig. 4-13 Golden Gate Bridge
Source: https://dy.163.com/article/DQCO0I1B0524SS65.html.

cold, the bridge is not completely stationary and can rise or fall up to 18cm.

The Golden Gate Bridge is a suspension bridge spanning the Golden Gate, the strait between San Francisco and Marin County to the north (Fig. 4-13). It was initially designed by engineer Joseph Strauss in 1917. At the time of its opening in 1937, it was both the longest and the tallest suspension bridge in the world, with a main span of 1280m and a total height of 227m. The famous red-orange color of the bridge was specifically chosen to make the bridge more easily visible through the thick fog that frequently shrouds the bridge. It has been declared one of the Wonders of the Modern World by the American Society of Civil Engineers.

The Normandy Bridge is a cable-stayed road bridge that spans the River Seine, linking Le Havre to Honfleur in Normandy, northern France (Fig. 4-14). Its total length is 2,143.21m, and the main span is 856m between the two towers. The span, 23.6m wide, is divided into four lanes for traffic and two lanes for pedestrians. The pylons, made of concrete, are shaped as upside-down Y. They weigh more than 20,000 tons and are 214.77m tall. More than 19,000 tons of steel and 184 Freyssinet cables were used in this bridge. Despite being a motorway toll bridge, there is a footpath as well as a narrow cycle lane in each direction allowing pedestrians and cyclists to cross the bridge free of charge.

The Akashi-Kaikyo Bridge, also known

Fig. 4-14 Normandy Bridge
Source: http://www.lovehhy.net/News/View/932669.

Fig. 4-15 Akashi-Kaikyo Bridge
Source: https://dy.163.com/article/DQCO0I1B0524SS65.html.

Fig. 4-16 Great Belt Bridge
Source: https://dy.163.com/article/DQCO0I1B0524SS65.html.

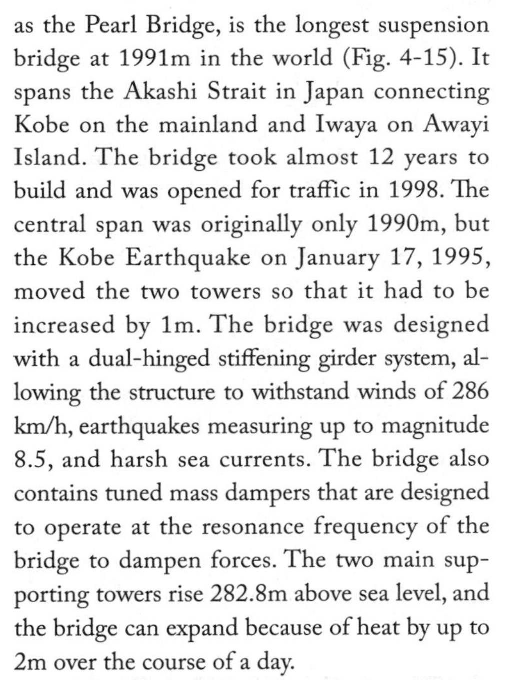

as the Pearl Bridge, is the longest suspension bridge at 1991m in the world (Fig. 4-15). It spans the Akashi Strait in Japan connecting Kobe on the mainland and Iwaya on Awayi Island. The bridge took almost 12 years to build and was opened for traffic in 1998. The central span was originally only 1990m, but the Kobe Earthquake on January 17, 1995, moved the two towers so that it had to be increased by 1m. The bridge was designed with a dual-hinged stiffening girder system, allowing the structure to withstand winds of 286 km/h, earthquakes measuring up to magnitude 8.5, and harsh sea currents. The bridge also contains tuned mass dampers that are designed to operate at the resonance frequency of the bridge to dampen forces. The two main supporting towers rise 282.8m above sea level, and the bridge can expand because of heat by up to 2m over the course of a day.

The Great Belt Bridge is actually two bridges—an Eastern and a Western section, split by the small island of Sprogø, as shown in Fig. 4-16. The East Bridge, a 1624m long suspension bridge crosses the deepest part of Storebælt between the island of Zealand and Sprogø. At 254m above the sea level, the East Bridge's two pylons are the highest points in Denmark. The West Bridge is a box girder bridge with a length of 6611m and has a vertical clearance for ships of 18m. It is actually two separate, adjacent bridges: the northern one carries rail traffic and the southern one carries road traffic. The pillars of the two bridges rest on common foundations below the sea level.

The Tatara Bridge is a cable-stayed bridge that is part of the Nishiseto Expressway (Fig. 4-17). The bridge has a center span of 890m. The Tatara Bridge was originally planned as a suspension bridge in 1973. In 1989, the design was changed to a cable-stayed bridge with the same span. By building a cable-stayed bridge, a large excavation for an anchorage would not be needed, thereby lessening the environmental impact on the surrounding area. The steel towers are 220m high and shaped like an inverted Y. The side-spans are 164.5m and 257.5m respectively, and there are also three very small cable spans.

The Millau Viaduct is a multi-span cable-

Fig. 4-17 Tatara Bridge
Source: https://www.zhihu.com/question/37395748.

Fig. 4-18 Millau Viaduct
Source: https://dy.163.com/article/DQCO0I1B0524SS65.html.

stayed road bridge that spans the valley of the River Tarn near Millau in southern France (Fig. 4-18). Each of the seven pylons is supported by four deep shafts with 15m deep and 5m in diameter. It is the tallest vehicular bridge in the world, with the highest pylon's summit at 343m—slightly taller than the Eiffel Tower. The abutments are concrete structures that provide anchorage for the road deck to the ground. The metallic road deck with total mass of around 40,000 tons, is 2460m long and 32m wide. It comprises eight spans. The six central spans measure 342m, and the two outer spans are 204m. The deck has an inverse airfoil shape, providing negative lift in strong wind conditions.

Donghai Bridge is the first sea-crossing bridge in China to connect Shanghai's Pudong New Area with the offshore Yangshan Deep-Water Port in Zhejiang's Shengsi County (Fig. 4-19). It has a total length of 32.5km and width of 31.5m. Most of the bridge is a low-level viaduct. There are also cable-stayed sections to allow for the passage of large ships, the largest with a span of 420m. Donghai Bridge is part of the S2 Hulu Expressway.

The Sutong Yangtze River Bridge is a cable-stayed bridge that spans the Yangtze River in China between Nantong and Changshu, a satellite city of Suzhou, in Jiangsu province (Fig. 4-20). With a span of 1088m, it was the cable-stayed bridge with the longest

Fig. 4-19 Donghai Bridge
Source: https://www.photophoto.cn.

Fig. 4-20 Sutong Yangtze River Bridge
Source: http://gov.eastday.com/renda/dfzw/ycbd/u1ai6180779.html.

Fig. 4-21 Chaotianmen Bridge
Source: http://cq.cqnews.net.

main span in the world in 2008~2012. Its two side spans are 300m each, and there are also four small cable spans. Two towers of the bridge are 306m high and thus the third tallest in the world. The total bridge length is 8206m. The tower is an inverted Y-shaped reinforced concrete structure with one connecting girder between tower legs. The bridge deck is a steel box girder with internal transverse and longitudinal diaphragms and fairing noses at both sides of the bridge deck. The total width of the bridge deck is 41m including the fairing noses.

The Chaotianmen Bridge is a road-rail bridge over the Yangtze River in the city of Chongqing, China (Fig. 4-21). The bridge, at the time of its opening in 2009, was the world's longest through arch bridge. The continuous steel truss arch bridge with tie girders has a height of 142m from middle supports to the arch top, main span of 552m and a total length of 1741m. It carries 6 lanes in two ways and a pedestrian lane on each side on the upper deck. The lower deck has 2 traffic lanes on each side with a reservation in the middle for the Chongqing Metro Loop Line.

The Tianxingzhou Yangtze River Bridge is a combined road and rail bridge across the Yangtze River in the city of Wuhan, the capital of Hubei Province of China, as shown in Fig. 4-22. The main span of the bridge is 504m. The rail and highway traffic run on separate levels of this bridge. The upper level carries six highway lanes while the lower deck accommodates four railway tracks. In order to reduce floorbeam span and increase cross-sectional stiffness, the bridge is designed with three truss planes of stay cables.

Jiaozhou Bay Bridge (or Qingdao Haiwan Bridge) is a 26.7km long roadway bridge in Shandong Province of eastern China, which is part of the 41.58km Jiaozhou Bay Connection Project (Fig. 4-23). The construction used 450,000 tons of steel and 2.3 million cubic meters of concrete. The bridge is designed to be able to withstand severe earthquakes, typhoons and collisions from ships. It is

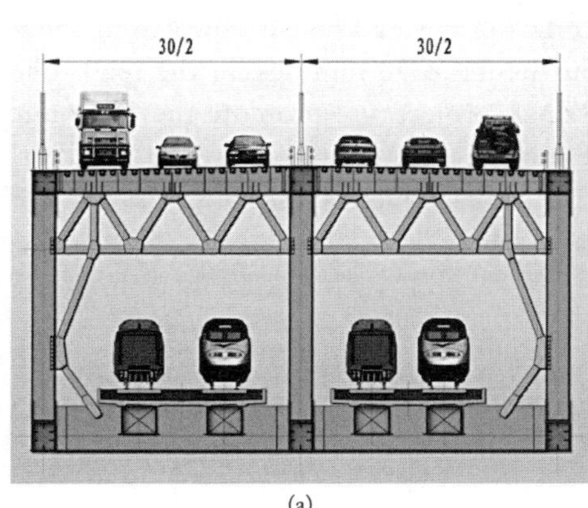

(a)

(b)

Fig. 4-22 Tianxingzhou Yangtze River Bridge
Sources: (a) https://bbs.zhulong.com/102020_group_200619/detail21071468/; (b) https://baike.baidu.com.

Fig. 4-23 Jiaozhou Bay Bridge
Source: http://qdtoqihpan1s.co.sonhoo.com/company_web/sale-detail-7481749.html.

(a)

(b)

(c)

Fig. 4-24 Hong Kong–Zhuhai–Macau Bridge
Sources: (a) http://k.sina.com.cn/article_3164957712_bca56c10042005cv6.html; (b) https://baike.baidu.com; (c) https://dy.163.com/article/DSIQP5SQ0524U85C.html.

supported by 5238 concrete piles. The cross section consists of two beams in total 35m wide, carrying six lanes with two shoulders.

Hong Kong-Zhuhai-Macau Bridge (HZMB) is the longest bridge-cum-tunnel sea crossing in the world. The length from Hong Kong Port to Zhuhai Port and Macao Port is approximately 42km, comprising the 12km Hong Kong Link Road and the 29.6km Main Bridge. The total length of HZMB is 55km when the approximately 13.4km Zhuhai Link Road is included. The Main Bridge, a dual 3-lane carriageway, is approximately 30km long comprising a sea viaduct section of approximately 23km and a tunnel section of approximately 7km. The viaduct section consists of three cable-stayed bridges, namely Qingzhou Channel Bridge, Jianghai Channel Bridge and Jiuzhou Channel Bridge, as shown in Fig. 4-24. Among the three bridges, the Qingzhou Channel Bridge has the longest main span of about 458m long. To minimize disturbance to marine life and to ensure the quality of works, precast and prefabrication techniques were widely adopted in the project. Pile caps, steel bridge decks and steel towers were manufactured off-site and transported to the construction site for erection. The HZMB links three major cities—Hong Kong, Zhuhai and Macau, which are geographically close but separated by water. With the bridge in place, travelling time between Zhuhai and Hong Kong would be cut down from about four hours to 30 minutes on the road. The HZMB project will promote the economic development of the whole area of the Pearl River Delta, which is also known as Greater Bay Area.

4.3 Components and Classification of Bridges

4.3.1 Components of Bridges

Bridge structures are composed of superstructure, substructure and accessories. The components of a typical bridge are shown in Fig. 4-25.

Superstructure is the portion of the bridge that supports the deck and connects one substructure element to another. It is composed of span structure and supports. Bridge span structure is used to bear actions brought by the dead load of structure and vehicle loads. Support system is applied to support the span structure and transmit the loads to the piers, so it should be suitable for the displacement restrictions under the actions of loads, temperature and other factors. The bigger the span of bridge is, the more complicated the structure is and the more difficult the construction is.

Substructure refers to the structure that is laid on the ground to support the bridge span structure and transmit loads to the ground, including piers, abutment, pier foundation and so on. It is a very important part of bridge structure. Bridge pier is located between the adjacent spans in the multi-span structure to support superstructure. Abutment is located on both ends of the bridge. One side of abutment connects with the embankment to prevent its slumping and the other one supports the superstructure of the bridge. Pier foundation protects the pier and transmits loads to the ground. It is the base of a bridge and has a

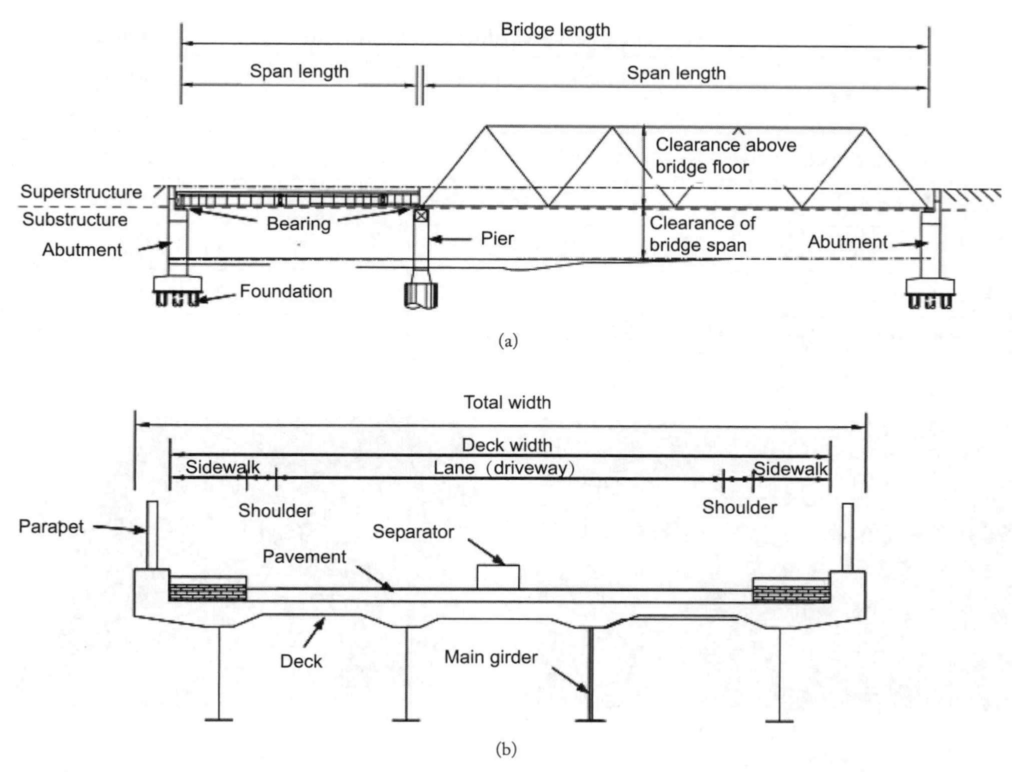

Fig. 4-25 Components of a bridge
(a) Longitudinal Direction; (b) Cross Section [1]

vital effect on the safety of the whole structure. Besides, it is the most complicated and difficult part during the bridge construction.

Accessories include bridge deck pavement, drainage and waterproof system, railing (or crash barrier), expansion joints and lighting. All of the above are related to the service functions of the bridge.

4.3.2 Classification of Bridges

Bridges can be categorized in several different ways. Common categories are listed as follows:

(1) According to the length of bridges, bridges can be classified as super-long span bridges, long span bridges, major bridges and minor bridges (Table 4-1).

(2) According to the construction material, bridges can be classified as steel bridges (Fig. 4-26), concrete bridges (Fig. 4-27), stone bridges (Fig. 4-28), wooden bridges (Fig. 4-29) and so on.

(3) According to the function, bridges can be classified as railway bridges (Fig. 4-30), highway bridges (Fig. 4-31), road-rail bridges (Fig. 4-32), footbridges (Fig. 4-33), aqueduct bridges (Fig. 4-34) and so on.

(4) According to the position of the bridge floor relative to superstructures, bridges can be classified as deck bridges (Fig. 4-35), half-through bridges (Fig. 4-36) and through bridges (Fig. 4-37).

(5) According to the horizontal alignment (Fig. 4-38), bridges can be classified as straight bridges, skewed bridges and curved bridges.

(6) According to the type of super structure, bridges can be classified as girder bridges (Fig. 4-39), arch bridges (Fig. 4-40), cable-stayed bridges (Fig. 4-41), suspension bridges (Fig. 4-42) and combination bridges (Fig. 4-43).

Types of bridges according to their lengths **Table 4-1**

Types	Total length of multi-span L (m)	Length of single-span L_k (m)
Super-long span bridge	$L>1000$	$L_k>150$
Long span bridge	$100\leqslant L\leqslant 1000$	$40\leqslant L_k\leqslant 150$
Major bridge	$30<L<100$	$20\leqslant L_k<40$
Minor bridge	$8\leqslant L\leqslant 30$	$5\leqslant L_k<20$

Date Source: Technical Standard of Highway Engineering JTG B01—2014.

Fig. 4-26 Steel bridge: Chaotianmen Bridge
Source: https://www.sohu.com.

Fig. 4-27 Concrete bridge
Source: https://baike.baidu.com.

Fig. 4-28 Stone bridge: Marco Polo Bridge
Source: https://dy.163.com/article/FCH6961R05438PKW.html.

Fig. 4-29 Wooden bridge: Kintaikyo Bridge
Source: https://baike.baidu.com.

Fig. 4-30 The former Amarube Bridge[1]

Fig.4-31 Sanshui River Grand Bridge, Xunyi County, Shanxi Province, China
Source: https://www.meipian.cn/3zhhs4m.

Fig. 4-32 Wuhan Yangtze River Bridge
Source: http://www.hubeilyw.com/raiders/show_103.html.

Fig.4-33 Lagan Weir Footbridge, Belfast, United Kingdom[1]

Fig. 4-34 Guardian Aqueduct, France
Source: https://www.sohu.com.

Fig. 4-35 Wanzhou Bridge, Wanzhou, Chongqing, China
Source: https://m.sohu.com/a/209174216_99953977.

Fig. 4-36 Wushan Yangtze River Bridge, Wushan, Chongqing, China
Source: https://clickme.net/40698.

Fig. 4-37 Jiefang Bridge, Guangzhou, Guangdong Province, China
Source: https://baike.baidu.com.

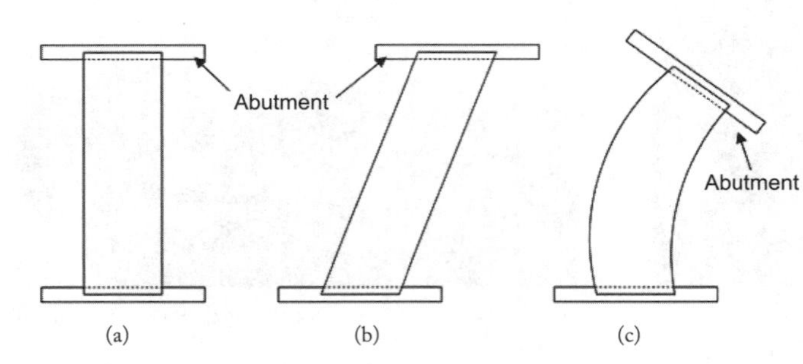

Fig. 4-38 Classification of bridges by the horizontal alignment[1]
(a) Straight bridge; (b) Skewed bridge; (c) Curved bridge

Fig. 4-39 Qingdao Jiaozhou Bay Bridge, Qingdao, China
Source: https://mp.weixin.qq.com/s/X-Jju7NTU3_EpIAG7tXXTw.

Fig. 4-40 New River Gorge Bridge, Fayette County, West Virginia, United States
Source: https://gamecrate.com/fallout-76-and-real-west-virginia/20000.

Fig. 4-41 Russky Bridge, Vladivostok, Russia
Source: https://www.tfzx.net/article/10730841.html.

Fig. 4-42 Xihoumen Bridge, Zhejiang Province, China
Source: https://baike.baidu.com.

Fig. 4-43 Seri Saujana Bridge, Lebuh Sentosa, Putrajaya, Malaysia
Source: https://www.tripadvisor.cn.

4.4 Bridge Structure Systems

4.4.1 Girder Bridges

A girder bridge is a basic, common type of bridge where the bridge deck is built on top of such supporting beams, that have in turn been placed on piers and abutments that support the span of the bridge. The types of beams used for girder bridges could be I-beam girders, T-beam girders, trusses, or box girder beams that are made of steel or concrete. Girder bridges are most commonly used for straight bridges that are 10~200m long, such as light rail bridges, pedestrian overpasses, or highway flyovers.

The superstructures of girder bridges generally vary by support type (simply supported, continuous, or superstructure and substructure rigidly connected). The main girders of a simply supported bridge are supported by a movable hinge at one end and a fixed hinge or roller at the other end. A continuous girder bridge has more than one span and it is carried by several supports (minimum of three supports). The moments are important in simply supported bridges as they determine both stresses and the girder curvature and deflection. Furthermore, the values of shearing force and bending moment will usually vary along a girder.

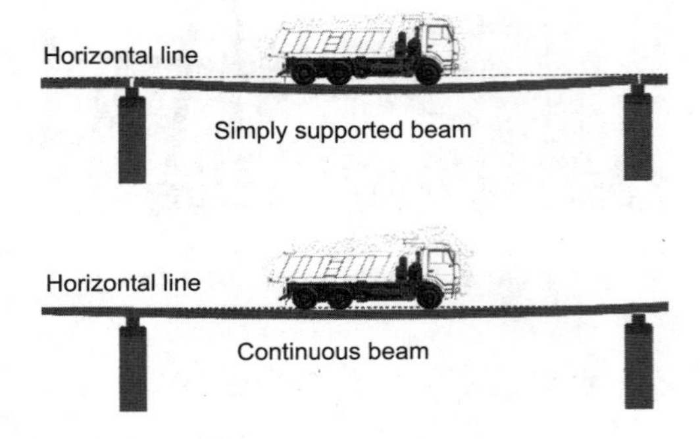

Fig. 4-44 Comparison of deflections between continuous beam and simply supported beam
Source: https://mp.weixin.qq.com/s/X-Jju7NTU3_EplAG7tXXTw.

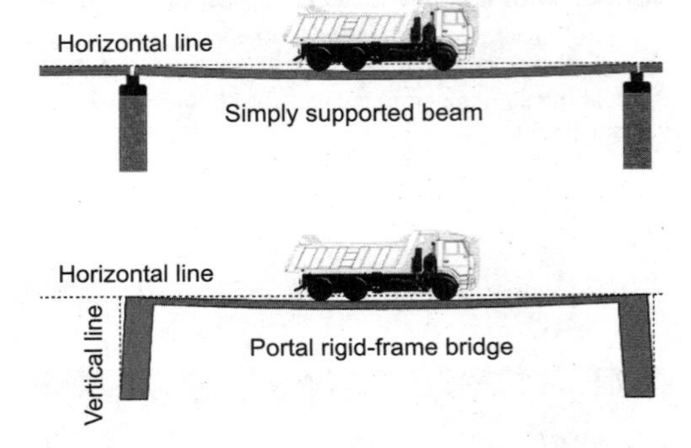

Fig. 4-45 Comparison of deflections between simply supported girder bridge and rigid-frame bridge
Source: https://mp.weixin.qq.com/s/X-Jju7NTU3_EplAG7tXXTw.

The girders of continuous span bridges are supported by two or more supports. The section of the girder at the support has hogging moments and the spans has sagging. Hence, the tension zone near the supports lies at the top fibers and in sagging at the bottom fibers. The stiffness of the continuous girder is more than that of the simply supported girder. The comparison of deflections between continuous beam and simply supported beam with the same main span and load is shown in Fig. 4-44.

A rigid-frame bridge consists of superstructure supported on vertical or slanted monolithic legs (columns), in which the superstructure and substructure are rigidly connected to act as a unit. This type of girder bridge is economical for moderate medium-span lengths. The use of rigid-frame bridges began in Germany in the early 20th century. The connections between superstructure and substructure are rigid connections which transfer bending moment, axial forces and shear forces. A bridge design consisting of a rigid frame can provide significant structural benefits but can also be difficult to design and construct. Moments at the center of the deck of a rigid-frame bridge are smaller than the corresponding moments in a simply supported deck. Therefore, a much shallower cross section at mid-span can be used. Additional benefits are that less space is required for the approaches and structural details for where the deck bears on the abutments are not necessary. However, as a statically indeterminate structure, the design and analysis are more complicated than that of simply supported or continuous bridges. The comparison of deflections between portal rigid-frame bridge and simply supported beam with the same span and load is shown in Fig. 4-45.

4.4.2 Arch Bridges

An arch is a curved structure that supports the loads parallel to its axis of symmetry, and a bridge with an arch as its load carrying system is called an arch bridge. The arch bridge

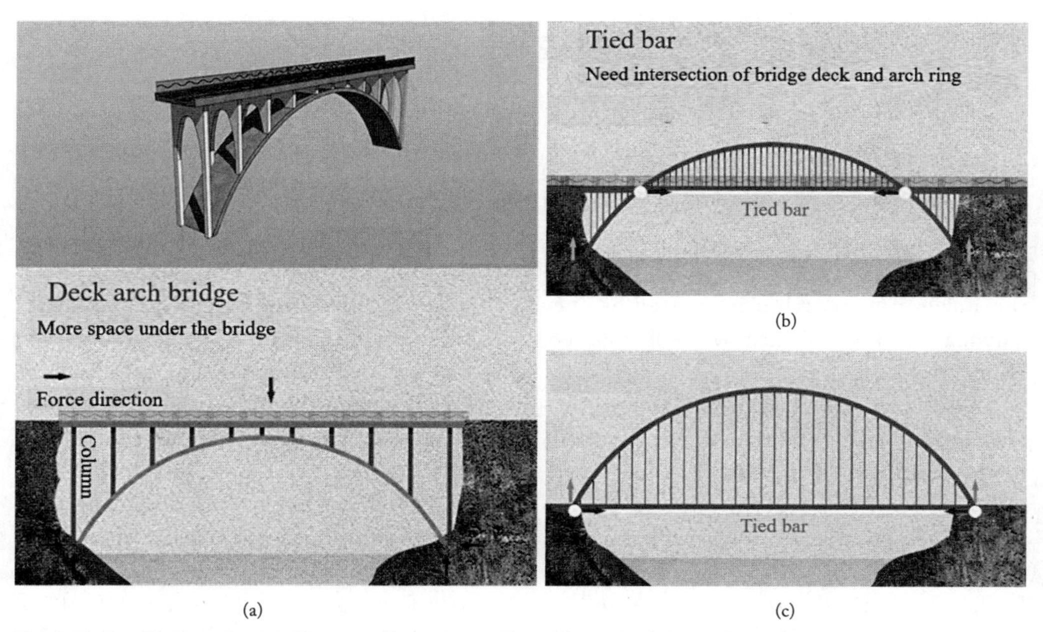

Fig. 4-46 Classification of arch bridge according to the position of the arch relative to the deck
(a) Deck arch bridge; (b) Half-through arch bridge; (c) Through arch bridge
Source: https://mp.weixin.qq.com/s/X-Jju7NTU3_EplAG7tXXTw.

carries the load primarily by compression, which exerts on the foundation both vertical and horizontal forces. Arch foundations must therefore prevent both vertical settling and horizontal sliding. In spite of the more complicated foundation design, the structure itself normally requires less material than a beam bridge of the same span. Concrete, stone, brick and other such materials which are strong in compression and poor in tension are usually used to construct arch bridges.

The arch bridges can be classified according to the position of the arch relative to the deck, including deck arch, through arch, and half-through arch bridges as shown in Fig. 4-46. The deck arch bridge comprises an arch where the deck is completely above the arch. The area between the arch and the deck is known as the spandrel. If the spandrel is solid, usually the case in a masonry or stone arch bridge, the bridge is called a closed-spandrel deck arch bridge. If the deck is supported by a number of vertical columns rising from the arch, the bridge is known as an open-spandrel deck arch bridge. Through or half-through arch bridge has an arch whose base is at or below the deck, but whose top rises above it, so the deck passes through the arch. The central part of the deck is supported by the arch via suspension cables or tied bars, as with a tied-arch bridge. The ends of the bridge may be supported from below, as with a deck arch bridge. Any part supported from arch below may have spandrels that are closed or open.

An arch system can also be grouped as a fixed-fixed arch, two-hinged arch, or three-hinged arch according to the support type (Fig. 4-47).

In the early part of the 19th century, three-hinged arches were commonly used for the long span structures as the analysis of such arches could be done with confidence. However, with the development in structural analysis, for long span structures starting from late 19th century, engineers adopted two-hinged and fixed-fixed (or hingeless arch) arches.

Two-hinged arch is the statically indeterminate structure to degree one. Usually,

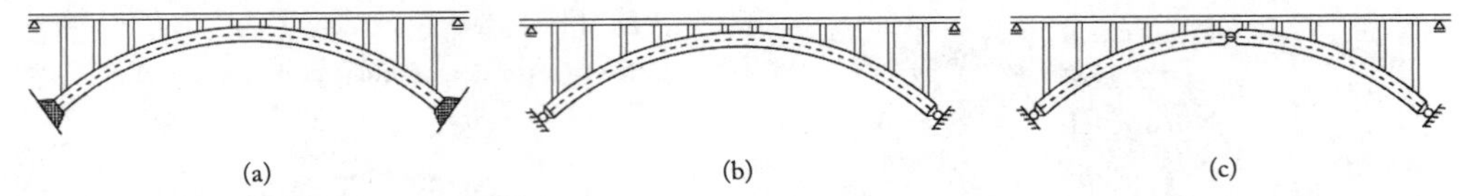

Fig. 4-47 Types of arch bridges
(a) Fixed-fixed arch; (b) Two-hinged arch; (c) Three-hinged arch[1]

the horizontal reaction is treated as the redundant and is evaluated by the method of least work.

The fixed-fixed arch is fixed at the abutments so that moment is transmitted to the abutment. The fixed arch has three redundancies, no rotation allowed at the foundations. Fixed arch is a very stiff structure and suffers less deflection than other arches. In the case of fixed–fixed arch, there are six reaction components: three at each fixed end.

4.4.3 Cable-stayed Bridges

In cable-stayed bridges, the superstructure is supported by cables or stays that are anchored to towers or pylons located at the main piers (Fig. 4-48). The weight of the deck is supported by a number of nearly straight diagonal cables in tension running directly to one or more vertical towers. The towers transfer the cable forces to the foundations through vertical compression. The tensile forces in the cables also put the deck into horizontal compression. The cables are typically in two planes separated by the width of the roadway, though numerous bridges have been built with a central plane of stays between the two opposing lanes of traffic. This requires a torsionally resistant superstructure. The cables are straight, resulting in greater stiffness than a suspension bridge. Anchoring the cables to the deck can be problematic to replace the deck. In general, a cable-stayed bridge is less efficient in carrying dead load than a suspension bridge but is more efficient in carrying live load. The most economical span length for a cable-stayed bridge is 150~600m. There have been some problems with cable excitation during rain/wind events, particularly on the longer stays. A cable-stayed bridge is very modern and pleasing in appearance and fits extremely well in almost any environment. Currently, the Russky Bridge in Russia is the largest cable-stayed bridge with the world's longest span at 1104m.

The cable-stayed bridge can be designed as single span, two spans, three spans, or multiple spans. However, cable-stayed bridges having either three or two cable-stayed spans are more widely used, because the stay cables and the anchor piers are important for the stability of the pylon. The cable-stayed bridges can be built with only one tower, like the Erasmus Bridge in Rotterdam, the Toyosato-Ohashi Bridge in Osaka, and the Chuoohashi

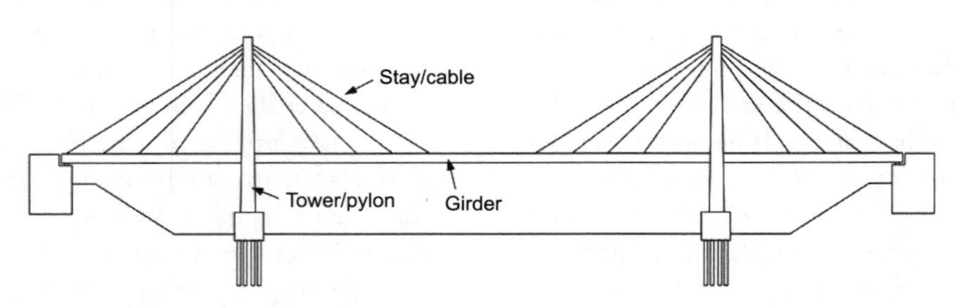

Fig. 4-48 A typical cable-stayed bridge[1]

Bridge in Tokyo. A multi-span cable-stayed bridge consists of more than three spans without intermediate anchor blocks, which would break the continuity. The problem with this type of bridge is its capacity to withstand alternated live loads. As shown in Fig. 4-49, this problem can be solved in the following ways: 1) Increasing the stiffness of pylons, for example, by using a frame braced pylon as shown in Fig. 4-49(a); 2) Using additional horizontal cables on tower tops, which can directly transfer any out-of-balance forces to the anchor stays in the end spans; 3) Using additional cables to connect the top of the internal pylons to the adjacent pylon at deck level so that any out-of-balance forces are resisted by the stiffness of the pylon below deck level; 4) Using additional tie-down piers at span centers; or 5) Adding additional cables at the mid-spans. These cables cross each other and extend for approximately 20% of span length beyond the span center. This method is efficient in reducing the bending moments in the girder and the towers to an acceptable level and still retains the slender look of a conventional cable-stayed bridge.

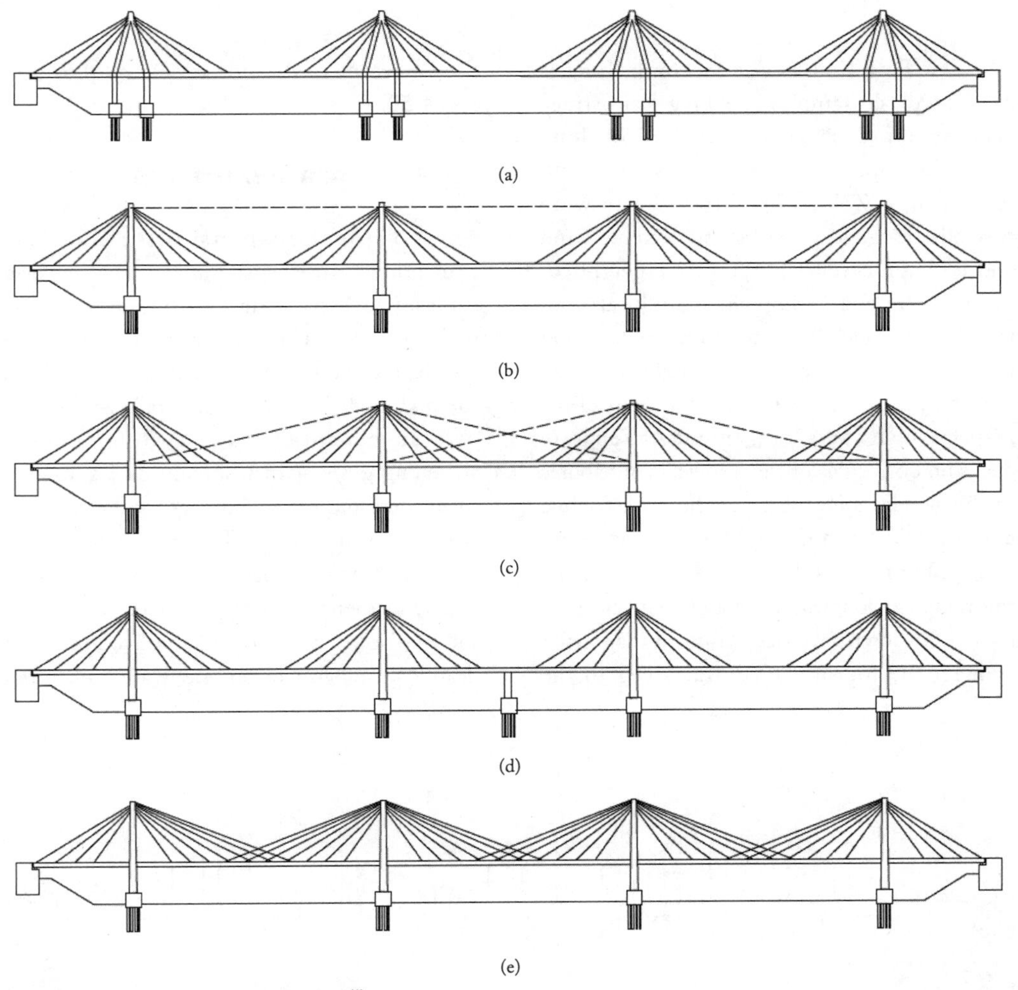

Fig. 4-49 Multi-span cable-stayed bridges[1]
(a) Aframe braced pylon; (b) Additional cable system between pylon tops; (c) Additional cable system between intermediate pylon and girders; (d) Additional tie-down piers at span centers; (e) Additional cables at the mid-spans

Fig. 4-50 The Ting Kau Bridge in Hong Kong [1]

The Ting Kau Bridge in Hong Kong is such a typical example. The Ting Kau Bridge and the approach viaduct are 1875m long while the triple-tower bridge has an overall length of 1177m. The three towers were specially designed to withstand extreme wind and typhoon conditions, and have heights of 170m, 194m and 158m, which are located on the Ting Kau headland, on a reclaimed island in Rambler Channel (which is 900m wide), and on the north-west Tsing Yi shoreline, respectively. The arrangement of separate decks on both sides of the 3 towers contributes to the slender appearance of the bridge while helping it act favourably under heavy wind and typhoon loads. In order to stabilize the central tower, longitudinal stabilizing cables up to 464.6m long have been used to diagonally connect the top of the central tower to the deck adjacent to the side towers, as shown in Fig. 4-50.

4.4.4 Suspension Bridges

A typical suspension bridge is a continuous girder suspended by suspension cables, which pass through the main towers with the aid of a special structure known as a saddle, and end on big anchorages that hold them (Fig. 4-51). The structural components of a suspension bridge system include stiffening girders/trusses, main cables, towers, suspenders and the anchorages for the cables at each end of the bridge. The main cables made of high-strength steel strands or wires are a crucial element for overall structural safety of the bridge because most of the live loads and dead loads of suspension bridge are transferred

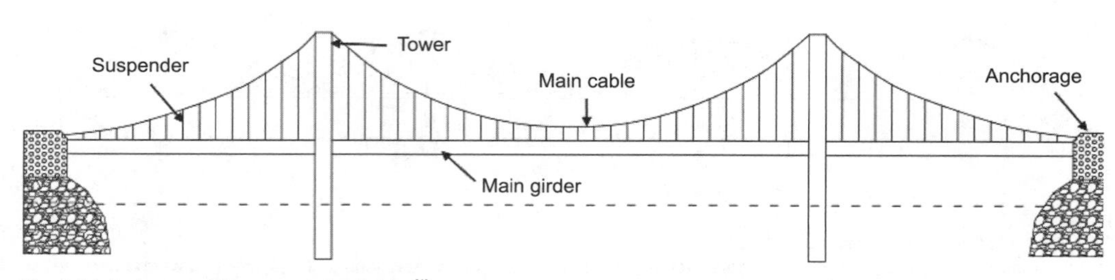

Fig. 4-51 An image of the suspension bridge[1]

Fig. 4-52 Rainbow Bridge
Source: https://baike.baidu.com.

Fig. 4-53 Liede Bridge
Source: https://baike.baidu.com.

to the anchorage systems by the main cables. Suspension bridges also have smaller cables called suspenders which run vertically from the deck up to the main supporting cables. The suspenders move the deck's compression forces to the towers through the main supporting cables. This creates graceful arcs between the towers and down to the ground. The towers of a suspension bridge can be fairly thin. That's because the forces at work are carefully balanced on each side of the towers. The force of the deck pulls inward on the towers. At the same time, the main support cables extend beyond the towers to anchor each end. These are usually solid rock or heavy concrete blocks secured underground. The anchors pull outward on the towers with an equal force to that of the deck. The weight of the bridge is thus transferred on the tower. Today's suspension bridges can span distances as great as 2000m or more.

Based on the number of towers, there are single-tower, two-tower, or multi-tower suspension bridges. Among them, two-tower suspension bridges with three spans are the most commonly used in engineering practice, like the Rainbow Bridge in Tokyo (Fig. 4-52). The Liede Bridge (Fig. 4-53) in Guangdong Province is a typical single-tower suspension bridge. For multi-tower suspension bridges with more than two towers, the horizontal displacement of the tower tops due to live loads can be a concern. It is necessary to control the horizontal displacement of the middle tower

Fig. 4-54 Tamate Bridge
Source: https://structurae.net/en/structures/tamate-bridge.

top. The Tamate Bridge (Fig. 4-54) built in 1928 in Japan is a typical multi-tower suspension bridge, which is still in use now. Since then, several bridges were built in France (Pont de Châteauneuf-sur-Loire, 1932; Chatillon Bridge, 1951; and Bonny-sur-Loire Bridge, etc.), Switzerland (Giumaglio Footbridge), Mozambique (Samora Machel Bridge, 1973), and Nepal (Dhodhara-Chandani Suspension Bridges, 2005). These bridges are generally built in a relatively short span except the Taizhou Yangtze River Bridge in China, which has three main towers and two main spans with a span length of 1080m, currently is the largest multi-tower suspension bridge.

4.4.5 Combination Bridges

A combination bridge consists of two or more structural systems (i.e. arch bridge combined with cable-stayed bridge, girder bridge combined with arch bridge, cable-stayed bridge combined with suspension bridge).

The Liancheng Bridge, also called the Fourth Xiangjiang River Bridge, links the east and west of Xiangtan (a city in Hunan Province, China), which is divided into two parts by the Xiangjiang River. As shown in Fig. 4-55, this cable-stayed concrete-filled steel tubular arch bridge has a unique configuration, which combines the features of an arch bridge and a two-pylon cable-stayed bridge. The main span of the bridge is 400m in length, and the side spans are 120m in length. The width of the bridge is 27m. The bridge has two parallel arch ribs, each of which has a rectangle cross section consisting of six steel tubes. The six steel tubes have an outer diameter of 850mm and a thickness that varies between 20mm and 28mm depending on the position of the arch rib. The main arch floor system, supported by two rows of 39 steel wire rope hangers with intervals of 8m, is composed of the deck, I-shaped transverse girders and longitudinal stringers. The stay cables are anchored on the bridge deck and arch ribs with intervals of 10m and 8m, respectively.

Tied-arch bridge (also called bowstring-arch or bowstring-girder bridge) is a type of bridge that has an arch rib on each side of the roadway (deck), and one tie beam on each arch to support the deck. Vertical ties connected to the arches support the deck from above. Tips of the arch of this bridge are tied together by a bottom chord. This allows the bridge to be constructed with less robust foundations because force on the abutments is low. Tied-arch bridges can be built on elevated piers or in areas of unstable soil. One more positive attribute of this type bridge is that it does not depend on horizontal compression forces for its integrity which allows them to be built off-site and then transported into place. Shenzhen Beizhan Bridge (Fig. 4-56) located in Guangdong Province is a rigid-frame tied-arch bridge. It has a span of 150m and clear width of 23.5m. The rise to span ratio of the arch is 1/4.5 and the axis is a centenary curve. The truss arch rib is 3m high and 2m wide, composed of four concrete filled steel tube (CFST) chords. The piers of the bridge are also CFST columns. The floor system is steel-composite structure, composed of prestressed steel transverse box beams and precast prestressed concrete hollow slabs.

Changzhou Longcheng Bridge is a single-tower self-anchoring cable-stayed bridge combined with suspension bridge (Fig. 4-57). The bridge is 664m long and 40m wide with 6 lanes. The main bridge and the tower of Longcheng Bridge are made of steel, with a total steel of more than 2000 tons. The two sides of the tower adopt the cooperative force mode of cable-stayed and suspension respectively. The bridge has become a landmark due to its unique shape and novel structure.

Fig. 4-55 The Liancheng Bridge
Source: https://sina.cn.

Fig. 4-56 Shenzhen Beizhan Bridge
Source: https://www.sohu.com/a/315429373_355786.

Fig. 4-57 Changzhou Longcheng Bridge
Source: http://gc.zbj.com/20151022/n32074.shtml.

4.5 Bridge Construction

4.5.1 Construction of Bridge Superstructures

In the 20th century, bridge construction technology evolved and was fueled by the Industrial Revolution. At the turn of the century, steel bridges were riveted together, not bolted; concrete bridges were cast in place, not precast; and large bridge members were built from lacing bars and smaller sections, not rolled in one piece. Plastic had not yet been invented. Bridge construction is changing as the new millennium begins. New construction techniques and new materials are emerging. The construction methods for different types of bridges are listed in Table 4-2.

(1) Cast-in-situ Method

The case-in-situ bridges are constructed fully in its final location, thus have relatively good structural integrity, and can be constructed into different shapes. However, building bridges with cast-in-situ concrete today suffers to some extent of inefficiency and less developed production methods. The construction work is time and labor consuming, expensive and often consists of poor working environment. A highway concrete bridge built in this method is shown in Fig. 4-58.

(2) Precast Method

The concrete bridges can be precast, which means the bridges are built at other locations and then transported to the construction site for placement in the whole bridge structure, as shown in Fig. 4-59. This method can have higher construction speed because the piling and member fabrication can be performed simultaneously and more efficient in suitable and well-organized site (factory), and is less affected or not affected from season to season. Also, the quality of the precast members can be better controlled because of the established procedures by “factory production”. However, as there is a distance from fabrication yard to bridge location, transportation could be a

Construction methods for different types of bridges

Table 4-2

Construction methods	Applicable span (m)	Girder bridge			Rigid-frame bridge	Arch bridge			Cable-stayed bridge	Suspension bridge
		Simply supported beam	Cantilever beam	Continuous beam		Masonry arch	Standard and composite arch	Truss arch		
Cast-in-situ method	20~60	√	√	√	√	√	√		√	
Precast method	20~50	√	√	√	√		√	√	√	√
Span-by-span method	20~60	√	√	√	√					
Balanced cantilever method	50~320		√	√	√		√	√	√	
Rotation method	20~140		√	√	√		√	√	√	
Incremental launching method	20~70			√	√				√	
Slide-in bridge Construction method	30~100	√	√	√					√	
Lift-slab method	10~80	√	√	√			√			

Fig. 4-58 Posttensioned concrete bridge, cast-in-situ[1]

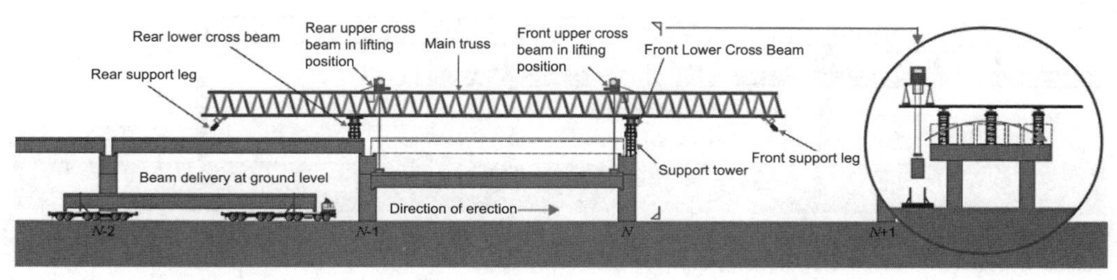

Fig. 4-59 Precast method
Source: https://structuraltechnologies.com/bridge-construction-systems/.

problem. The transportation facilities, lifting equipment and element weight/size must be considered.

In the last few decades, this construction method (Fig. 4-60) has been widely used around the world due to the reduction of costs, construction time, environmental impacts and the maintenance of traffic. They also offer additional structural advantages of durability, fire resistance, deflection control, insensitivity to fatigue and other redundancies. Precast segmental erection techniques for concrete bridges include the erection on falsework, erection by gantry, erection by crane, erection by lifting frame, full span erection techniques, etc.

(3) Span-by-span Method

Span-by-span (Fig. 4-61) method is a relatively new construction technique historically associated with cantilever construction but the advancement in external prestressing has enabled its own potential use to grow. Today it is considered to be the most economic and rapid method of construction available for long bridges and viaducts with individual spans up to 60m.

Decks are begun at one abutment and constructed continuously by placing segments to the other end of the bridge. Segments can be positioned by a temporary staying mast system through more commonly using an

Fig. 4-60 Precast concrete segmental construction
Source: https://structuraltechnologies.com/bridge-construction-systems/.

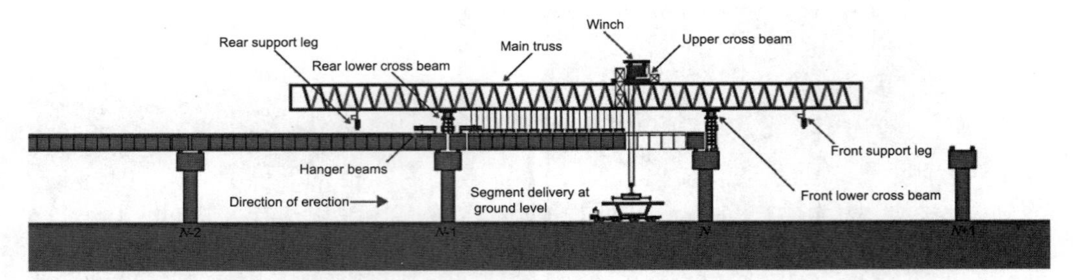

Fig. 4-61 Span-by-span erection with launching gantries
Source: https://structuraltechnologies.com/bridge-construction-systems/.

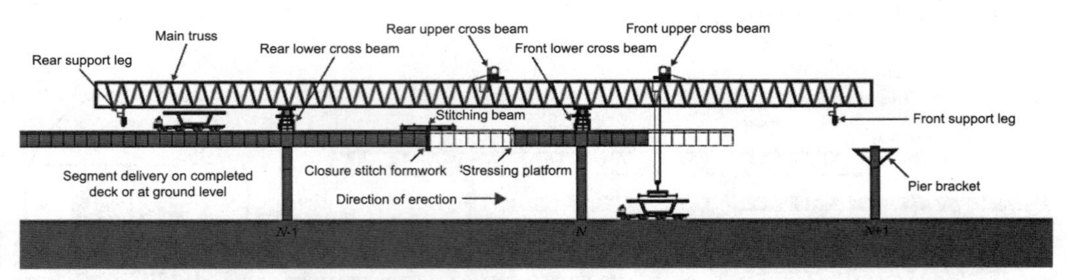

Fig. 4-62 Balanced cantilever erection with launching gantries
Source: https://structuraltechnologies.com/bridge-construction-systems/.

assembly truss. Before segments are placed, the truss with sliding pads is braced over two piers. Depending on the bridge location, the segments are then transported by lorry or barge to the span under construction. Each segment is then placed on the sliding pads and slid into its position. Once all segments are in position, the pier segment is then placed. The final stage is then begun by running longitudinal prestressing tendons through segments' ducts and prestressing the entire span. Deck joints are then cast and closed, and ducts are grouted. When a span is constructed, the assembly truss is lowered and moved to the next span where construction cycle begins until the bridge is completely constructed.

(4) Balanced Cantilever Method

Balanced cantilever method (Fig. 4-62) is an economical method when cast-in-situ with formworks is expensive or access below the bridge is practically impossible. Construction starts from the top of a pier, and the concrete segment is normally fixed to the pier either permanently or temporarily during the construction. This method can be used for both cast-in-situ and precast concrete bridge constructions. After a segment is cast, the formwork moves for the construction of the next segment. The segments on both sides can be built simultaneously so that the unbalanced moment is kept to a minimum. Also, rapid construction becomes possible by doing this. A temporary support may be used to improve the stability of the bridge during construction. A concrete girder bridge constructed by using balanced cantilever method is shown in Fig. 4-63.

This method is more frequently used for

Fig. 4-63 Balanced cantilever method of a girder bridge[1]

modern cable-stayed bridge construction, in which bridge deck is directly supported by the cable during construction. In this case, the cable-stayed bridge is always in a cantilever condition before the deck erection is completed. The construction can be carried out in the following stages: 1) Pylons and the deck units above the main piers are erected and fixed to the piers; 2) New deck segments are erected by free cantilevering from the pylon, either symmetrically in both directions or only into the main span. Simultaneously, the stay cables are installed and tensioned initially to relieve the bending moments in the deck; and 3) the stage 2) is repeated until the decks at midspan are connected, as shown in Fig. 4-64. For this

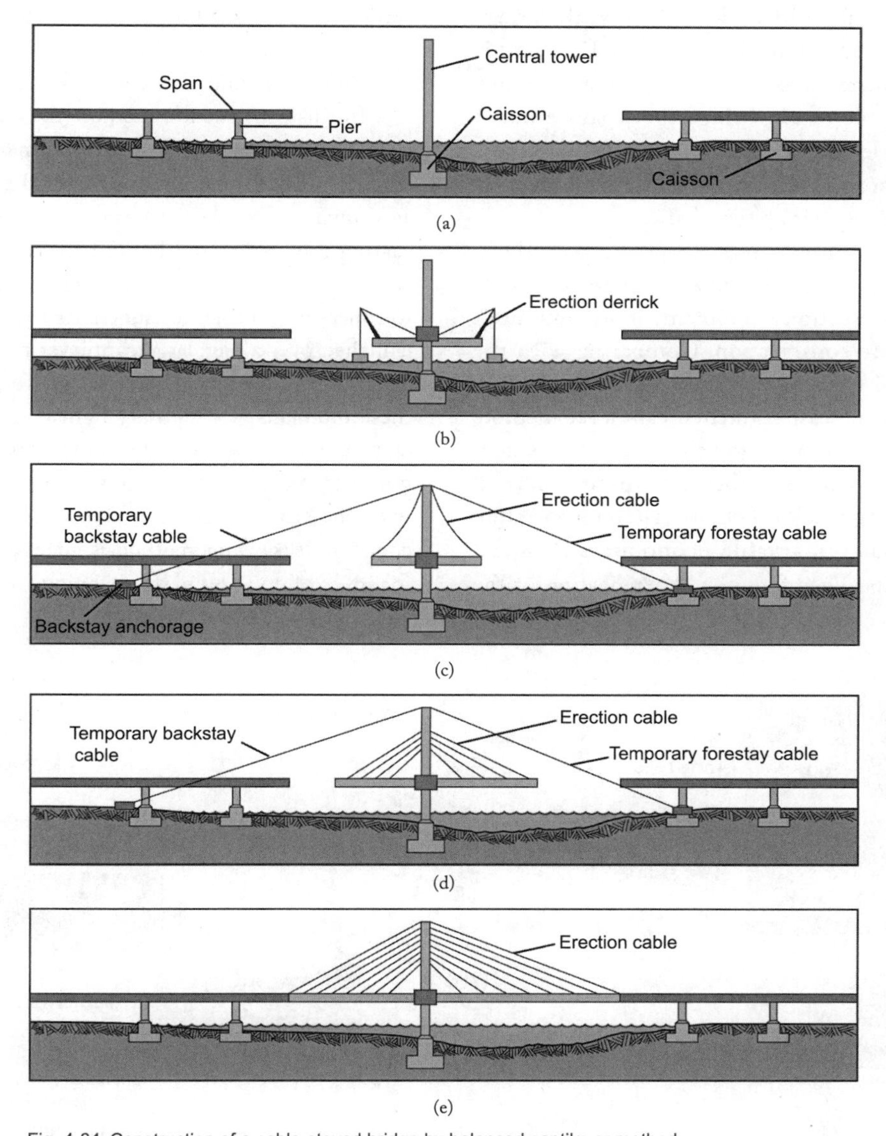

Fig. 4-64 Construction of a cable-stayed bridge by balanced cantilever method
(a) Erection of piers and support spans; (b) Construction of work station on central tower; (c) Installation of temporary stay cables and first erection cables; (d) Extension of central span; (e) Completion of central span and removal of temporary cables
Source: https://www.britannica.com/technology/cable-stayed-bridge.

method, the construction safety, especially of the final stage before the connection of the deck at midspan (largest cantilever condition), should be carefully confirmed.

(5) Rotation Method

The rotation method refers to that the main bridge span (girder, truss or arch) is divided into two main parts. Each part is built at one side of the obstacle—normally at a 90° to its final position, depending on the bridge layout and surrounding landform. Each part is then rotated into its final position. It is mainly applicable to any bridge that spans a major obstacle with challenging traffic or access conditions, such as a navigational water body or a railway with heavy traffic.

It is an efficient method especially in the gorge, cliff or arduous landform for its simple erection equipment, short time limit and safe construction. Comparing with the normal construction methods of the bridge, such as precast segmental cantilever method, site concreting method and so on, the rotation method shortens the construction period, does not break off the traffic transportation and has remarkable economic and social benefits. Therefore, the generalization of the bridge rotation construction has profound significance in reality.

When an arch bridge is constructed by this method (Fig. 4-65), the procedure consists of building two semi-arches in a quasi-vertical position over the abutments and rotating them later by means of a back stay, until closing them with a keystone. This method can be further divided into plane rotating method, vertical rotating method, and combined plane and vertical rotating method.

(6) Incremental Launching Method

In this construction method, the concrete is cast in segments behind the abutment, the deck is pushed or pulled out over the piers, as shown in Fig. 4-66. A specially prepared casting area is located behind the abutment with sections to assemble the reinforcement, to concrete, and then to launch. As the deck is launched over a pier, large cantilever moments occur until the next pier is reached. To reduce these moments, a temporary lightweight steel launching nose can be fixed to the front of the girder (typically 60% of the span length and with the stiffness of about 10%~15% of the concrete deck). This method is generally used for spans of up to 60m, the technique has been

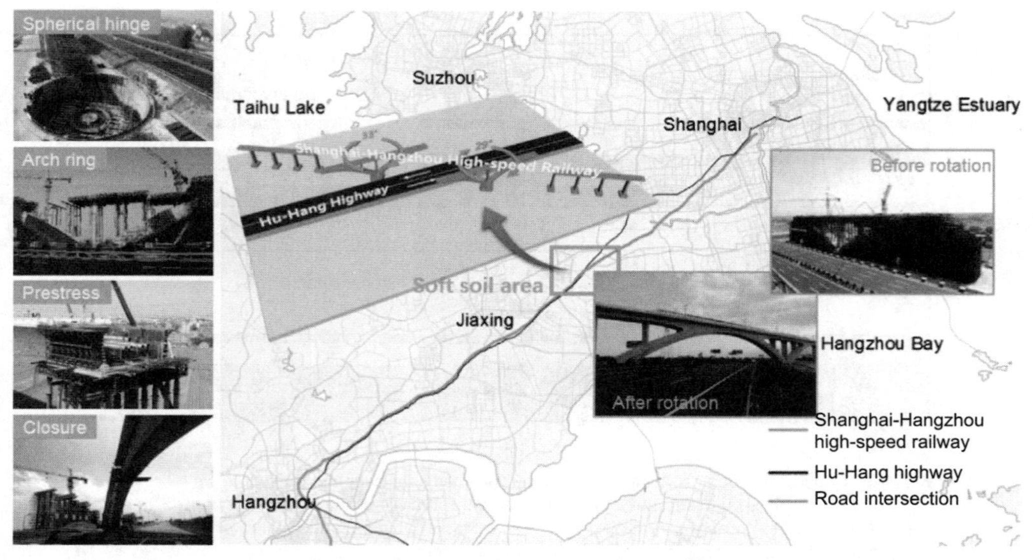

Fig. 4-65 Rotation construction of heavy swivel arch bridge for high-speed railway[22]

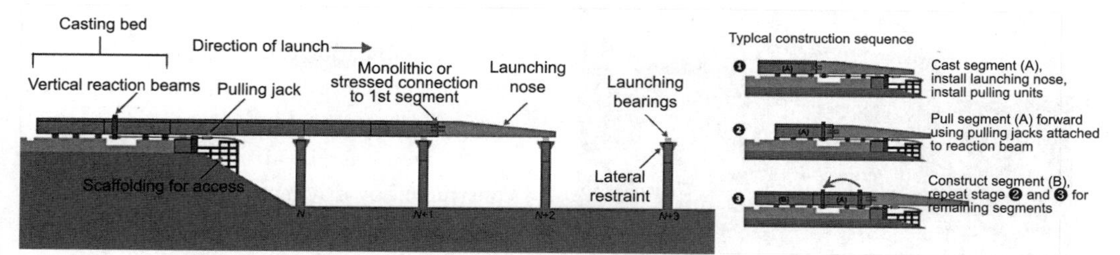

Fig. 4-66 Incremental launching method
Source: https://structuraltechnologies.com/bridge-construction-systems/.

used for longer spans up to 100m with the help of temporary piers placed to reduce the effective span length during launch.

(7) Slide-in Bridge Construction Method

Slide-in bridge construction (SIBC) allows for construction of a new bridge while maintaining traffic on the existing bridge. The new superstructure is built on temporary supports adjacent to the existing bridge (Fig. 4-67). Once construction is complete, the road is closed, the existing bridge structure is demolished or slid to a staging area for demolition, and the new bridge is slide into its final, permanent location. Once in place, the roadway approach tie-ins to the bridge are constructed. The replacement time ranges from overnight to a week or several weeks. A variation of this method is to slide the existing bridge to a temporary alignment, place traffic on the temporary alignment, and construct the new bridge in place.

(8) Lift-slab Method

Also known as the Youtz-Slick method, the lift-slab method (Fig. 4-68) ensures time efficiency and safety. Basically, the concrete slabs are cast on ground level, and are then lifted through hydraulic jacks into the designated placement. This method not only saves time, but also does not require workers to be creating and working with formwork on high ground levels.

4.5.2 Construction of Bridge Foundation

A foundation of the bridge is the part constructed under the pier/abutment and over the underlying soil or rock. Foundation is the important structural part of bridges. The construction quality of the bridge foundation

Fig. 4-67 Overhead view of slide-in bridge construction[25]
Note: NB represents Northbound; SB represents Southbound.

Fig. 4-68 Lift-slab method
Source: https://www.heavyequipmentguide.ca/article/21105/vacuum-lifting-makes-road-and-bridge-work-more-efficient.

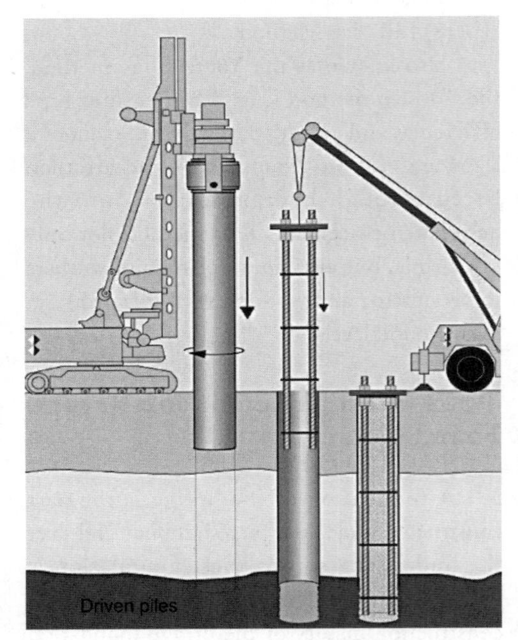

Fig. 4-69 Construction of driven cast-in-situ concrete piles
Source: https://www.kellerme.com/expertise/techniques/driven-cast-situ-piles.

directly decides the load carrying capacity, stiffness, stability, durability and safety of the bridge.

Selection of foundation types shall be based on an assessment of the magnitude and direction of loading, depth to suitable bearing materials, evidence of previous flooding, potential for liquefaction, undermining or scour, swelling potential, frost depth, and ease and cost of construction.

(1) Pile Foundations

Pile foundation can be classified into cast-in-situ concrete pile foundation and precast concrete pile foundation.

Driven cast-in-situ concrete piles are constructed by driving a closed-ended hollow steel or concrete casing into the ground and then filling it with concrete. The casing may be left in position to form part of the pile, or withdrawn for reuse as the concrete is placed. The details of driven cast-in-situ piles are shown in Fig. 4-69. The concrete is then rammed into position by a hammer as the casing is withdrawn ensuring firm contact with the soil and the compaction of concrete. Care must be taken to see that the concrete is not over-rammed or the casing is not withdrawn too quickly. There is a danger that as the liner tube is withdrawn, it will lift the upper portion of the cast-in-situ concrete, thus leaving a void or necking in the upper portion of the pile. This can be avoided by good quality control of the concrete and slow withdrawal of the casing. Driven cast-in-situ concrete piles can prove to be economic for sand, loose gravels, soft silts and clays, particularly when large numbers of piles are required. For small numbers of piles, the on-site costs can prove expensive.

Driven precast concrete piles are constructed by hammering the piles into the soil to the designed depth by an adjustable hydraulic or diesel hammer. Driven precast concrete piles are widely used because of their versatility and suitability for most ground conditions. These piles can be used for the foundation of all types of engineering structures under nearly every soil condition. Driven precast concrete piles (Fig. 4-70) are particularly suited where the founding stratum is overlain by soft deposits and aggressive or contaminated soils. Piles are manufactured in factories under high-quality control, and consist of segmental lengths of reinforced concrete sections of lengths between 3m and 15m with required or standard cross-section.

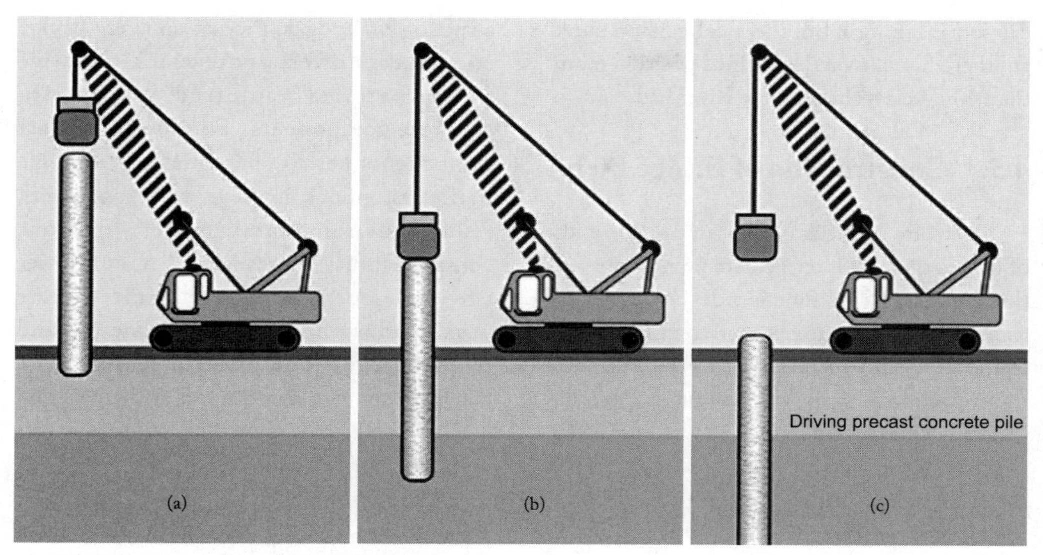

Fig. 4-70 Construction of precast concrete piles
(a) Placement of a pile; (b) Installation of a pile; (c) Repetition of the process
Source: https://basiccivilengineering.com/2016/11/type-of-pile-foundation-in-construction.html.

(2) Caisson Foundations

A caisson foundation also called foundation is a watertight retaining structure used as a bridge pier, in the construction of a concrete dam, or for the repair of ships. It is a prefabricated hollow box or cylinder sunk into the ground to some desired depth and then filled with concrete thus forming a foundation.

Caisson foundations are mostly used in the construction of bridge piers and other structures that require foundation beneath rivers or oceans. This is because that caissons can be floated to the job site and sunk into place. Caisson foundations are similar in form to pile foundations, but are installed using a different method. It is used when soil of adequate bearing strength is found below surface layers of weak materials such as fill or peat. It is a form of deep foundations which are constructed above ground level, then sunk to the required level by excavating or dredging material from the caisson.

Changzhou Taizhou Yangtze River Bridge has currently the world's largest caisson foundation (Fig. 4-71). A large floating crane was slowly hoisting a half-arc behemoth weighing 800 tons from the transport ship. Under the combined action of the hoist and the tug, the floating crane was accurately hoisted to the last high section of the steel caisson, completing the last hoisting construction of the steel caisson. Being the foundation of the main tower of ChangzhouTaizhou Yangtze River Bridge, the steel caisson is the largest of its kind in the world with an area equivalent to 13 standard basketball courts. With the height of the steel

Fig. 4-71 Construction of caisson foundation of Changzhou Taizhou Yangtze River Bridge
Source: https://www.yangtse.com/zncontent/798158.html.

caisson reaching 64m, the steel caisson sunk another 23m to a predetermined position more than 40m below the Yangtze River bed.

4.5.3 Construction of Bridge Piers

Piers are substructures located at the ends of bridge spans at intermediate points between the abutments. The function of the piers is to transfer the loads from superstructure to the foundation and to resist all horizontal and transverse forces acting on the bridge. Piers are generally constructed of masonry or reinforced concrete. Since piers are one of the most visible components of a bridge, the piers contribute to the esthetic appearance of the structure. They are found in different shapes, depending on the type, size and dimensions of the superstructure and also on the environment in which the piers are located. Underwater piers construction is shown in Fig. 4-72.

Increasing traffic volumes and a deteriorating transportation infrastructure have stimulated the development of new systems and methods to accelerate the construction of bridges. Precast concrete bridge components offer a potential alternative to conventional reinforced, cast-in-situ concrete components. The use of precast components has the potential to minimize traffic disruptions, improve work zone safety, reduce environmental impacts, improve constructability, increase quality, and lower life-cycle costs. The prefabricated piers usually adopt high-strength concrete and reinforcement. The prestressed steel bars penetrate the pier and the top of the pier, and are tensioned and anchored to connect the pier body with the pile cap (Fig. 4-73).

The Hong Kong Link Road (HKLR) is a prominent structure that serves to connect the main bridge of HongKong-Zhuhai-Macao Bridge (HZMB) at the HKSAR Boundary and the tunnel portal at Scenic Hill in Airport Island. Precast segmental technology technique was selected as the predominant form in the design of the viaducts in this project. In the western waters, tall piers except the twin-blade piers in long-span decks were designed as precast prestressed concrete

Fig. 4-72 Underwater piers construction
Source: http://www.wzrb.com.cn/article389731show.html.

(a) (b) (c)

Fig. 4-73 Construction of prefabricated piers[28]

Fig. 4-74 Precast piers of Hong Kong–Zhuhai–Macao Bridge
Source: https://www.sohu.com/a/270812207_795558.

structures (Fig. 4-74). The typical pier section is hollow with typical external dimension of 5.0m×3.2m and internal dimension of 3.0m×1.5m. A typical precast column unit is 6m and an in-situ stitch of 400mm is used to connect the in-situ pier base and the first precast unit. U-shape internally prestressed tendons were used to connect the precast units. All prestressed structures in this project were designed as Class 1 structure under the service load combinations as stipulated in the *Structures Design Manual for Highways and Railways* (SDMHR) (3rd edition) published by Highways Department, the Government of the Hong Kong Special Administrative Region. For the formation of monolithic connection between the pier/diaphragm segment and the pier, prestressing system was used instead of traditional reinforced concrete approach in the connection design.

4.6 Inspection, Monitoring and Assessment

4.6.1 Bridge Inspection

Bridge inspection is an essential element of the bridge management system particularly for aged and deteriorated bridges and a path way to condition rating. The validity of condition assessment relies heavily on the quality of the inspection. A variety of bridge inspections may be required on a bridge during its service life (Fig. 4-75).

Firstly, the inspector should conduct a cursory visual inspection of the entire bridge looking for indications of problems. Next, the inspector should conduct a hands-on visual inspection of the bridge parts taking into account any indications of problem found during the cursory inspection. During the hands-on visual inspection (Fig. 4-76), the inspector should look for signs of deterioration that will need a physical examination. In general, the inspector should start at the top of the bridge and work one's way down the load paths. This will help the inspector avoid missing any parts of the structure.

One inspection order could be (Fig. 4-77): (1) Observation of the whole bridge; (2) Observation of railings; (3) Observation of curbs; (4) Observation of the road surface; (5) Observation of the bridge deck; (6) Observation of the girder/cross-girder; (7) Observation of bearings; (8) Observation of abutments; (9) Observation of the whole bridge (final check).

The first step of bridge inspection is to perform a preliminary investigation of the bridge, which includes a visual inspection of the bridge (Fig. 4-78). It is important to observe anything out of ordinary that can influence the bridge behavior, such as concrete deterioration, beam deformations, support

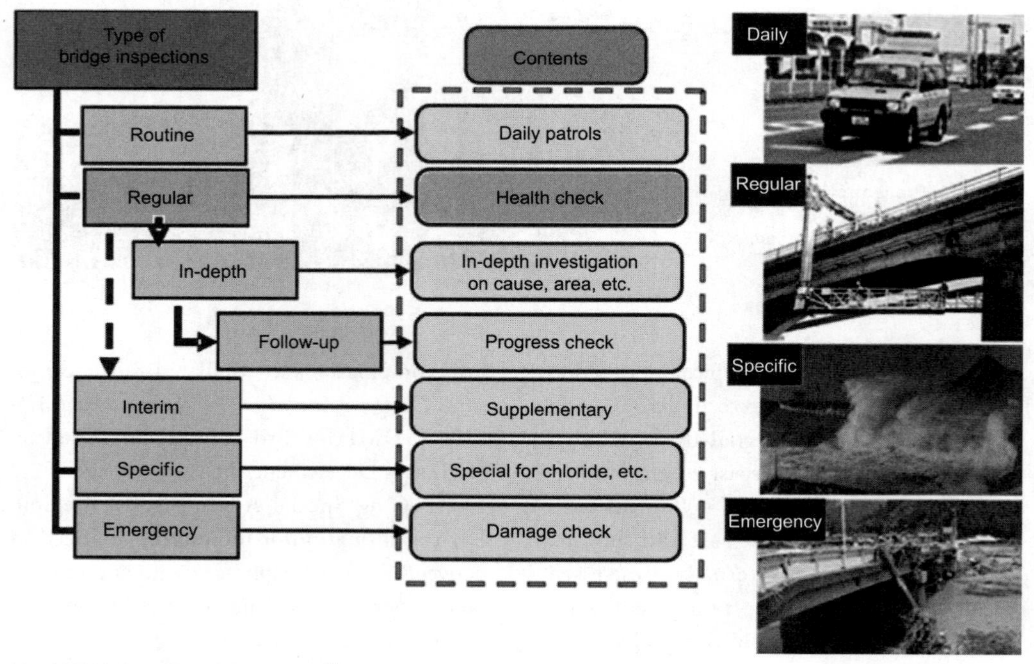

Fig. 4-75 Variety of bridge inspections[1]

Fig. 4-76 Hands-on inspection[1]

conditions, etc. In addition, if possible, previous maintenance and inspection reports should also be reviewed.

There are two types of nondestructive load testing for the purpose of bridge load rating: diagnostic and proof. Diagnostic load testing involves loading the bridge in question with a known truck load at set positions and measuring the bridge response. The results of a diagnostic test would typically be used to facilitate rating calculations. Proof load

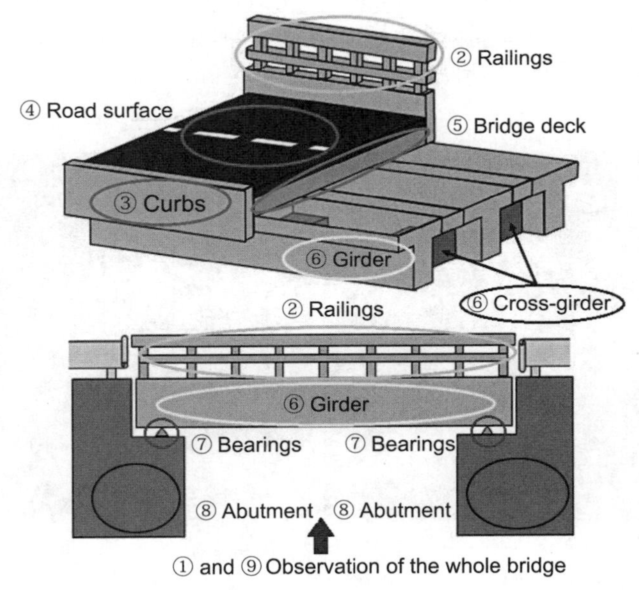

Fig. 4-77 Inspection order[1]

(a) (b)

Fig. 4-78 Visual inspection

testing involves setting a limit or goal for the bridge and gradually increasing the vehicle load until the limit or goal is reached. Both types of load tests can yield knowledge of a particular structure's behavior and can be used to generate more accurate load ratings.

The Structural Testing System (STS) is the field component of the testing system. The main purpose of using the STS is to collect bridge behavior data. Specifically, strain data is collected as a truck with known dimensions and weight is driven over the bridge (Fig. 4-79). It is common to position the truck in at least three different transverse positions (Fig. 4-80): the outer wheel line placed at two feet from each curb and the truck centered on the bridge. Additional positions may also be included if needed. Typically, the truck will be driven in each lane twice to verify that the recorded strains are consistent. If any strain asymmetry is determined (by comparing data from symmetric load paths), the analytical model must be developed accordingly.

4.6.2 Bridge Health Monitoring

Bridge health monitoring is performed through a set of activities that includes observation, data acquisition, transfer and

(a) (b)

Fig. 4-79 Data collecting as a truck driven over the bridge[1]

(a) (b)

Fig. 4-80 Positions of the trucks for load testing

Sources: (a) http://roll.sohu.com/20121208/n359872617.shtml; (b) http://k.sina.com.cn/article_3881380517_e7592aa502000gpbw.html?cre=tianyi&mod=pcpager_focus&loc=35&r=9&doct=0&rfunc=100&tj=none&tr=9&wm=3049_0005235218359.

analysis of data acquired by long time measurement during the bridge's exploitation. The goal for monitoring the bridge's structural health is to form a database for tracking the behavior of the bridge's structure in order to avoid any potential deterioration in the bridge's safety and performance (the bearing capacity, stiffness, serviceability and durability).

Health monitoring can be subdivided into multiple types of categories. Both the time frame of monitoring and the scale of monitoring are necessary considerations that need to be addressed before choosing a type of monitoring system. A bridge owner may want to monitor the bridge health for a period of a year or a few months, while in other cases only a one-time short-term solution may be necessary. Conversely, a new structure may have an expected lifetime of 50 years and the owner would like a monitoring system that would last an extended period of the time as well. Regarding the scale of monitoring, a specific joint or member in a bridge that has been problematic in the past may be the focus of the monitoring. On the other hand, an overall assessment of bridge response to loading may be the goal.

The overall framework of monitoring should include: (1) networked sensor arrays, (2) a high-performance database, (3) computer vision applications, (4) tools of data analysis and interpretation in light of physics-based models, (5) visualization allowing flexible and efficient comparison between experimental and numerical simulation data, (6) probabilistic modeling, structural reliability and risk analysis, and (7) computational decision theory.

Data analysis includes tasks aimed at evaluating, calibrating and applying several appropriate approaches for detecting small structural changes or anomalies and quantifying their effects up to the decision-making process (Fig. 4-81).

Up to now, no evaluation method can accurately evaluate the bridge state under various conditions. There is still a long way to go to study the theory and technology of bridge condition monitoring and evaluation.

4.6.3 Structural Assessment

Assessing bridges in attempts to identify the optimal sustainable solutions is far from

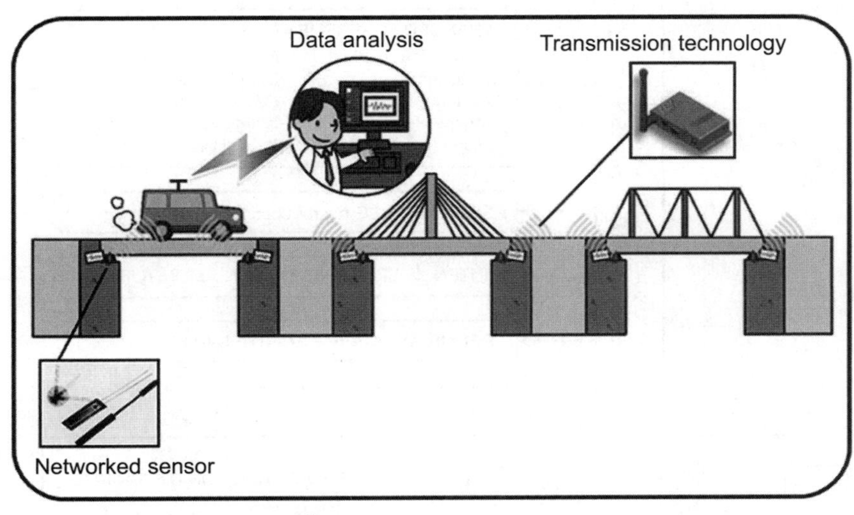

Fig. 4-81 Bridge monitoring system[1]

straightforward, so it can be useful to follow a predefined systematic approach. One such approach is illustrated by the flow-chart in Fig. 4-82. The requirement for a bridge assessment is usually associated with doubts arising from changing specifications for the structure, deterioration and damage, or reconstruction. In initial assessment of a bridge, traditional and standardized methods are used, similar to those used when designing the new structure. If the requirements of the bridge are not proven to be fulfilled by the initial assessment, different available and technically feasible options must be identified. To find the most sustainable solution, economic, societal and environmental aspects should be taken into consideration with an acceptable level of safety for the user. As suggested in Fig. 4-82, a risk-based decision-making process, taking into account the above-mentioned aspects, should be followed in the assessment, either leading to the bridge being kept in service or demolished and replaced. However, the approach in Fig. 4-82 highlights the importance of risk-based

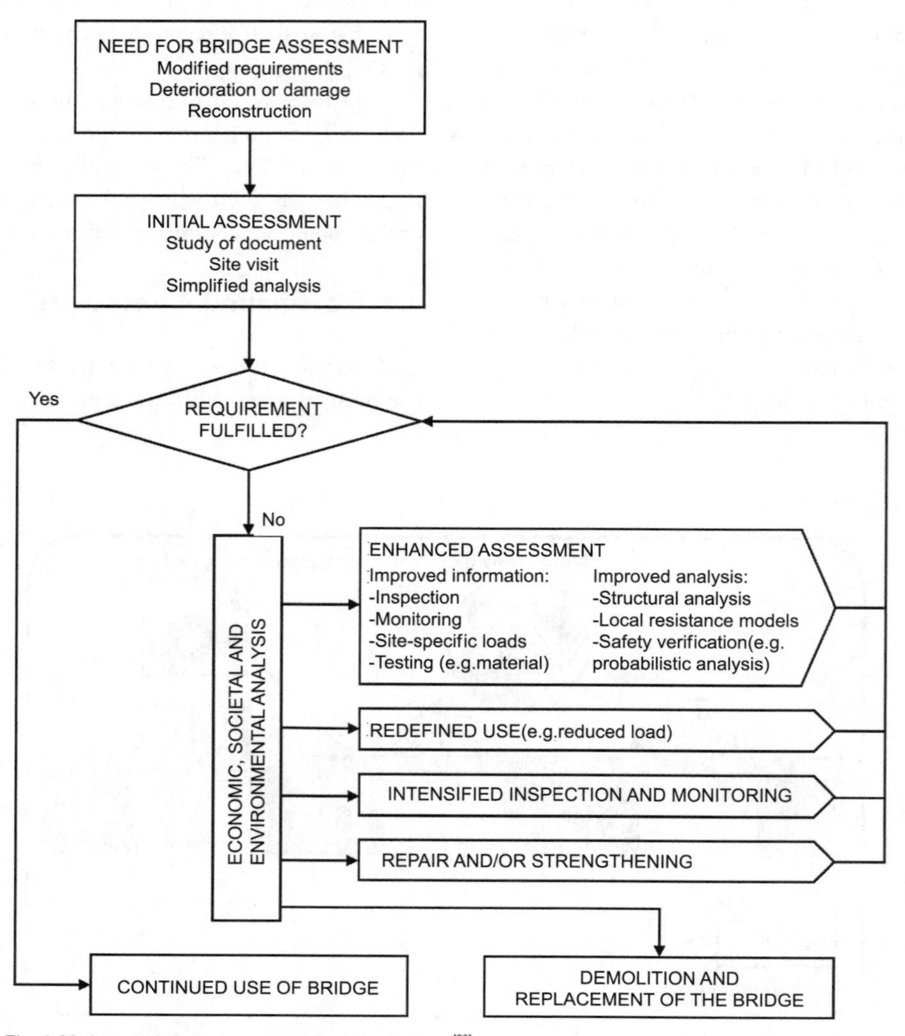

Fig. 4-82 Approach for assessment of existing bridges[30]

decision-making throughout the assessment process in order to find the most sustainable solution from the available options.

If the initial assessment cannot demonstrate that actual requirements are fulfilled, there is a range of further measures to take into consideration. They can be categorized as: (1) enhanced assessment; (2) redefined use of the bridge; (3) intensified inspection and monitoring; (4) repair and/or strengthening; and (5) demolition and replacement.

The enhanced assessment involves improvements to the assessment with regard to updated information and/or analysis. Informative improvements for model updating can be provided by inspection, monitoring, evaluation of site-specific loads and testing (e.g. material testing and proof loading). Thus, the current state of the bridge is further investigated to improve knowledge about, for instance, the actual materials, geometry, possible degradation and defects, residual prestressing, loading conditions and boundary conditions. In order to improve the analysis, refined structural analysis and local resistance models can be used. Moreover, refined concepts for verification of the structural safety can also be useful (e.g. probabilistic analysis). Successive improvements with an increasing level of complexity are fundamental elements of the enhanced assessment, together with other available measures (the loop-like procedure shown in Fig. 4-82 with risk-based decision-making). Consequently, several steps at the enhanced level may be needed to meet the requirements. The successive improvements proposed are based on causes of uncertainty identified in the assessment, mainly focusing on those of highest relevance. Moreover, to provide reliable results, both informative and analysis-oriented improvements may be needed, since (for instance) only refining the analysis will not necessarily provide a more accurate and reliable representation of the assessed bridge.

4.7 Conclusions

(1) Bridge engineering is a field of engineering (particularly a significant branch of structural engineering) dealing with the surveying, plan, design, analysis, construction, management and maintenance of bridges that support or resist loads. This variety of disciplines requires knowledge of the science and engineering of natural and man-made materials, composites, metallurgy, structural mechanics, statics, dynamics, statistics, probability theory, hydraulics and soil science, among other topics.

(2) A bridge consists of the superstructure, the substructure and bearings. Thus, a bearing is a component of a bridge which typically locates between bridge substructures (piers or abutment) and bridge superstructures, playing an important role in the force transmission and in accommodating the deformation caused by temperature variation and the earthquake. A bridge bearing carries the loads or movement in both vertical and horizontal directions from the bridge superstructure and transfers those loads to the bridge piers and abutments. The loads can be live load and dead load in vertical directions, or wind load, earthquake load, etc., in horizontal directions.

(3) Aged bridges are facing an increasing risk of failures, due to the deterioration of structural members caused by corrosion, fatigue, cracks, concrete alkali-silica reaction, and the rise of traffic load (occasional

overloading), etc. In addition, apart from these inevitable service reasons, other extreme events resulting from accidents or natural disasters such as ship-collision, flood, hurricane and earthquake, etc., also threaten bridges' safety. In order to increase reliability of aged bridges, engineers are considering different ways to settle these issues associated with aged bridges.

(4) Bridge construction tends to involve huge projects that encompass the utilization of skills related to several engineering disciplines including geology, civil, electrical, mechanical and computer sciences. Therefore, integrating the efforts of all involved must be meticulous.

Exercises

4-1 Define a bridge and describe the main components of a bridge.

4-2 Briefly describe the history of bridge engineering.

4-3 Classify the bridges according to the length, the construction material, the function, the position of the bridge floor relative to superstructures, the horizontal alignment, and the type of superstructure, respectively.

4-4 Describe the structural characteristics of girder bridge, arch bridge, cable-stayed bridge, suspension bridge and combination bridge, respectively.

4-5 Describe the characteristics and scope of application of main construction methods of bridge superstructures.

4-6 Classify the bridge foundations and describe their characteristics.

4-7 What are the bridge inspection, monitoring and assessment?

References

[1] LIN W, YODA T. Bridge Engineering: Classifications Design Loading, and Analysis Methods [M]. Oxford Butterworth-Heinemann, 2017.

[2] HE X L, CHEN Z Q, YU Z W, et al. Fatigue damage reliability analysis for Nanjing Yangtze River bridge using structural health monitoring data [J]. Journal of Central South University of Technology, 2006, 13: 200-203.

[3] Portland Cement Association. Analysis of Rigid Frame Concrete Bridges [M]. Portland Cement Association, Chicago, 1936.

[4] QIN Q, MEI G, XU G. Chapter 20: bridge engineering in China [M]. //CHEN W F, DUAN L. Handbook of International Bridge Engineering. Boca Raton CRC Press, 2013.

[5] WANG H L, XIE C L, LIU D, et al. Continuous reinforced concrete rigid-frame bridges in China [J]. Practice Periodical on Structural Design and Construction, 2019, 24(2): 05019002.

[6] XU G, YANG B, CHEN C, et al. A study on analysis of long span continuous rigid frame bridge [J]. IOP Conference Series: Materials Science and Engineering, 2019, 611(Conference 1): 012045.

[7] MA Z D, LIU A S. Technical measures for control of excessive deflection of girders of long span continuous rigid-frame bridges [J]. Bridge Construction, 2015, 45(2): 71-76.

[8] XANTHAKOS P P. Theory and Design of Bridges [M]. New York: Wiley-

Interscience, 1993.

[9] MA B, LIN Y, ZHANG J, et al. Decade review: bridge type selection and challenges of Lupu Bridge [J]. Structural Engineering International, 2013, 23(3): 317-322.

[10] TANG M C. Multispan cable-stayed bridges [C]. International Bridge Conference Bridges into the 21st Century, Hong Kong, 1995.

[11] NI Y Q, ZHOU H F, CHAN K C, et al. Modal flexibility analysis of cable-stayed Ting Kau Bridge for damage identification [J]. Computer-Aided Civil and Infrastructure Engineering, 2008, 23: 223-236.

[12] KHAN M A. Bridge and Highway Structure Rehabilitation and Repair [M]. New York: The McGraw-Hill Companies Inc., 2010.

[13] WANG H, TAO T Y, LI A Q, et al. Structural health monitoring system for Sutong Cable-stayed Bridge [J]. Smart Structures and Systems, 2016, 18(2): 317-334.

[14] MARCUSSEN J B. Design and construction of composite bridges [J]. ce/papers, 2017, 1(2-3): 4246-4255.

[15] KADIR A, MURSIDI B. Design concept long span bridge with floating foundation [J]. IOP Conference Series: Materials Science and Engineering, 2020, 797: 012021.

[16] SARGSYAN A N, SARGSYAN G G, SARGSYAN A A. Bridge health monitoring: the Davtashen Bridge example in Yerevan [J]. IOP Conference Series: Materials Science and Engineering, 2019, 698: 077009.

[17] XU Y Y, TURKAN Y. BrIM and UAS for bridge inspections and management[J]. Engineering. Construction and Architectural Management, 2020, 27(3): 785-807.

[18] COVIAN E, CASERO M, MENÉNDEZ M, et al. Application of HDS techniques to bridge inspection [J]. Nondestructive Testing and Evaluation, 2018, 33(3): 301-314.

[19] ELGAMAL A W, CONTE J P, FRASER M, et al. Health monitoring for civil infrastructure [C]. 9th Arab Structural Engineering Conference (9ASEC), Abu Dhabi, United Arab Emirates, 2003.

[20] RASHIDI M, GIBSON P. A methodology for bridge condition evaluation [J]. Journal of Civil Engineering and Architecture, 2012, 6 (9): 1149-1157.

[21] SUN Q S, GUO X G, ZHANG D P, et al. Research on the application of horizontal rotation construction method with flat hinge in cable-stayed bridge construction [J]. Advanced Materials Research, 2011, 255-260: 856-860.

[22] FENG Y, QI J, WANG J, et al. Rotation construction of heavy swivel arch bridge for high-speed railway [J]. Structures, 2020, 26: 755-764.

[23] WANG W, YAN W C, DENG L, et al. Dynamic analysis of a cable-stayed concrete-filled steel tube arch bridge under vehicle loading [J]. Journal of Bridge Engineering, 2014, 20(5): 04014082.

[24] YANG Y, CHEN B. Rigid-frame tied through concrete filled steel tubular arch bridge [C]. ARCH'07 – 5th International Conference on Arch Bridges, 2007: 863-867.

[25] Project #F-ST99(232). Slide-in bridge construction implementation guide: Planning and executing projects with the lateral slide method [R]. Federal Highway Administration, 2013.

[26] HIEBER D G, WACKER J M, EBERHARD M O, et al. Precast concrete pier systems for rapid construction of bridges in seismic regions [R]. Bridge Piers, 2005.

[27] OU Z, XIE M, LIN S, et al. The practice and development of prefabricated bridges [J]. IOP Conference Series: Materials Science and Engineering, 2018, 392: 062086.

[28] CHAN M, CHAN D S H, NG P W H, et al. Innovative solutions in Hong Kong - Zhuhai - Macao Bridge (HZMB) Hong Kong Link Road Project [C]. ICE HKA Annual Conference 2015: Thinking out of the box in infrastructure development and retrofitting, 2015: 53-64.

[29] GASTINEAU A, JOHNSON T, SCHULTZ A. Bridge health monitoring and inspections–a survey of methods [R]. Minnesota: Minnesota Department of Transportation, 2009.

[30] BAGGE N. Structural assessment procedures for existing concrete bridges: experiences from failure tests of the Kiruna Bridge [D]. Luleå: Luleå University of Technology, 2017.

Chapter 5
Road and Transportation Engineering

5.1 Overview

Transportation engineering is the application of technology and scientific principles to planning, functional design, operation and management of facilities for any mode of transportation in order to provide the safe, efficient, rapid, comfortable, convenient, economical and environmentally compatible movement of people and goods transport. The planning and designing aspects of transportation engineering relate to elements of urban planning and the design of road and infrastructure. Therefore, there are many intersections between transportation engineering and civil engineering, which are concentrated in the sub-disciplines such as road and railway engineering. This chapter focuses on road engineering mainly from the view point of civil engineers and traffic engineers, starting from transportation planning to the geometric design and ending with pavement structure and materials design.

5.1.1 Transportation Engineering

The transportation industry is the social production sector in the national economy that is engaged in transporting goods and passengers. It can be called the fourth sector of material production after extractive industry, processing industry and agriculture. Unlike other economic sectors, it does not directly produce new products, but transfers goods and passengers from one place to another. It organically links the various links of social production, distribution, exchange and consumption. It is a prerequisite to ensure the normal progress and development of social and economic activities, and it acts as a link in the entire social mechanism.

Transportation engineering deals with planning, design, operations, control, management, maintenance and rehabilitation of transportation systems, services and components. The subareas that are parts of transportation engineering are transportation planning, traveler behavior, design and analysis of transportation networks, traffic flows analysis, analysis and control of traffic operations, queueing analysis, vehicle routing and scheduling, logistics and supply chain management, etc. Within transportation engineering, technology, mathematics, physics, computer science, social sciences and cultural heritage converge, transportation engineering methods and techniques have a high impact on transportation system performances, in regard to the level of service, capacity, safety, reliability, resource consumption, environment, economics, etc.

The basic elements of transportation are vehicles, guideways, terminals and control policies. Transportation engineering primarily involves planning, design, construction, maintenance and operation of transportation facilities. The facilities support air, highway, railroad, pipeline, water and even space transportation. The design aspects of transportation engineering include the sizing of transportation facilities (how many lanes or how much capacity the facility has), determining the materials and thickness used in pavement, designing the geometry (vertical and horizontal alignment) of the roadway (or rail track).

A range of control systems in transportation were originally created, mainly to improve traffic safety. Later on, engineers started with the development of control systems, intending to reduce traffic congestion. This congestion is an outcome of many decisions that different users make. Traffic and transportation systems are composed of decentralized individuals and each individual acts together with other individuals in accordance with localized knowledge. Occasionally, individuals collaborate, and at other times they are in conflict. They interact simultaneously with transportation infrastructure and the environment. Through the aggregation of the individual interactions, the global picture of the transportation system emerges.

These control systems of transportation engineering are associated closely with traffic engineering. Traffic engineering is a branch of civil engineering that uses engineering techniques to achieve the safe and efficient movement of people and goods on roadways. It focuses mainly on research for safe and efficient traffic flow, such as road geometry, sidewalks and crosswalks, cycling infrastructure, traffic signs, road surface markings and traffic lights. Traffic engineering deals with the functional part of transportation system, except the infrastructures provided. Traffic engineering is closely associated with other disciplines, such as transport engineering, pavement engineering, bicycle transportation engineering, highway engineering, transportation planning, urban planning and human factors engineering. Typical traffic engineering projects involve designing traffic control device installations and modifications, including traffic signals, signs and pavement markings. However, traffic engineers also consider traffic safety by investigating locations with high crash rates and developing countermeasures to reduce crashes. Traffic flow management can be short-term (preparing construction traffic control plans, including detour plans for pedestrian and vehicular traffic) or long-term (estimating the impacts of proposed commercial/residential developments on traffic patterns). Increasingly, traffic problems are being addressed by developing systems for intelligent transportation systems, often in conjunction with other engineering disciplines, such as computer engineering and electrical engineering.

Transportation and traffic systems are, in essence, different from other technical systems. Their performances depend a great deal on the users' behavior. A good understanding of the human decision-making mechanism is one of the key factors in the transportation planning process, as well as in developing appropriate real-time traffic control. There are continuous construction and expansion of traffic networks. Before the construction of a new bridge, road expansion, or development of a toll road, it is necessary to study how the potential users of the facility will react. A proper understanding of the human decision-making mechanism is of high importance, since it has been shown that building additional roads, in some cases, does not automatically produce a reduction in total travel time in the transportation network.

Transportation science and transportation engineering offer various techniques related to transportation modeling, transportation planning and traffic control. These techniques should be used for predicting travel and freight demand, planning new transportation networks and developing traffic control strategies. The range of engineering concepts and methods should be used to make future transportation systems safer, more cost-effective and greener.

5.1.2 Highway Engineering

Highway engineering is an engineering discipline branch from civil engineering that involves planning, design, construction, operation, and maintenance of roads, bridges and tunnels to ensure safe and effective transportation of people and goods. Highway engineering became prominent towards the latter half of the 20th century after World War II. Standards of highway engineering are continuously being improved. Highway engineers must take into account future traffic flows, design of highway intersections/interchanges, geometric alignment and design, highway pavement materials and design, structural design of pavement thickness, and pavement maintenance.

The beginning of road construction could be dated to the time of the Romans. With the advancement of technology from carriages pulled by two horses to vehicles with power equivalent to 100 horses, road development had to follow suit. The construction of modern highways did not begin until the late 19th to early 20th century. The development of the road can be roughly divided into four stages as follows.

(1) The Stage of Trails for Pedestrians, Cattle, Horses and Other Animals to Walk and Carry Goods

The Chinese writer Lu Xun once wrote, "I thought: Hope cannot be said to exist, nor can it be said not to exist. It is just like roads across the earth. For actually the earth had no roads to begin with, but when many men pass one way, a road is made". The second half of this famous quote describes exactly the origin of trails.

(2) The Stage of Cart Way that for Vehicles and Pedestrians to Pass Through

In China, there are names such as "Kang-Qu" (康衢), "Chi-Dao" (驰道) and "Post Way" (驿道). In Europe, Roman roads are very famous, and there is a saying that "All roads lead to Rome". The long straight roads built by the Romans wherever they conquered have, in many cases, become just as famous names in history as their greatest emperors and generals. Building upon more ancient routes and creating a huge number of new ones, Roman engineers were audacious in their plans to join one point to another in as straight a line as possible whatever the difficulties in geography and the costs in manpower. Consequently, roads used bridges, tunnels, viaducts and many other architectural and engineering tricks to create a series of breathtaking but highly practical monuments which spread from Portugal to Constantinople. The network of public Roman roads covered over 120,000km, and it greatly assisted the free movement of armies, people and goods across the empire. Roads were also a very visible indicator of the power of Rome, and they indirectly helped unify what was a vast melting pot of cultures, races and institutions.

(3) The Stage of Highway for Motor Vehicles to Drive on

A highway is any public or private road or other public way on land. It is used for major roads, but also includes other public roads and public tracks: It is not an equivalent term to controlled-access highway, or a translation for high-speed, access-controlled motorway. According to *Merriam Webster*, the use of the term predates the 12th century. According to *Etymonline*, "high" is in the sense of "main". In North American and Australian English, major roads such as controlled-access highways or arterial roads are often state highways, while in Canada, are provincial highways. Other roads may be designated "county highways" in the U.S. and Ontario Province of Canada. These classifications refer to the level of government (state, provincial, county) that maintains the roadway.

(4) The Stage of High-speed Access-controlled Highway

The high-speed access-controlled highway is also called motorway, freeway or expressway. The idea for the construction of the Autobahn (German name for motorway) was first conceived in the mid-1920s during the days of the Weimar Republic, but the construction was slow, and most projected sections did not progress much beyond the planning stage due to economic problems and a lack of political support. One project was the private initiative HaFraBa which planned a "car-only road" crossing Germany from Hamburg in the north via central Frankfurt am Main to Basel in Switzerland. Parts of the HaFraBa were completed in the late 1930s and early 1940s, but the construction eventually was halted by World War II. The first public road of this kind was completed in 1932 between Cologne and Bonn and opened by Konrad Adenauer (Lord Mayor of Cologne and future Chancellor of the Federal Republic of Germany) on August 6, 1932. Today, that road is the Bundesautobahn 555. This road was not yet called Autobahn and lacked a central median like modern motorways, but instead was termed a Kraftfahrstraße (motor vehicle road) with two lanes that each direction does not have intersections, pedestrians, bicycles, or animal-powered transportation.

5.1.3 Highway System in China

Nowadays, roads can be very multi-functional. A road is a thoroughfare, route, or way on land between two places that has been paved or otherwise improved to allow travel by foot or by some form of conveyance (including a motor vehicle, cart, bicycle or horse). Other names for a road include: parkway; avenue; freeway, motorway or expressway; tollway; interstate; highway; thoroughfare; or primary, secondary and tertiary local road.

In China, road is normally divided as airport runways, highways, urban roads, roads for factory and mine, forest road, and sometimes even railways.

(1) Runway is a special type of road. According to the International Civil Aviation Organization (ICAO), a runway is a "defined rectangular area on a land aerodrome prepared for the landing and takeoff of aircraft". Runway may be a man-made surface (often asphalt, concrete, or a mixture of both) or a natural surface (grass, dirt, gravel, ice, sand or salt). Runways, as well as taxiways and ramps, are sometimes referred to as "tarmac", though very few runways are built using tarmac. Runways made of water for seaplanes are generally referred to waterways.

(2) Highway is a public road on which cars and other vehicles can be driven, and/or pedestrians, bicycles, rickshaws, horses, etc. can walk.

(3) Urban roads are the roads which are accessible to all areas of the city, for transportation and pedestrian use and convenience for residents to live, work, and for cultural and entertainment activities, as well as connection with roads outside the city carrying external traffic. Urban roads are characterized by more lanes, good lightening and greening conditions as well as complicated substructures, which is therefore the topic of municipal engineering.

(4) Roads for factory and mine are roads that serve factories, mines, oil fields, ports, warehouses and other enterprises. They are divided into roads outside the factory, roads inside the factory, and open-pit mine roads. Roads outside the factory are external roads connecting factories and mining enterprises with highways, urban roads, stations, port raw material bases, and other factories and mining enterprises; roads inside the factory are internal roads in the factory (field) area, reservoir area, station area, port area, etc. The open-pit mine road is the road between the mining field and the unloading point, between the factory (field) areas within the mining area; or the road leading to the affiliated factories and auxiliary facilities. The grades and main technical indicators of roads in factories and mines should be determined comprehensively according to the scale of factories and mines, types of enterprises, road properties and usage requirements (including road service life, traffic volume, vehicle models, etc.), local topography and geology, with extra consideration of future development.

(5) Forest roads are built in forest areas and are mainly used for the passage of various forestry vehicles. Forest roads include four types. The first is the skidding road. The simple road opened from the timber harvesting point to the loading yard is exclusively for skidding. Generally, the line is short and there is no strict standard. The second is the timber transportation road. As the main body of forest roads, it directly undertakes the task of transporting timber from the loading yard to the wood storage yard. The form and standard of road construction differ greatly according to the difference of the material transportation tools and the transportation volume. The third is the forestry road which is regular road built according to the needs of afforestation, forest

protection, etc. Usually the traffic volume is very small. To ensure long-term use, there are certain technical standards. The timber transportation road and the forestry road are often integrated. The fourth is the fire prevention road. Under normal circumstances, the width, thickness and strength of the road surface can meet the needs of forest protection and fire protection.

Within these categories of roads, highways are the most common type of roads and have the largest number and the longest mileage. Research and practice in highways are the most active, the planning and design methods of roads are mostly from the highway experiences. In the following, we will focus on highways.

According to the *Technical Standard of Highway Engineering* JTG B01—2014 by the Ministry of Transport of the People's Republic of China (abbreviated as *Technical Standard* in the following), the highway is divided into five technical grades: motorway, 1st-class highway, 2nd-class highway, 3rd-class highway and 4th-class highway. There are still some rural highways below the 4th-class.

(1) The motorway is a multi-lane highway exclusively for motor vehicles to drive in different directions and lanes, and is access-controlled. The design annual average daily traffic volume (*AADT*) of motorways should be more than 15,000 passenger cars.

(2) The 1st-class highway is a multi-lane highway where motor vehicles can drive in different directions and lanes, and access can be controlled as needed. The design annual average daily traffic volume of 1st-class highways should be more than 15,000 passenger cars. That is to say, the 1st-class highway has an equivalent design traffic volume to motorway.

(3) The 2nd-class highway is a two-lane road for motor vehicles. The design annual average daily traffic volume of 2nd-class highways should be 5000~15,000 passenger cars.

(4) The 3rd-class highway is a two-lane highway for mixed driving of motor vehicles and non-motor vehicles. The annual average daily traffic volume of 3rd-class highways should be 2000~6000 passenger cars.

(5) The 4th-class highways are two-lane or single-lane highways for mixed driving of motor vehicles and non-motor vehicles. The annual average daily traffic volume of two-lane 4th-class highways should be less than 2000 passenger cars; the annual average daily traffic volume of single-lane 4th-class highways should be less than 400 passenger cars.

The following principles should be followed in the selection of highway technical grades:

(1) The selection of highway technical grades should be determined according to the highway network planning and function in combination with the prognosis of traffic volume.

(2) Primary arterial highways should be selected as motorways.

(3) Secondary arterial highways should be selected as 2nd-class or higher grade highways.

(4) Primary distribution highways should be selected as 1st-class or 2nd-class highways.

(5) Secondary distribution roads should be selected as 2nd-class or 3rd-class highways.

(6) Branch highways should be classified as 3rd-class or 4th-class highways.

From the view of political and economic importance, highways in China can be classified as national trunk highways, provincial trunk highways, county highways, township highways and village highways. The former two categories are also called trunk highways, and latter three are rural highways. The technical grade has little to do with the administrative category of the highways.

(1) National trunk highways refer to major arterial highways of national political and economic significance, including important international highways, national defense highways, highways connecting the capitals of provinces, autonomous regions and

municipalities, as well as connecting major economic centers, port hubs, commodity production bases, and highways in strategic locations. The construction, maintenance and management of highways that span across provinces in the national highway shall be undertaken by special agencies approved by the Ministry of Transport.

(2) Provincial trunk highways are the highways that has a provincial political, economic and national defense significance, and has been designated as a provincial-level trunk highway by the unified planning of the provincial government. Provincial highway authorities are responsible for the construction, maintenance and management of provincial trunk highways.

(3) County highways refer to the main highways with political and economic significance of counties or county-level cities, connecting counties and main towns in the county and other main places, functioning as major commodity production and distribution centers, as wells as inter-county highways that are not national or provincial trunk highways. County highways are constructed, maintained and managed by the county or city highway authorities.

(4) Township highways refer to the highways that mainly serve the internal economic, cultural and administrative functions of the townships, the highways connecting townships that are not county highways or above, as well as the highways that connect the township to the outside.

China has a large highway network. Up to the end of 2019, the total mileage of highways in China was 5.0125 million km. The highway density is relatively low, about 52.21km per 100km^2. The highway maintenance mileage is 4,953,100km, accounting for 98.8% of the total highway mileage. Highways of 4th-class and higher grades had a mileage of 4,698,700km. The mileage of 2nd-class and higher grades highways was 672,000km, accounting for 13.4% of the total highway mileage. The mileage of all motorways was 149,600km; the mileage of motorway lanes was 669,400km. The mileage of national motorway network was 108,600km. The national highway mileage was 366,100km and the provincial highway mileage was 374,800km. The mileage of rural highways is 4.205 million km, including 580,300km of county highways, 1,198,200km of township highways, and 2,422,200km of village roads.

According to the *National Highway Network Planning* (2013~2030), the total planned scale of China's national highway network is 401,000km. The national motorway network has been adjusted from the originally planned "7 Shots, 9 Verticals, 18 Horizontals" to "7 Shots, 11 Verticals, 18 Horizontals" as well as 6 regional loop lines, 16 parallel lines and 104 tie lines. The total scale is about 118,000km, and another planned long-term prospect line is about 18,000km. The ordinary national highway network has been adjusted from the originally planned 70 routes of "12 Shots, 28 Verticals, and 30 Horizontals" to "12 Shots, 47 Verticals, and 60 Horizontals", as well as 81 connecting lines, with a total of 200 routes and a total scale of about 26.5 million km.

5.2 Highway Planning Process

Highways are of vital importance to a country's economic development. The construction of a high-quality road network directly increases a nation's economic output by reducing journey times and costs, making a region more attractive economically. The actual construction process will have the added effect of stimulating the construction market.

However, the building and maintenance of highways costs a lot, and most important of all, highways as transportation infrastructure occupy much land. Therefore, a scientific planning process is necessary.

5.2.1 Introduction

Transportation planning is the process of defining future policies, goals, investments and designs to prepare for future needs to move people and goods to destinations. As practiced today, it is a collaborative process that incorporates the input of many stakeholders including various government agencies, the public and private businesses. Transportation planners apply a multi-modal and/or comprehensive approach to analyzing the wide range of alternatives and impacts on the transportation system to influence beneficial outcomes.

To be specific, transportation planning refers to the form of transportation network that is formulated to predict the demand for transportation under the conditions of future population, socio economic development and land use after a survey of current transportation conditions, and to draw up this transportation planning plan. The work involves process of preparing the implementation proposal, schedule and budget.

Urban transportation planning and highway network planning are two important aspects in traffic planning. In urban transportation planning, road system planning is the main aspect. However, with the emergence of modern transportation facilities such as urban subways, elevated roads and rapid light rails, urban transportation planning is no longer limited to the layout of a simple urban flat road network system, but a comprehensive plan for various forms of transportation. For road traffic planning, due to the different nature and functions of urban roads and highways, different environments, and different concentrations of population, industry and agriculture, the planning of highways and urban roads is also different. But whether it is highway network planning or urban road network planning, it must involve various factors such as people, vehicles, roads and the environment, and it is also closely related to land use and development, social and economic development, transportation policies and traffic management. Therefore, transportation planning is essentially an important part of the overall planning for socio economic development of a certain region or city.

The purpose of transportation planning is to design a reasonable transportation system in order to serve various land use patterns that are compatible with the social and economic development in the future. Specifically, the significance of transportation planning is manifested in following aspects.

(1) Transportation planning is an important means to establish and improve the transportation system because transportation planning coordinates the links between the five modes of transportation (roads, railways, water transportation, aviation and pipelines), and puts forward tasks and requirements for roads to make them compatible with other transportation. The methods cooperate closely, complement each other and jointly complete the transportation task. At the same time, certain prejudices in the planning of a single and isolated road system in the past can be eliminated, such as focusing only on the form of the road network and not focusing on the internal connections between various modes of transportation.

(2) Transportation planning is the fundamental measure to solve the current road traffic problems. As traffic problems are a holistic and comprehensive problem, increasing investment in road construction or improving the level of traffic management alone cannot solve the problem fundamentally. It must be related to the social economy. The fundamental measures are the rational layout of industrial, agricultural, commercial, cultural service facilities and

population distribution, and a comprehensive and scientifical based transportation plan.

(3) Transportation planning is an effective way to obtain the best benefits of transportation. Because the size of road construction investment, the choice of vehicle transportation methods and routes, the level of vehicle operating costs and the level of traffic management are all closely related to transportation planning, only formulating reasonable transportation planning can form a safe and unobstructed transportation network, so as to use the shortest distance, the least time and cost to complete the scheduled transportation tasks and obtain the best transportation benefits.

As for highways, the basic requirements for the planning and layout of the network are as follows:

(1) Highway network planning must work closely with other transportation networks to form a coordinated comprehensive transportation. Since the layout of highway routes is less objectively restricted and more flexible than railways and water transportation, it is necessary to create convenient conditions for the connection and development of railways and water transportation as much as possible.

(2) The technical grade of the trunk highway network, traffic facilities along the lines and their construction sequences should be planned according to the importance of the passing area and the size of the traffic volume.

(3) It should make full use of the original roads and local roads, and gradually improve to meet the road network level and technical standard requirements through improvement.

(4) It should conform to the principles of phased construction and engineering economy.

(5) It should strive to meet the requirements of low road network density, short transportation lines, high transportation efficiency and low transportation costs.

(6) Highway network planning should also pay attention to the needs of local farmland water conservancy construction and development of local resources.

(7) The highway grade should be determined comprehensively according to the planning of the highway network and the long-term traffic volume, starting from the overall situation and combining the tasks and nature of the highway.

(8) The environmental protection is required. Whether in the process of road construction, environmental protection, or vehicle exhaust gas, noise, and road sewage discharge and diversion during operation, it should be fully considered in the planning process.

The form of the highway network generally depends on the following factors: (1) the transportation needs between administrative and economic centers, (2) the size and direction of passenger and cargo transportation flows, (3) the natural conditions of the planned area, especially the distribution of mountains, the direction of large rivers and negative engineering geological conditions, (4) special requirements for national defense, etc.

Regarding the trunk highway system, most countries in the world are centered on the capital and provincial capitals. The national trunk highway network and the provincial trunk highway network are arranged radially. The radical trunk lines cross with circular trunk roads, forming a trunk network system of intersecting radial and circular lines. In addition, there are also grid-like arrangements. China's highway network adopts the principle of combining vertical and horizontal networks and radiation to connect the capitals and military regions of provinces, cities, autonomous regions, important port hubs, industrial and agricultural bases, and large and medium-sized cities with a population of more than 500,000. For the sub-level and local roads between the trunk lines, they are arranged according to the regional topography and the distribution of transportation hubs. Dendritic and square grid patterns are generally adopted. Nevertheless, the road network layout should be selected in consideration of the social, natural and economic conditions of that region.

In short, the planning and design of the road network should not be limited to one point or one line, but should focus on the entire road network system. The layout of the road network has a great influence on the efficiency of the entire transportation system. A good road network layout can greatly improve the efficiency of the transportation system, increase the accessibility of the road network, save a lot of investment, and save transportation time and transportation costs, meanwhile achieve good economic, social and environmental benefits.

The process of transportation planning entails developing a transportation plan for an urban region. It is an ongoing process that seeks to address the transport needs of the inhabitants of the area, and with the aid of a process of consultation with all relevant groups, strives to identify and implement an appropriate plan to meet these needs.

The process takes place at a number of levels. At an administrative/political level, a transportation policy is formulated and politicians must decide on the general location of the transport corridors/networks to be prioritized for development, on the level of funding to be allocated to the different schemes and on the mode or modes of transport to be used within them.

Below this level, professional planners and engineers undertake a process to define in some details the corridors/networks that comprise each of the given systems selected for development at the higher political level. This is the level at which what is commonly termed a "transportation study" takes place. It defines the links and networks, and involves forecasting future population and economic growth, predicting the level of potential movement within the area and describing both the physical nature and modal mix of the system required to cope with the region's transport needs, be they road, rail, cycling or pedestrian-based.

5.2.2 Highway Planning Strategies

When the highway planning process takes place within a large urban area and other transport options such as rail and cycling that may be under consideration alongside car-based ones, the procedure can become quite complex and the workload involved in data collection can become immense. In such circumstances, before a comprehensive study can be undertaken, one of a number of broad strategy options must be chosen:

(1) The Land Use Transportation Approach

Within this method, the management of land use planning is seen as the solution to controlling the demand for transport. The growing trend that many commuters live in suburbs of a major conurbation or in small satellite towns while working within or near the city center has resulted in many using their private cars for their journey to work. This has led to congestion on the roads and the need for both increased road space and the introduction of major public transport improvements. Land use strategies such as the location of employment opportunities close to large residential areas and actively limiting urban sprawl which tends to increase the dependency of commuters on the private cars, are all viable land use control mechanisms.

(2) The Demand Management Approach

The demand management approach entails planning for the future by managing demand more effectively on the existing road network rather than constructing new road links. Demand management measures include the tolling of heavily trafficked sections of highway, possibly at peak times only, and car-pooling, where high occupancy rates within the cars of commuters is achieved voluntarily either by the commuters themselves, in order to save money, or by employers in order to meet some target stipulated by the planning authority. Use of car-pooling can be promoted by allowing private cars with multiple occupants to use bus-lanes during peak hour travel or by allowing them reduced parking charges at their destination.

(3) The Car-centered Approach

The car-centered approach has been favored by a number of large cities within the U.S., most notably in Los Angeles. It seeks to cater for future increases in traffic demand through the construction of bigger and better roads, be they inter-urban or intra-urban links. Such an approach usually involves prioritizing the development of road linkages both within and between the major urban centers. Measures such as in-car information for drivers regarding points of congestion along their intended route and the installation of state-of-the-art traffic control technology at all junctions, help maximize usage along the available road space.

(4) The Public Transport-centered Approach

In the public transport-centered approach, the strategy will emphasize the importance of bus and rail-based improvements as the preferred way of coping with increased transport demand. Supporters of this approach point to the environmental and social advantages of such a strategy, reducing noise and air pollution and increasing efficiency in the use of fossil fuels while also making transport available to those who cannot afford to run a car. However, the success of such a strategy depends on the ability of transport planners to induce increasing numbers of private car users to change their mode of travel during peak hours to public transport. This will minimize highway congestion as the number of peak hour journeys increase over the years. Such a result will only be achieved if the public transport service provided is clean, comfortable, regular and affordable.

5.2.3 Travel Data Survey

As mentioned above, transportation planning studies a transportation system that can make people and goods run safely, efficiently and economically, and make people travel comfortable, convenient, and the environment is not disturbed. It generally includes the following content and work steps.

The planning process commences with the collection of historical traffic data covering the geographical area of interest. Growth levels in past years act as a strong indicator regarding the volumes that one can expect over the chosen future time, be it 15, 20 or 30 years. If these figures indicate the need for new/upgraded transportation facilities, the process then begins with considering what type of transportation scheme or suite of schemes is most appropriate, together with the scale and location of the scheme or group of schemes in question. The demand for highway schemes stems from the requirements of people to travel from one location to another in order to perform the activities that make up their everyday lives. The level of this demand for travel depends on a number of factors: (1) The location of people's work, shopping and leisure facilities relative to their homes; (2) The type of transport available to those making the journey; (3) The demographic and socio economic characteristics of the population in question.

Characteristics such as population size and structure, number of cars owned per household and income of the main economic earner within each household tend to be the demographic/socio economic characteristics having the most direct effect on traffic demand. These act together in a complex manner to influence the demand for highway space.

High levels of residential and employment growth will inevitably result in increased traffic demand as more people link up to greater employment opportunities, with the higher levels of prosperity being reflected in higher levels of car ownership. Increasing numbers of jobs, homes, shopping facilities and schools will inevitably increase the demand for traffic movement both within and between centers of population.

On the assumption that a road scheme is selected to cater for this increased future demand, the design process requires that the traffic volumes for some year in the future, termed the design year, can be estimated. The design year is generally taken as 10~15 years after the highway has commenced operation. The basic building block of this process is the current level of traffic using the section of highway at present. An estimate for the normal traffic growth must be added onto this figure, which is due to the year-on-year annual increases in the number of vehicles using the highway between the present and the design year. Additionally, the generated traffic, that is the extra trips brought about directly from the construction of the new road, must be added onto these two constituents of traffic volume. Computation of these three components enables the design-year volume of traffic to be estimated for the proposed highway. Within the design process, the design volume will determine directly the width of the traveled pavement required to deal with the estimated traffic levels efficiently and effectively.

Traffic survey is a necessary means to provide basic data and information for traffic planning, and it is also one of the main contents. For highway planning, traffic survey is an important part of the feasibility study of a highway construction project. The purpose is to understand the traffic characteristics and composition in the project area, to grasp the data of traffic volume, flow direction and vehicle composition. It provides not only basic data for predicting the future traffic volume, but also a reliable basis for the economic evaluation and highway design. It consists of the person travel survey (PT), cargo flow survey, origin-destination (OD) survey of motor vehicles as well as the traffic flow survey through a certain road cross section.

Initially, the responsible transport planners decide on the physical boundary within which the study will take place. Most transport surveys have at their basis the land-use activities within the study area and involve making an inventory of the existing pattern of trip making, together with consideration of the socio economic factors that affect travel patterns. Travel patterns are determined by compiling a profile of the origin-destination (OD) of all journeys made within the study area, together with the mode of travel and the purpose of each journey. For those journeys originating within the study area, household surveys are used to obtain the OD information. These can be done with or without an interviewer assisting. In the case of the former, termed a personal interview survey, an interviewer records answers provided by the respondent. With the latter, termed a self-completion survey, the respondent completes a questionnaire without the assistance of an interviewer, with the usual format involving the questionnaire being delivered/mailed out to the respondent who then mails it back/has it collected when all questions have been answered. For those trips originating outside the study area, traversing its external “cordon” and ending within the study area, the OD information is obtained by interviewing trip makers as they pass through the “cordon” at the boundary of the study area. These are termed intercept surveys where people are intercepted in the course of their journey and asked where their trip started and where it will finish.

A transportation survey should also gather information on the adequacy of existing infrastructure, the land use activities within the study area and details on the socio economic classification of its inhabitants. Traffic volumes along the existing road network together with journey speeds, the percentage of heavy goods vehicles using it and estimates of vehicle occupancy rates are usually required. For each designated zone within the study area, office and factory floor areas and employment figures will indicate existing levels of industrial/ commercial activity, while census information and recommendations on housing densities will indicate population size. Some form of personal household-based survey will be required within each zone to determine household incomes and their effect on the frequency of trips made and the mode of travel used.

5.2.4 Prediction of Demands

At this point, having gathered all the necessary information, models are developed to translate the information on existing travel patterns and land-use profiles into a profile of future transport requirements for the study area. The four stages in constructing a transportation model are trip generation and attract, trip distribution, modal split and traffic assignment, illustrated in Fig. 5-1.

The first stage estimates the number of trips generated by each zone based on the nature and level of land-use activity within it. The second distributes these trips among all possible destinations, thus establishing a pattern of trip making between each of the zones. The mode of travel used by each trip maker to complete their journey is then determined and finally the actual route within the network taken by the trip maker in each case. Together they form the process of transportation demand analysis which plays a central role within highway engineering. It attempts to describe and explain both existing and future travel behavior in an attempt to predict demand for both car-based and other forms of transportation modes.

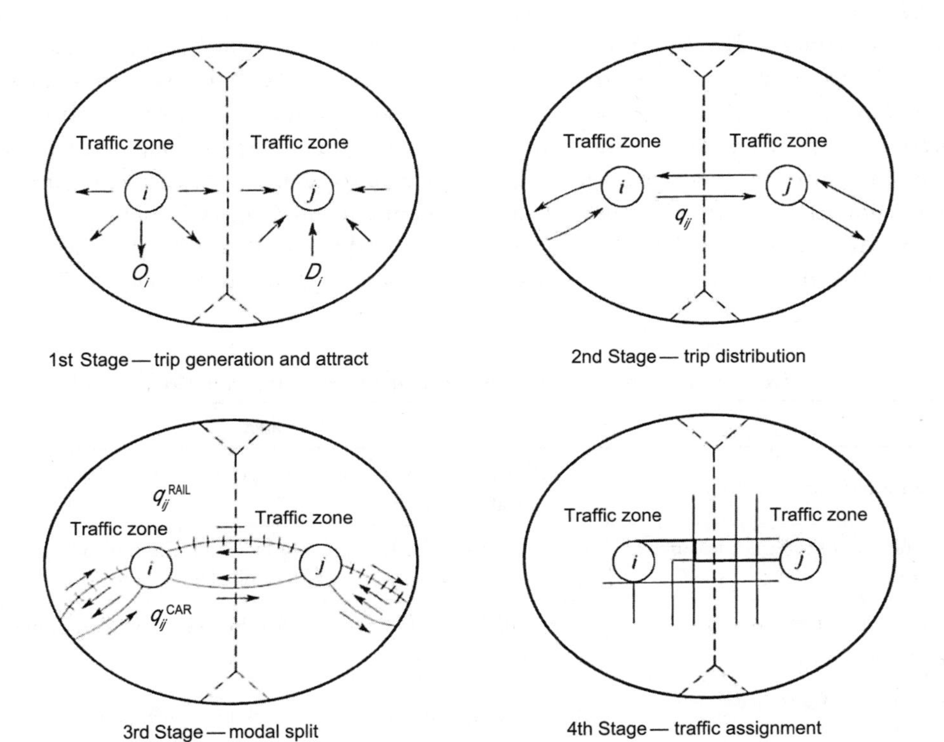

Fig. 5-1 Illustration of the four-stage-model
Source: Shao C F. Traffic Planning (2nd Edition). Beijing: China Railway Press, 2014.

5.2.5 Decision-making and Assessment

Highway and transportation planning can be described as a process of making decisions which concerns the future of a given transport system. The decisions relate to the determination of future demand, the relationships and interactions which exist between the different modes of transport, the effect of the proposed system on each existing land uses and those proposed for the future, the economic, environmental, social and political impacts of the proposed system and the institutional structures in place to implement the proposal put forward. Transport planning is generally regarded as a rational process, i.e. a rational and orderly system for choosing between competing proposals at the planning stage of a project. It involves a combined process of information gathering and decision making. The five steps in the rational planning process are summarized in Table 5-1.

In the main, transport professionals and administrators subscribe to the values underlying rational planning and utilize this process in the form detailed below. The rational process is, however, a subset of the wider political decision-making system, and interacts directly with it both at the goal-setting stage and at the point in the process at which the preferred option is selected. In both situations, inputs from politicians and political/community groupings representing those with a direct interest in the transport proposal under scrutiny are essential in order to maximize the level of acceptance of the proposal under scrutiny.

Assuming that the rational model forms a central part of transport planning and that all options and criteria have been identified, the most important stage within this process is the evaluation/appraisal process used to select the most appropriate transport option. Broadly speaking, there are two categories of appraisal process. The first category consists of a group of methods that require the assessments to be solely in money terms. They assess purely the economic consequences of the proposal under scrutiny. The second category consists of a set of more widely-based techniques that allow consideration of a wide range of decision criteria—environmental, social and political as well as economic, with assessments allowable in many forms, both monetary and non-monetary. The former group of methods are termed economic evaluations, with the latter termed multi-criteria evaluations.

Evaluation of transport proposals requires various procedures to be followed. These are

Steps in the rational planning process for a transportation project Table 5-1

Step	Purpose
Definition of goals and objectives	To define and agree the overall purpose of the proposed transportation project
Formulation of criteria/measures of effectiveness	To establish standards of judging by which the transportation options can be assessed in relative and absolute terms
Generation of transportation alternatives	To generate as broad a range of feasible transportation options as possible
Evaluation of transportation alternatives	To evaluate the relative merit of each transportation option
Selection of preferred transportation alternative/group of alternatives	To make a final decision on the adoption of the most favorable transportation option as the chosen solution for implementation

ultimately intended to clarify the decision relating to their approval.

It is a vital part of the planning process, be it the choice between different location options for a proposed highway or the prioritizing of different transport alternatives listed within a state, regional or federal strategy. As part of the process by which a government approves a highway scheme, in addition to the carrying out of traffic studies to evaluate the future traffic flows that the proposed highway will have to cater for, two further assessments are of particular importance to the overall approval process for a given project proposal: cost-benefit analysis (CBA) and environmental impact assessment (EIA).

Layered on top of the evaluation process is the need for public participation within the decision process. Although a potentially time-consuming procedure, it has the advantages of giving the planners an understanding of the public's concerns regarding the proposal and also actively draws all relevant interest groups into the decision-making system. The process, if properly conducted, should serve to give the decision-makers some reassurance that all those affected by the development have been properly consulted before the construction phase proceeds.

The program evaluation of transportation planning mainly includes four aspects: the evaluation of technology, economy, social environment and traffic operation effect.

(1) The technical evaluation of transportation planning is based on the construction level and technical performance of the transportation network. Adaptability of its construction scale to social and economic development and the internal structure and functions of the transportation network will be analyzed. Certain technologies will be introduced into transportion planning. A comprehensive qualitative and quantitative study on the social impact of the coming transportation plan will be performed, so as to make a comprehensive evaluation of its pros and cons. The economic evaluation of transportation planning is an analysis of the economic benefits of the transportation network as a whole. One of the fundamental purposes and important principles of transportation planning is to obtain the best economic benefits of the transportation system with the least investment.

(2) The economic evaluation of the transportation planning is to analyze and demonstrate the economic rationality of the plan by comparing the construction and operating costs with benefits of the planning plans, and combining the forecast of future funds during the planning period. The social environment assessment of transportation planning is to analyze the role and influence of transportation network system on the social environment of the planned area, including the promotion of the development and utilization of land and natural resources, the improvement of water and soil conservation and environmental protection conditions, and the impact on regional politics, economy, cultural heritage and scenic spots. Compared with technical evaluation and economic evaluation, social environmental evaluation has the characteristics of macroscopic, long-term, multi-objective, multiple indirect benefits and difficult quantitative indicators. Starting from the requirements of quantitative analysis, social environmental evaluation is a more difficult type of evaluation. Further exploration is needed.

(3) Environmental impact assessment is essentially a strategic environmental impact assessment, which is forward-looking. It helps to resolve conflicts that cannot be resolved at the project level for a long time, and can analyze the cumulative environmental impact of a large number of projects. And it requires detailed discussion of strategic countermeasures for environmental protection and economic development from many aspects. The planned environmental impact assessment must be carried out before the

detailed planning of construction activities, and a reasonable planning plan should be formulated to maximize economic, social and environmental benefits.

(4) The evaluation of traffic operation effect takes safety, smoothness and speed as the target value, from a broader perspective, comprehensively and systematically evaluates traffic operation characteristics, including the use of road infrastructure, traffic demand characteristics, the supply of urban parking facilities, the usage and the efficiency of traffic management facilities in the urban traffic system, etc., so as to get corresponding improvement plans.

Each of the above is an evaluation sub-system, which uses single indicator corresponding to multiple factors to make quantitative or qualitative analysis and judgment on the performance and value of the transportation network from different aspects. Finally, a comprehensive evaluation of the overall transportation plan is required, which is called the overall goal evaluation. The overall goal evaluation is to obtain the optimization of the overall function of the planning system on the basis of the evaluation of each part, stage and level of sub-systems. At the same time, in the overall optimization process of the system, it continuously provides various related information to the decision maker. Modern science and technology theories, especially the theoretical development of systems engineering, provide a basis for comprehensive evaluation. The purpose of program evaluation is to conduct a comprehensive technical review of several programs, use mathematical tools to judge the overall effect of the program, and select the best program among multiple programs to provide a scientific basis for decision makers' approval.

5.2.6 Summary

Highway engineering involves the application of scientific principles to the planning, design, construction, maintenance and operation of a highway project or system of projects. The aim of this section is to give students an understanding of the analysis and design techniques that are fundamental to the topic. This section has briefly introduced the context within which highway projects are undertaken, and details the frameworks, both institutional and procedural, within which the planning, design, construction and management of highway systems take place. During the transportation planning process, some important design standards are already determined, such as the function of a highway in the network and its technical grade. The starting point and the end point determine the orientation of a highway line. Meanwhile the designing process starts. This mainly includes the geometric design and the pavement structural and material design. The remaining of this chapter will deal specifically with the geometric design, pavement structural and material design.

5.3 Geometric Design of Highway

The geometric design of highway is the branch of highway engineering concerned with the positioning of the physical elements of the roadway according to standards and constraints. The basic objectives in geometric design are to optimize efficiency and safety while minimizing cost and environmental damage. Geometric design also affects an emerging fifth objective called "livability", which is defined as designing roads to foster broader community goals, including providing access to employment, schools, businesses and residences, accommodate a range of travel modes such as walking, bicycling, transit and vehicles, and minimizing fuel use, emissions and environmental damage. As a result, every country has developed its own design codes or specifications, of which the basic principles are more or less the same. In China, the corresponding code is *Design Specification for Highway Alignment* JTGD20—2017 for highways and *Code for Design of Urban Road Engineering* CJJ37—2012 for urban roads. In the following, the geometric design principles and details will be introduced in the background of the *Design Specification for Highway Alignment*, which will be abbreviated as *Design Specification*.

The proper geometric design of a highway ensures that drivers use the facility with safety and comfort. The process achieves this by selecting appropriate vertical and horizontal curvature along with physical features of the road such as sight distances and super elevation. The ultimate aim of the procedure is a highway that is both justifiable in economic terms and appropriate to the local environment. Geometric design of highway can be divided into three main parts: alignment, profile and cross-section. Together, they provide a three-dimensional layout for a roadway. The alignment is the route of the road, defined as a series of horizontal tangents and curves. The profile is the vertical aspect of the road, including crest and sag curves, and the straight grade lines connecting them. The cross section shows the position and number of vehicle and bicycle lanes and sidewalks, along with their cross slope or banking. Cross sections also show drainage features, pavement structure and other items outside the category of geometric design.

5.3.1 Design Controlling Factors

Road design begins with the establishment of basic controlling factors for the design. These include environmental factors (such as terrain, specific location on the road, climate), driver and pedestrian characteristics, traffic elements, and so on. The above controlling factors are selected or determined by the designer, they determine the grade of the road, and at the same time provide the basis for the linear design (longitudinal gradient, curvature, width, sight distance, etc.). The design control factors discussed below are an important basis for road geometric design. In engineering practice, various control factors such as economy, safety and beauty should also be considered.

(1) Design Vehicle

The design vehicle refers to the representative vehicle model used for the geometric design. Its external dimensions, weight and operating characteristics are the basis for the road geometric design, which has a decisive effect on the road geometric design. The size of the vehicle directly affects the lane width, minimum turning radius, curve widening, sight distance and road construction limits, and the dynamic characteristics affect the longitudinal section design, climbing lanes, etc. There are many types of vehicles running on the road, with different shapes and great differences in power. Therefore, the characteristics of the various vehicles running on the road should be combined. They are divided into various types according to the day of use, structure or engine. Vehicles with a representative weight size, and running characteristics from the various types should be selected as the design vehicles. There are five types of design vehicles selected for highway design: cars, buses, articulated buses, trucks, and articulated trucks. The outline dimensions are shown in Table 5-2 and Fig. 5-2. The front overhang refers to the front of the car body to the center of the front wheel axle. The wheel base refers to the distance from the center of the front wheel axle to the center of the rear wheel axle. And the rear overhang refers to the distance from the center of the rear wheel axle to the rear end of the vehicle body. Along

Dimensions of design vehicles (m) Table 5-2

Vehicle type	Length	Width	Height	Front overhang	Wheel base	Rear overhang
Car	6	1.8	2	0.8	3.8	1.4
Bus	13.7	2.55	4	2.6	6.5+1.5	3.1
Articulated bus	18	2.5	4	1.7	5.8+6.7	3.8
Truck	12	2.5	4	1.5	6.5	4
Articulated truck	18.1	2.55	4	1.5	3.3+11	2.3

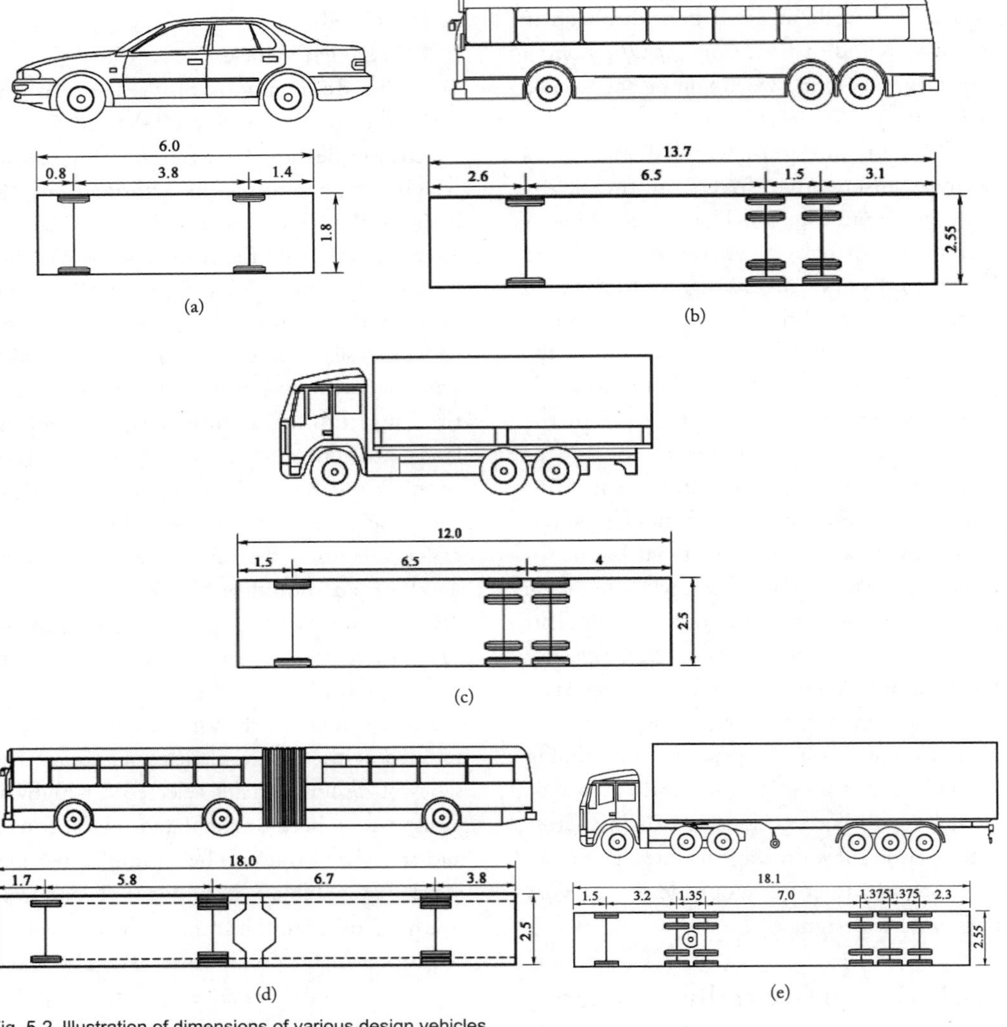

Fig. 5-2 Illustration of dimensions of various design vehicles
(a) Car; (b) Bus; (c) Ttruck; (d) Articulated bus; (e) Articulated truck
Source: Ministry of Transport of the People's Republic of China. Technical Standard of Highway Engineering JTG B01—2014. Beijing: China Communications Publishing & Media Management Co., Ltd., 2014.

with clearance, length of overhangs affects the approach and departure angles, which measures the vehicle's ability to overcome steep obstacles and rough terrain. The longer the front overhang is, the smaller the approach angle will be, and thus reduces the car's ability to climb or descend steep ramps without damaging the front bumpers.

(2) Design Speed

The concept of design speed lies at the center of the geometric design process. The design speed of a highway serves as a guide in the selection of the geometric features of a highway. Selection of the correct design speed ensures that issues of both safety and economy in the design process are addressed. The chosen design speed must be consistent with the anticipated vehicle speeds on the highway under consideration.

When roads are planned, the selected design speed may be based on or influence several factors, including the geometric design of road features, the planned operating speed, the legislated speed limit caps, the anticipated traffic volume and the road's functional classification. An excerpt of the specification as to the design speed and the influenced geometric design indicators according to the standard of China is shown in Table 5-3.

(3) Traffic Volume and Capacity

1) Planned Traffic Volume

Traffic volume refers to the number of vehicles passing through a certain section of the road in a unit time. The general measurement unit is the annual average daily traffic volume (*AADT*), which is the total annual traffic volume divided by 365 days. The planned traffic volume (also called the design traffic volume) refers to the achievable annual average daily traffic volume (pcu/d) within the time from the planned road to the predicted age. It is calculated based on the forecast of traffic observation data in history. When this data is not available, it will be calculated according to an annual average growth rate with the following equation:

$$AADT = ADT \times (1+\gamma)^{n-1} \quad (5\text{-}1)$$

Where *AADT* — design traffic volume (pcu/d), where pcu (passenger car unit) is a summary of various vehicle types through a conversion factor according to Table 5-4;

ADT — initial traffic volume (pcu/d);

γ — annual average growth rate, which is based on certain

Design speed and geometric design indicators of different highway technical grades

Table 5-3

Technical grades of the highway		**Motorway**			**1st-class**			**2nd-class**		**3rd-class**		**4th-class**
Design speed (km/h)		120	100	80	100	80	60	80	60	40	30	20
Number of lanes		≥4			≥4			2	2	2	2	1 or 2
Lane width (m)		3.75	3.75	3.75	3.75	3.75	3.5	3.75	3.5	3.5	3.25	3
Stopping sight distance (m)		210	160	110	160	110	75	110	75	40	30	30, 20
Minimum Radius of circular curves (m)	I_{max} = 4%	810	500	300	500	300	150	300	150	65	40	20
	I_{max} =6%	710	440	270	440	270	135	270	135	60	35	15
	I_{max} = 8%	650	400	250	400	250	125	250	125	60	30	15
	I_{max}=10%	570	360	220	360	220	115	220	115	—	—	—
Maximum longitudinal gradient (%)		3	4	5	4	5	6	5	6	7	8	9

Conversion factors of different vehicles on highways **Table 5-4**

Types	Conversion coefficient	Illustration
Passenger car	1	Passenger car of less than 19 seats; truck of axle-load less than 2t
Medium vehicle	1.5	Passenger car of more than 19 seats; truck of axle-load between 2t and 7t
Large vehicle	2.5	Truck of axle-load between 7t and 20t
Trailer	4	Truck of axle-load more than 20t

prediction models or according to experiences of similar highway project;

n — predicting period, which is selected according to the function of the highway in the network, for important national and provincial arterial highways, n = 20 years; for national and provincial arterial highways, n = 15 years; and for county highways, n = 10 years.

The forecasting year of the planned traffic volume should be the planned opening year in the project feasibility study report. When the feasibility study report is submitted to the road opening year for more than 5 years, the planned traffic volume should be checked before the preliminary design is prepared. The planned traffic volume plays an important role in determining the road grade, demonstrating the planned cost of the road, and carrying out various structural designs, but it should not be used directly in the road design. Because the monthly, daily and hourly traffic volume of the year is changing, it may be several times higher than the annual average daily traffic volume in certain seasons and certain periods, so it should not be used as a basis for specific design.

AADT is a simple, but useful, measurement of how busy the road is. Newer advances from GPS traffic data providers are now providing *AADT* counts by side of the road, by day of week and by time of day. *AADT* is the standard measurement for vehicle traffic load on a section of road, and the basis for most decisions regarding transport planning, or to the environmental hazards of pollution related to road transport.

2) Hourly Traffic Volume

Hourly traffic volume is the traffic volume calculated in hours, which is the basis for determining the number of lanes and lane width or evaluating service levels. A large number of traffic statistics show that the hourly traffic volume varies considerably during a day and throughout the year. If the maximum peak hour traffic volume in a year is used as the design basis, it will cause waste, but if the daily average hourly traffic volume is used, the traffic demand cannot be met, causing traffic congestion. In order to make the value of the designed traffic volume not only ensure the traffic safety and smooth flow, but also make the project cost economical and reasonable, it is necessary to use the change curve of the traffic volume per hour in the year to guide the determination of the hourly traffic volume that meets the requirements of use.

The method is as follows: The traffic volume (two-way) of each hour among all 8760 hours in a year is listed in descendent order. A typical distribution is shown in Fig. 5-3, where the curve descends exponentially. The curve on the right side drops slower, while the gradient on the left side is larger. There is a reasonable value of the designed hourly traffic volume, normally in the 30th hour position. If the traffic volume of the 30th hour is used as the design basis, it means that only 29 hours of traffic in a year exceed the design value, and congestion will occur, accounting for 0.33%

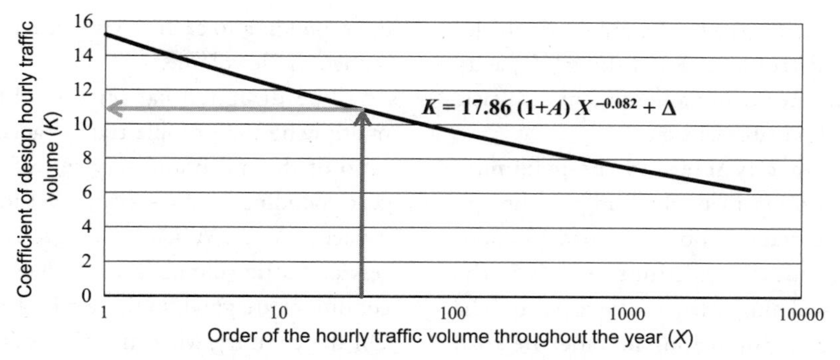

Fig. 5-3 Distribution of hourly traffic volume of the year (drawn according to Sun 1991)

of the annual hours. On the contrary, within 99.67% of the whole year smooth traffic can be ensured.

When determining the design hourly traffic volume, the traffic volume change curve of each route should be drawn based on the observation data. The road sections without observation data can be determined by referring to the observation data of other roads with similar nature and similar traffic conditions. The design hourly traffic volume is calculated as follows:

$$N_h = N_d \times D \times K \quad (5-2)$$

Where N_h — directional design hourly traffic volume;

N_d — annual average design traffic volume;

D — direction unevenness factor, normally D = 0.5~0.6;

K — coefficient of design hourly traffic volume (%), which is determined with the observation data according to Fig. 5-3. When no data is available, empirical value will be given, which is related to the economic territory and the technical grade of the highway.

3) Standard Vehicle Model and Conversion

There are many types of vehicles driving on the road, and their speed, driving law and the headroom are quite different, but the designed traffic volume of a road should be converted into a certain standard vehicle type. The standard vehicle type specified in the *Technical Standard* is a small passenger car, and the motor vehicle conversion coefficient used for road planning and technical grade classification is adopted according to Table 5-4. For mixed traffic roads where non-motor vehicles account for a large proportion, bicycles, pedestrians and animal-drawn vehicles are no longer involved in traffic conversion as lateral interference factors. Each tractor driving on the road can be converted into 4 small passenger cars. Various vehicles on urban roads can be converted according to the corresponding urban road design regulations, as shown in Table 5-5.

4) Traffic Capacity and Level of Service

Road capacity refers to the maximum amount of traffic that a certain road section can withstand, also called road capacity, expressed in terms of the maximum number of vehicles passing through a unit time (vehicles/hour). For a multi-lane road, it is the number of vehicles passing through one lane, and for a

Conversion factors of different vehicles on urban roads **Table 5-5**

Vehicle type	Passenger car	Bus	Truck	Articulated vehicle
Conversion factor	1	2	2.5	3

two-lane road, it is the total number of vehicles in the round-trip lane. Road design capacity is obtained after many amendments to basic capacity and possible capacity.

Basic capacity refers to the maximum number of passenger cars that can pass through a lane or a certain section of a road in a unit time under ideal conditions, which is the basis for calculating other traffic capacity. The so-called ideal conditions include the road itself and traffic: the road itself should have sufficient lane width, lateral clear width, and good horizontal, vertical geometric conditions and sufficient sight distances; in terms of traffic, there are only small passenger cars on the lanes, no other vehicle models are mixed in and the speed is not restricted. Even on motorways, no ideal conditions exist, and the number of passing vehicles is generally lower than the basic capacity. The calculation of basic capacity can be obtained by "time headway" or "space headway". The time headway is the time interval between two consecutive vehicles passing through the lane or the same place on the road. The space headway is the distance between two consecutive cars in the traffic flow.

Possible traffic capacity is the traffic capacity through amendment of the basic capacity, considering the factors that affect the capacity, such as lane width, lateral clear width and large vehicles. The designed capacity is the maximum number of vehicles that can pass through a certain section of the road in unit time when the operating state of the road traffic is maintained at a certain designed service level. In China, the traffic flow is divided into six service levels according to the operating status of the traffic flow, from the free flow with only little traffic volume to the restricted flow where the traffic volume reaches the highest possible state. The traffic volume corresponding to each service level is called the service traffic volume.

The designed capacity is obtained by multiplying the possible traffic capacity by the ratio of the maximum service traffic volume corresponding to the service level to the basic capacity (V/C). When V/C value is small, the service traffic volume is small, the traffic flow conditions are good, so the service level is high. On the contrary, when the V/C value is large, the service traffic volume is high, the traffic conditions are poor and the service level will be low. When the design hourly traffic volume exceeds the design capacity, congestion will happen.

Service level or level of service (LOS) is a quality indicator for drivers to experience the operating state of the traffic flow. It is usually characterized by indicators such as average driving speed, driving time, traffic delay, degree of freedom of maneuverability and safety. The service level of highways at all levels should not be lower than those specified in Table 5-6.

When a 1st-class road is used as a distribution road, the design service level can be reduced by one level. Long and extra-long tunnel sections, non-motorized and pedestrian-intensive sections, as well as the converging, diverging and weaving sections of an interchangeable grade separated junction and the design service level can be reduced by one level.

In the *Technical Standard* a description for the various level of service is given:

① First level of service: The traffic flow is in a completely free flow state. The traffic volume is small, the speed is high and the driving density is low. The driver can freely choose the required speed according to his own wishes. The driving vehicle is not or basically not affected by other vehicles in the traffic flow. The freedom of driving in the traffic

Service levels adopted in highways **Table 5-6**

Highway technical grade	Motorway	1st-class	2nd-class	3rd-class	4th-class
Allowable service level	3	3	4	4	No requirement

flow is great, and the comfort and convenience provided for drivers, passengers or pedestrians are very superior. The impact of minor traffic accidents or traffic obstacles is easy to eliminate, and there will be no stagnation and queuing on the accident section, and it will soon be restored to the first service level.

② Second level of service: The state of traffic flow is relatively free flow. Drivers can basically choose the driving speed according to their own wishes, but they should start to notice that there are other users in the traffic flow, and drivers have a high level of physical and mental comfort. The impact of minor traffic accidents or traffic obstacles is easy to eliminate, and the operation and service conditions in the accident section are worse than the first level.

③ Third level of service: The traffic flow state is in the upper half of the steady flow, the mutual influence of the vehicle brakes becomes greater, the selection speed is affected by other vehicles, the driver must be extra careful when changing lanes, and minor traffic accidents can still be eliminated. However, the service quality of the road section where the accident occurred was greatly reduced, and the traffic flow in line was formed behind the severe blockage and the driver was nervous.

④ The fourth level of service: The traffic flow is at the lower limit of the stable flow range, but the vehicle operation is obviously affected by the mutual influence of other vehicles in the traffic flow, and the speed and driving freedom are obviously restricted. A slight increase in traffic volume will result in a significant reduction in service levels and a reduction in the level of physical and mental comfort of drivers. Even minor traffic accidents are difficult to eliminate, and a long queue of traffic will form.

⑤ The fifth level of service: The traffic flow is at the upper half of the traffic congestion, and the lower part is the operating state when the maximum capacity is reached. For any disturbance to the traffic flow, such as the flow of traffic entering from the road or the change of lanes, it will produce a disturbance wave in the traffic flow. The traffic flow cannot eliminate it. Any traffic accident will form a long queue of traffic, and the flow of traffic will be flexible. Extremely restricted, the driver's physical and mental comfort level is very poor.

⑥ The sixth level of service: The traffic flow is at the lower half of the congestion, which is the forced flow or blocked flow in the usual sense. Under this service level, the traffic demand of the transportation facility exceeds its allowable capacity. The traffic flow is queued, and the vehicles in the queue appear to repeatedly stop and go. The operating state is extremely unstable, and sudden changes may occur between different traffic flow states.

An approximate illustration of different traffic flow states can be seen in Fig. 5-4.

5) Driver Characteristics

The road is mainly for vehicles, and the vehicle is manipulated by the driver. Whether the road design is appropriate should be judged from the effect of the driver to meet the safety and effectiveness. If the designed road is compatible with the driver's ability, the road will help improve driving efficiency. If it is not compatible with the driver's ability, the driver's error probability will increase and this will cause traffic accidents. Driver characteristics include many aspects, such as driver's reaction time, visual characteristics, driver's driving age, etc., which all have a certain impact on the geometric design. The reaction time directly affects the sight distance and traffic safety facilities design. In the alignment and landscape design, the visual characteristics of the driver need to be considered.

When a vehicle is driving, it will encounter various emergencies at any time. The driver needs to respond quickly to the situation encountered and deal with it in a timely and proper manner to avoid accidents. It is worth noting that the driver's reaction time to different events is different, and the reaction

Fig. 5-4 Visual appearances of various levels of service of highway
(a) Level 1; (b) Level 2; (c) Level 3; (d) Level 4; (e) Level 5; (f) Level 6
Source: Transportation Research Board. Highway Capacity Manual: A Guide for Multimodal Mobility Analysis (6th Edition), 2016.

time to judge more complicated or unexpected events will be even longer. According to the composition of drivers, it is also necessary to consider the situation of elderly drivers. Research on the reaction time of drivers to anticipated events and emergencies in the United States shows that the average reaction time of most drivers to anticipated events is 0.6s, a few are as long as 2s. The reaction time to emergencies is 35% longer. For some drivers, this time reaches 2.75s. Therefore, from the perspective of safety, the design involving the driver's response time is generally considered as 3s.

6) Building Limit, Land Use and Red Lines

Road building limit is also called clear space, which consists of two parts: clear height and clear width. It is to ensure the normal passage and safety of various vehicles and people on the road, and no obstacles are allowed to invade within a certain height and width. The road building limit is an important basis for cross-sectional design. During the design, the interrelationship of the elements of the road width and the installation plan of various road facilities should be fully studied and reasonable arrangements should be made in the limited space. Bridge abutments, piers, lighting poles, guardrails, signals, signs, street trees, electric poles and other facilities are not allowed to invade into the road building limit.

The clear height is the height of the clearance, which refers to the vertical height that the road must meet to ensure safe passage within the cross section. The clear height should be determined according to factors such as vehicle loading height, safety height and road paving. The loading height of trucks in China is limited to 4.0m, plus a safety height of 0.5m, and a net height of 4.5m is generally adopted. Taking into account the transportation of large-scale equipment, snow and pavement thickening during maintenance, etc., it is stipulated that the clear height for motorways, 1st-class and 2nd-class highways is 5.0m, and the 3rd-class and 4th-class roads are 4.5m. For the 3rd-class and 4th-class roads with intermediate or low-level pavement types,

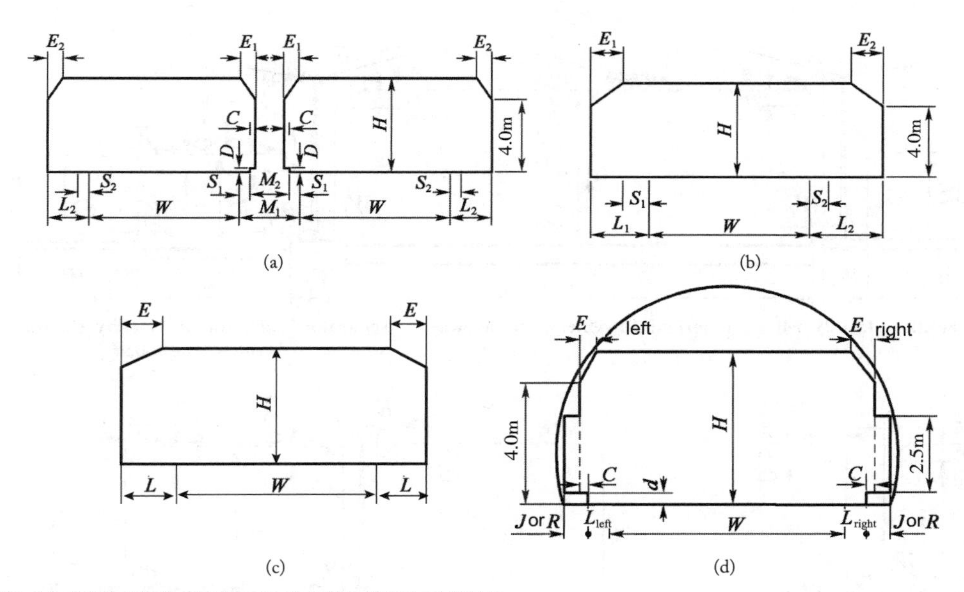

Fig. 5-5 Building limit of highways of various technical grades
(a) Motorway or 1st-class highway (integral); (b) Motorway or 1st-class highway (separate); (c) 2nd-class, 3rd-class and 4th-class highways; (d) Highway tunnel
Source: Ministry of Transport of the People's Republic of China. Technical Standard of Highway Engineering JTG B01—2014. Beijing: China Communications Publishing & Media Management Co., Ltd., 2014.

considering the requirements of pavement, an extra 20cm can be reserved for the clear height. The clear height along a highway should be consistent. When the structure is located above the concave vertical curve, the passage of large vehicles will form a suspension and reduce the effective clear height under the structure. Full consideration should be given to ensure the effective clear height requirement during the design. In the same way, when the road passes underneath, it should be ensured that any point between the road surface and the bottom of the structure should meet the clear height requirements. The minimum clear height of urban roads is 4.5m for various cars, 5.0m for trolleybuses, 5.5m for trams, 2.5m for bicycles and pedestrians, and 3.5m for other non-motorized vehicles.

The clear width refers to the horizontal width that the road must meet to ensure safe passage within the cross section. The clear width includes the width of the driving belt, road shoulder, central strip and green belt. The road shoulder is within the clearance range, so various facilities (signs, guardrails, etc.) on the road should be set up on the protective road shoulder beyond the hard shoulder, and the projected part must be above the clear height. The bridge piers or portal pillars set on the central strip and the road shoulder should not be set close to the building limit, and there should be room for the location of the protective fence (not less than 0.5m).

The headroom of bridges, tunnels and elevated roads should generally be the same as that of road sections. Sometimes when the headroom needs to be compressed in order to reduce the cost, the compressed part is mainly reflected in the lateral width. However, sidewalks are required in bridges and tunnels, and when the width of the sidewalk is greater than the lateral width, the increased width should be included in the clear width. When the sidewalks, bicycle lanes, inspection lanes and traffic lanes are set separately, the clear height is generally 2.5m.

The limits of highway construction at all levels are shown in Fig. 5-5, and the limits

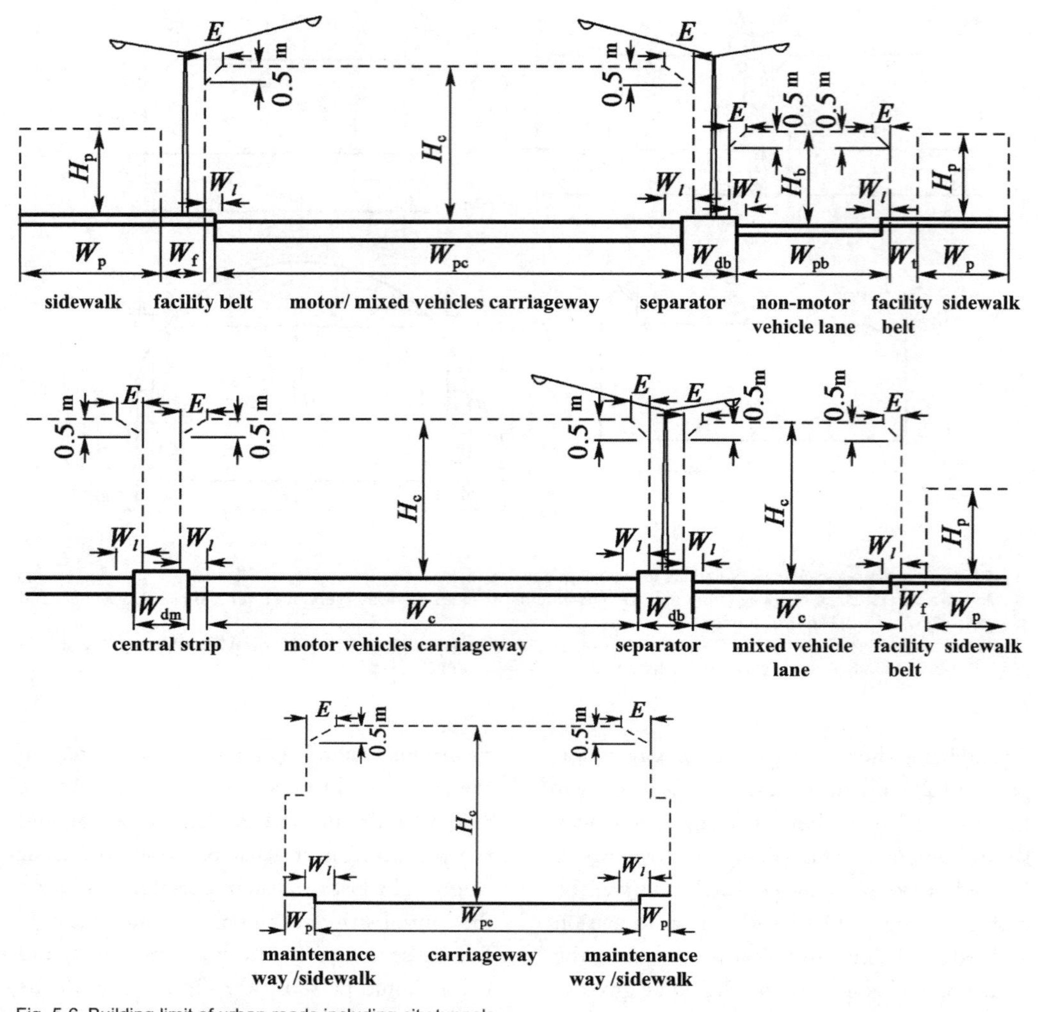

Fig. 5-6 Building limit of urban roads including city tunnels
Source: Ministry of Housing and Urban-Rural Development of the People's Republic of China. Code for Design of Urban Road Engineering CJJ 37—2012. Beijing: China Architecture & Building Press, 2012.

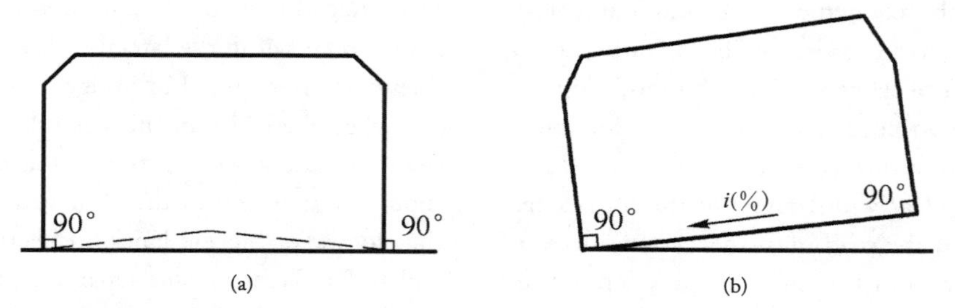

Fig. 5-7 Building boundaries of highways and urban roads
(a) With crown; (b) With superelevation
Source: Ministry of Housing and Urban-Rural Development of the People's Republic of China. Code for Design of Urban Road Engineering CJJ 37—2012. Beijing: China Architecture & Building Press, 2012.

of urban road construction are shown in Fig. 5-6. The building limit line regulations of the road construction boundary are shown in Fig. 5-7. For general crown sections, the upper boundary line is a horizontal line, and the boundary lines on both sides are perpendicular to the horizontal line; for sections with superelevation, the upper boundary line is a diagonal line parallel to the superelevation cross slope, and the boundary lines on both sides are parallel to the superelevation cross slope. The high cross slope line is vertical.

Highway land refers to the land that needs to be occupied for highway construction, maintenance and laying of various facilities along the line. Land acquisition procedures for highways must be handled in accordance with relevant national policies. Non-road buildings, such as digging channels, burying pipelines, cables, poles and other facilities, are not allowed within the highway land. When determining land use, it is necessary to not only meet the scope of land required for road construction, but also fully consider the precious characteristics of land resources in China. It is necessary to save every inch of land in terms of design and construction as much as possible, and do not occupy or occupy less high-yield land. It is advocated to use soil borrowing or discarding soil for preparing cultivated land. Specifications of the scope of the highway land as well as principles of highway land use are as follows:

①For newly built motorway, the land scope should be not less than 2m away from the outer edge of the drainage ditch on both sides (the embankment or the toe of the slope protection if there is no drainage ditch), or the outer edge of the intercepting ditch at the top of the cutting slope (top of the slope if there is no intercepting ditch). For highway of lower-grades, the land scope should be not less than 1m beyond the above-mentioned edge line.

②In the high-filling and deep-excavation section, in order to ensure the stability of the roadbed, the land use scope should be determined according to stability analysis.

③In zones with windy sand, snow damage and special geological conditions, the scope of land should be determined according to the needs of setting up shelter forests, planting sand-fixing plants, installing sand or snow fences, or loading berm.

④Street trees should be planted within the road land area outside the drainage ditch or interception ditch. The land scope should be determined according to the specific conditions, if multiple rows of forest belts needs to be planted according to environmental protection requirements.

⑤The land used for projects such as grade separation, intersection, service facilities, safety facilities, traffic management facilities, parking areas, maintenance management facilities, and stockyards and nurseries along the highway shall be determined according to actual needs.

⑥For road reconstruction, the scope of land can be determined by referring to the newly built road.

The road red line refers to the boundary control line of the urban road land. The width between the red lines is the road land range, which is called the road building red line width or roadway width. Planning the road red line is to determine the side line of the road or the width of the road building red line. Its purpose is to comprehensively stipulate the scope of roads, squares, intersections and other land at all levels, which is convenient for road design, construction and the arrangement and layout of buildings on both sides. It is also the main basis for engineering design, construction and adjustment. Once the road red line is determined, the land outside the red line must be constructed according to the plan, and various pipelines must be laid out according to the red line. Once completed, it is difficult to change. Therefore, planning the red line is very important.

The road red line is usually determined by the city planning department based on the city's overall plan to determine the form of the road network and the functions, nature, direction and location of each road. The main contents of the road red line planning and design are as follows:

①Determine the width of the road building red line. According to the nature and function of the road, consider the appropriate cross-sectional form, determine the reasonable width of each component such as motor vehicle lanes, non-motor vehicle lanes, sidewalks, green belts, etc., so as to determine a reasonable road red line width. In determining the red line width, in addition to considering special political needs, the factors that must be considered are the necessary width for traffic functions (including the number and width of the carriageways, partitions, non-motorized lanes, sidewalks, the green belts, etc.), sunshine and ventilation needs, width required by air defense, fire prevention and earthquake prevention, and the width required by architectural art. When the red line width plan is too narrow to meet the requirements of various influencing factors, it will bring difficulties to future reconstruction and expansion, and too wide will cause uneconomical urban land use. Therefore, when determining the width of the red line, the principle of "combination of near and far with emphasis of near" should be fully considered.

②Determine the position of the road red line. On the basis of the city master plan, for the roads in the new district, draw a red line according to the planned red line width and location of the centerline of the planned road. For road reconstruction in the old district, if it is planned to widen to the width of the red line in the near future, according to the principle of less demolition, it can be widened on one or both sides, and it is better to widen on one side. If it is a long-term controlled and gradually formed road, especially when the width of the red line is wider than the current road, it is advisable to keep the current center line in place and to retreat the buildings on both sides equally.

③Determine the type of intersection. According to the short-term and long-term planning and the specific conditions of the intersection, the form, land use range, specific location and main geometric dimensions of the intersection is determined and drawn on the plan together with the red line.

④Determine the coordinates and elevation of the control points. The turning point of the center line of the planned road and the intersection of each road are the control points. The plane coordinates of the control points can be directly measured on the spot, and the control elevation is determined by the vertical plan.

5.3.2 Cross Section Elements

The basic features of a highway are the carriage way itself, expressed in terms of the number of lanes used, the central reservation or center strip and the shoulders (including verges). Depending on the level of the highway relative to the surrounding terrain, side slopes may also be a design issue.

The road cross section refers to the normal tangent plane at any point on the centerline, which is composed of the cross-section design line and the ground line. The cross-section design line includes traffic lanes, non-motorized vehicle lanes, sidewalks, shoulders, partitions, side ditches, side slopes, intercepting ditches, slope protection, borrow pits, spoil piles, environmental protection facilities and other parts. The ground line is a line that characterizes the undulations of the ground. It is obtained through on-site measurements or large-scale topographic maps, aerial photographs, digital ground models, etc. The cross-sectional design is generally limited to the part of the road directly related to driving, that is, the width and cross slope of the components between the outer edges of the shoulders, for urban roads, the planned red lines. The design of side slopes, side ditch, intercepting ditch, slope protection and other facilities is part of the subgrade engineering, which is not topic here.

The composition of the highway cross section and the size of each part should be determined according to factors such as highway function, design traffic volume, traffic composition, design speed and terrain conditions. On the premise of ensuring highway functions, highway traffic capacity, traffic safety and smoothness, the designer should try to save land, so that economic and social benefits can be maximized.

For highways with high grades and heavy traffic, such as motorways and 1st-class highways, the up and down vehicles are usually separated. There are two ways to separate: one is to separate with equal width and the same height partition, the other is to separate the up and down lanes on different planes or different longitudinal sections. The former is called an integrated cross section, and the latter is called a separated cross section.

The integrated cross section of motorways and 1st-class highways is composed of carriage ways, central strip (central median zone, left marginal strip), road shoulder (right hard shoulder and soil shoulder) and other parts. The separated cross section is composed of carriage ways, shoulders (hard shoulders on both sides, and right soil shoulders). When there are additional lanes such as speed-change lanes, ramp lanes, emergency stop lanes and escape lanes, the cross section of motorways and 1st-class highways should also include these parts. Integrated cross sections without central strip, such as those of 2nd-class, 3rd-class and 4th-class highways, are composed of carriage ways, shoulders and other parts. When the proportion of trucks on the 2nd-class highways is high, overtaking lanes can be added locally as needed. When there are many slow vehicles on the 2nd-class highways, the slow lane can be set by widening the hard shoulder as needed. When a single-lane lane is adopted for the 4th-class highway, the turn-out lane should be set. For non-motorized vehicles, densely-pedestrian highways, and highways at urban entrances and exits, side dividers, non-motorized vehicle lanes and sidewalks can be set on the cross section of the highways as needed. The typical cross sections of highway are shown in Fig. 5-8, Fig. 5-9 and Fig. 5-10.

5.3.3 Sight Distance

Sight distance is defined as the length of carriage way that the driver can see in both the horizontal and vertical planes. Two types of sight distance are detailed: stopping distance and overtaking distance.

In order to ensure driving safety, the driver should be able to see a considerable distance ahead of the car at any time. Once an obstacle is found on the road ahead or an oncoming car can take timely measures to avoid collisions, this required shortest distance is called driving sight distance. Whether the sight distance is sufficient is directly related to the safety and speed of driving, and is one of the important

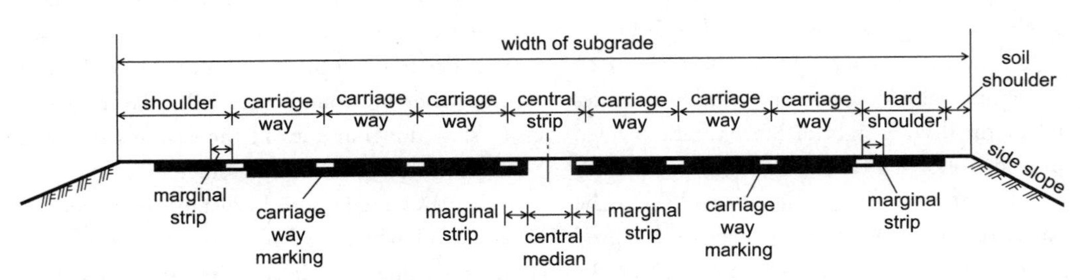

Fig. 5-8 Integrated cross section of motorway and 1st-class highway
Source: Ministry of Transport of the People's Republic of China. Design Specification for Highway Alignment JTG D20—2017. Beijing: China Communications Publishing & Media Management Co., Ltd., 2017.

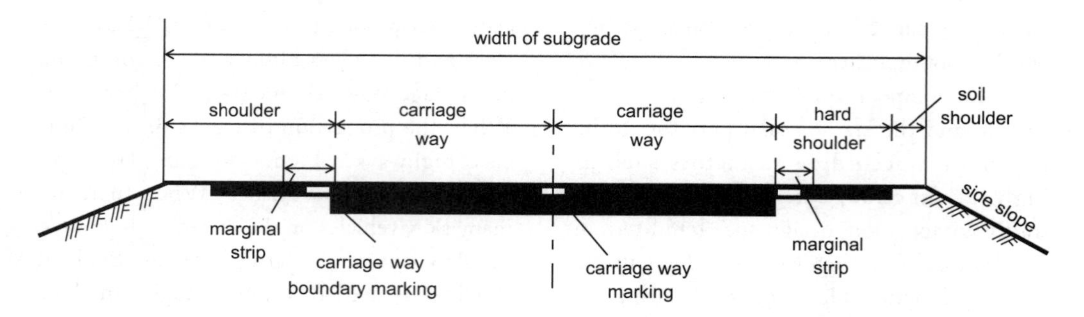

Fig. 5-9 Separated cross section of motorway and 1st-class highway
Source: Ministry of Transport of the People's Republic of China. Design Specification for Highway Alignment JTG D20—2017. Beijing: China Communications Publishing & Media Management Co., Ltd., 2017.

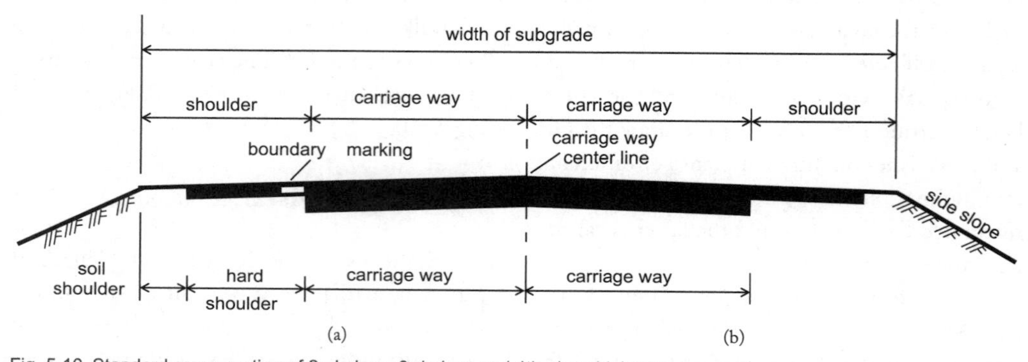

Fig. 5-10 Standard cross section of 2nd-class, 3rd-class and 4th-class highways
(a) Widening of hard shoulders for 2nd-class highway; (b) Normal case for 3rd-class and 4th-class highways
Source: Ministry of Transport of the People's Republic of China. Design Specification for Highway Alignment JTG D20—2017. Beijing: China Communications Publishing & Media Management Co., Ltd., 2017.

indicators of road quality. There may be insufficient sight distance on concealed bends (such as horizontal curves in excavated section or with obstacles on the inner side, central strip and tunnel wall on a horizontal curve), convex vertical curves on longitudinal section, and concave vertical curves on an underpass grade separation, as shown in the Fig. 5-11.

When the driver finds an obstacle or an oncoming vehicle, depending on the measures taken, the driving sight distance can be divided into the following types:

(1) Stopping sight distance—When the car is running, the shortest distance required for the driver to stop safely before reaching the obstacle from the moment he sees the obstacle ahead.

(2) Sight distance of meeting vehicles—The two vehicles are driving towards each other. The shortest distance required for the driver to stop the two vehicles from driving when he sees the vehicle in front of him to meet safely.

(3) Passing sight distance—On a two-lane road with no clearly demarcated lane lines, two cars driving in opposite directions meet, and since they are found, they will take measures to slow down and avoid the shortest distance required to make a safe passing.

(4) Overtaking sight distance—On a two-lane road, when the following vehicle overtakes the preceding vehicle, it is the shortest distance required to safely return to the original lane after the oncoming vehicle is visible and able

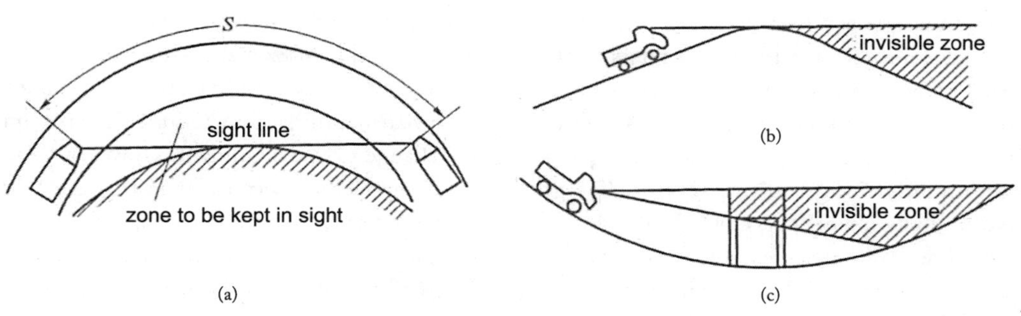

Fig. 5-11 Zones limiting sight distance
Source: Xu J L, et al. Road Survey and Design (4th Edition)[M]. Beijing: China Communications Publishing & Media Management Co., Ltd., 2016.

to pass the vehicle from the point where it starts to leave the original lane.

Among the above four types of sight distances, the first three are relevant for driving in opposite directions, and the fourth is for driving in the same direction. In the fourth condition, the longest distance is required and needs to be studied separately. Among the first three, the sight distance of meeting vehicles is the longest. As long as the sight distance of meeting vehicles can be guaranteed on the road, the stopping sight distance and the passing sight distance can be guaranteed. According to calculation and analysis, the sight distance of meeting vehicles is approximately equal to twice the stopping sight distance, so only the stopping sight distance is calculated.

The eye height and object height need to be determined in the sight distance calculation. "Eye height" refers to the height of the driver's eyes from the road surface. It is stipulated that a low-body passenger car is the standard, and 1.2m is used according to actual measurement. "Object height" refers to the height of obstacles on the road. In addition to the oncoming vehicles, there are also pedestrians crossing the road, goods dropped by vehicles in front, and stones dropped from the excavation slope. Considering that the minimum height of the car chassis from the ground is 0.14~0.20m, the object height is specified as 0.10m. Table 5-7 and Table 5-8 list the stopping sight distance for highways and urban roads.

5.3.4 Horizontal Alignment

During geometric design, the focus is the route of the road. The route refers to the spatial position of the center line of the road, and the projection of the route on the horizontal plane is called the route plane. By cutting vertically along the center line and then expanding it, the longitudinal section of the route is obtained. The normal tangent to any point on

Stopping sight distance for highways **Table 5-7**

Design speed (km/h)	120	100	80	60	40	30	20
Stopping sight distance (m)	210	160	110	75	40	30	20

Stopping sight distance for urban roads **Table 5-8**

Design speed (km/h)	80	60	50	45	40	35	30	25	20	15	10
Stopping sight distance (m)	110	70	60	45	40	35	30	25	20	15	10

the center line is the cross section of the road at that point. Geometric design refers to the work of determining the spatial location of the route and the geometric dimensions of each part, including route plan design (horizontal alignment), longitudinal section design (vertical alignment) and cross section design. The three tasks are interrelated, and must be dealt separately as well as comprehensively.

Horizontal alignment deals with the design of the directional transition of the highway in a horizontal plane. The alignment of the road is restricted by factors such as social economy, natural conditions and technical standards. Task of the alignment is to reasonably determine the geometric parameters of the route under the premise of comprehensively considering various constraints, to meet the requirements of technical standards, driving safety and engineering economy, and to coordinate with the terrain, environment and landscape. The geometric design of the highway runs roughly in the following order: Under the premise of taking into account the balance of the vertical and cross sections as much as possible, the horizontal alignment is firstly determined, and the elevation and cross section measurements are carried out along this alignment to obtain the ground line and geological, hydrological and other necessary information. The longitudinal section and the cross section are determined. In order to achieve the design goal, if necessary, the alignment can be modified and repeated many times to achieve a more satisfactory result. In the design of urban roads, the horizontal position and vertical elevation of roads are often strictly controlled by urban planning, and there is little room for change of the alignment. However, there are many factors to consider in the layout of cross-sections. Therefore, during urban road designing, layout of the cross section comes first, and then come the horizontal and vertical alignment.

When the horizontal alignment is affected by obstacles such as terrain and ground features, a curve should be set at the turning point. The curve is usually a circular curve. In order to make the line pattern more in line with the driving trajectory of the car and ensure the smoothness and safety of driving, a transition curve should be inserted between a straight line and a circular curve (an arc) or between two circular curves with different radii. There are three relationships between the rotating surface of the guide wheel and the longitudinal axis of the vehicle in a driving car, that is, the angle is zero, the angle is constant, and the angle is variable. The driving trajectory lines corresponding to the above three states are: straight lines with zero curvature, circular curves with constant curvature and transition curves with variable curvature. Therefore, the main components that constitute the horizontal alignment of the road are straight lines, circular curves and transition curves, as shown in Fig. 5-12.

When using a straight line for the

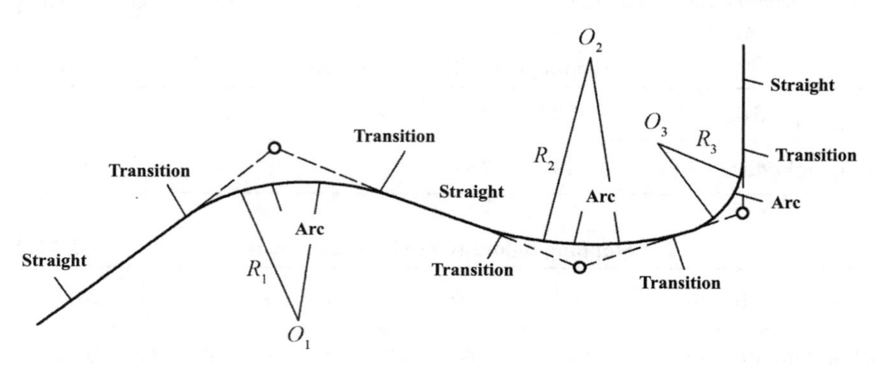

Fig. 5-12 Horizontal alignment

road's horizontal alignment, attention should be paid to the relationship between the alignment and the topography, and should comply with the limits of the maximum and minimum length of the straight line. Whether straight lines or curves are to be applied depends on the actual needs. Generally, long straight lines should be avoided as much as possible. Nevertheless, straight lines should be used (1) in plain or a wide valley between mountains with no restriction from topography or features; (2) in towns and sub-urban roads, or areas with straight lines as main planning element; (3) in sections of structures such as long bridges and tunnels; (4) around intersection and its vicinity; and (5) on sections of two-lane highways that provide overtaking.

When designing the road plan with a circular curve, a larger radius should be selected according to the terrain and features along the route to ensure safe and comfortable driving. When selecting the radius, it must be technically reasonable and economical. Neither blindly adopts high standards (large radius) and excessively increases the amount of work, nor does it adopt low standards only considering the current traffic requirements. The transition curve should be applied as the main element rather than just as a transition. Attention should be paid to the coordination and cooperation with the straight line and the circular curve. The linear combination and the linear appearance produce good driving and visual effects.

According to the *Methods for the Preparation of Design Documents for Highway Engineering Construction Projects*, the results of the horizontal alignment are mainly the drawings and tables of the alignment, including the straight line, curve and intersection angle table, stake-by-stake coordirate table and the plan drawing.

Plane drawing reflects the plane position, line pattern and geometric dimensions of the route. It also reflects the arrangement of artificial structures and important engineering facilities along the route, as well as the relationship between the highway and the topography, features and administrative divisions along the route. An example of plan drawing is shown in Fig. 5-13.

The plan drawing of the route should include the topography, features, line positions and mileage along the route, the relationship between the main stake positions of broken chainage and other traffic routes, and the boundaries of the county and so on. Level points, traverse points and coordinate grids or north-pointing diagrams should be marked in the drawing. Also, super-large bridges, large and medium bridges, tunnels, route intersections, etc. should be marked out. Horizontal curve elements and intersection coordinate tables, etc. should be listed together with the plan. The scale of the plan is generally 1 : 5000~1 : 2000.

5.3.5 Vertical Alignment

The ground surface on which the road is paved is undulating. When the undulation is not large, the road can be constructed to conform to the terrain. When the ups and downs are severe and cannot meet the vehicle's dynamic performance and stability requirements, it is necessary to fill and excavate the area that the road passes (including the construction of bridges, tunnels and other structures) to ensure the safety and speed of the vehicle, and also consider many requirements such as economical processing and no major damage to the environment.

These are the work of longitudinal section design. The longitudinal profile is the main result of the alignment. Combining the longitudinal section view (Fig. 5-14) with the plan view, the spatial position of the road can be accurately determined.

In the design documents of higher-class highways, in addition to the above-mentioned route plan drawing, an overall plan of the

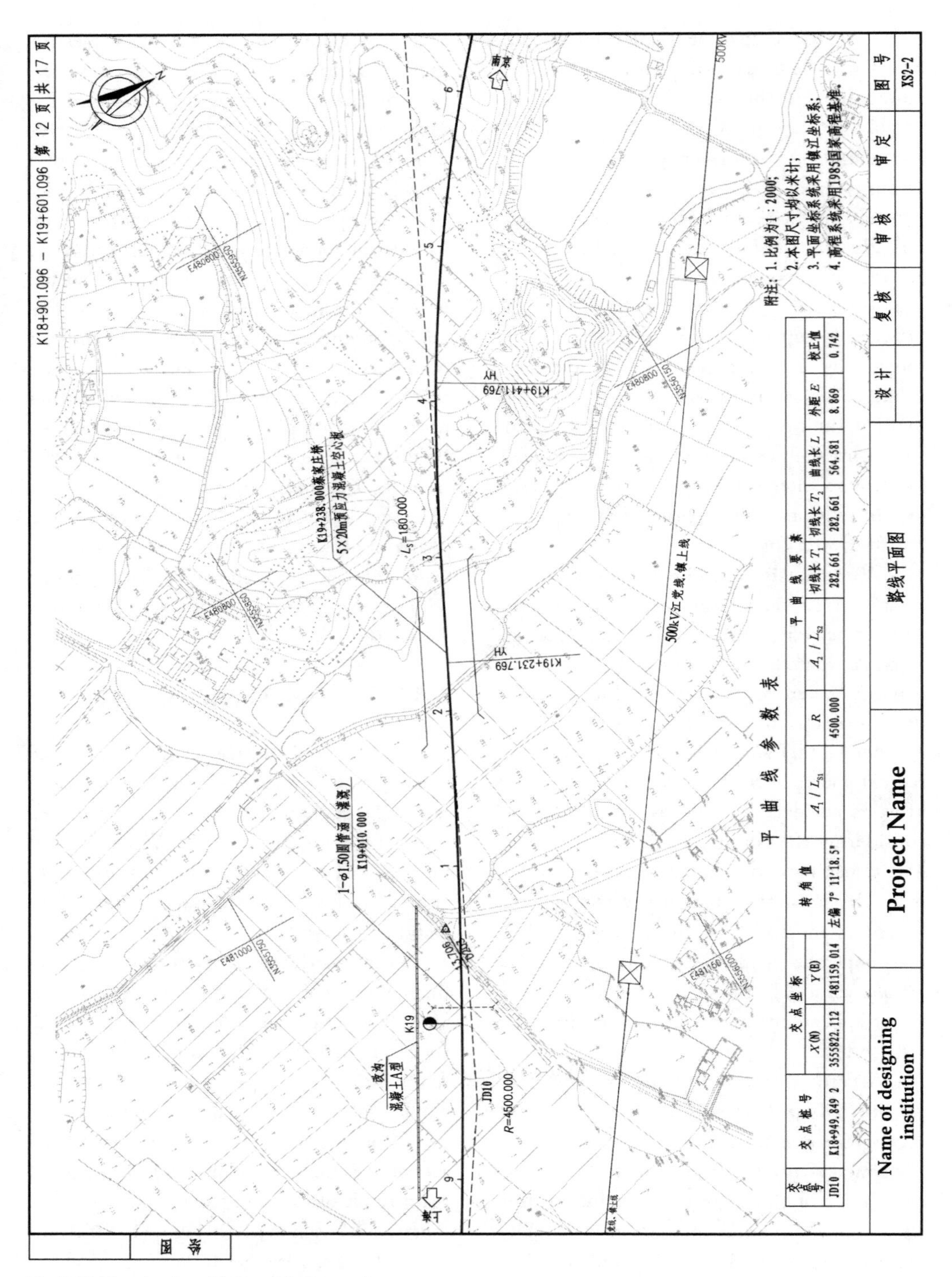

交点号	交点桩号	交点坐标 X(N)	交点坐标 Y(E)	转角值	A_1 / L_{S1}	R	A_2 / L_{S2}	切线长 T_1	切线长 T_2	曲线长 L	外距 E	校正值
JD10	K18+949.849 2	3555822.112	481159.014	左偏 7° 11′18.5″		4500.000		282.661	282.661	564.581	8.869	0.742

Fig. 5-13 Plan drawing of horizontal alignment

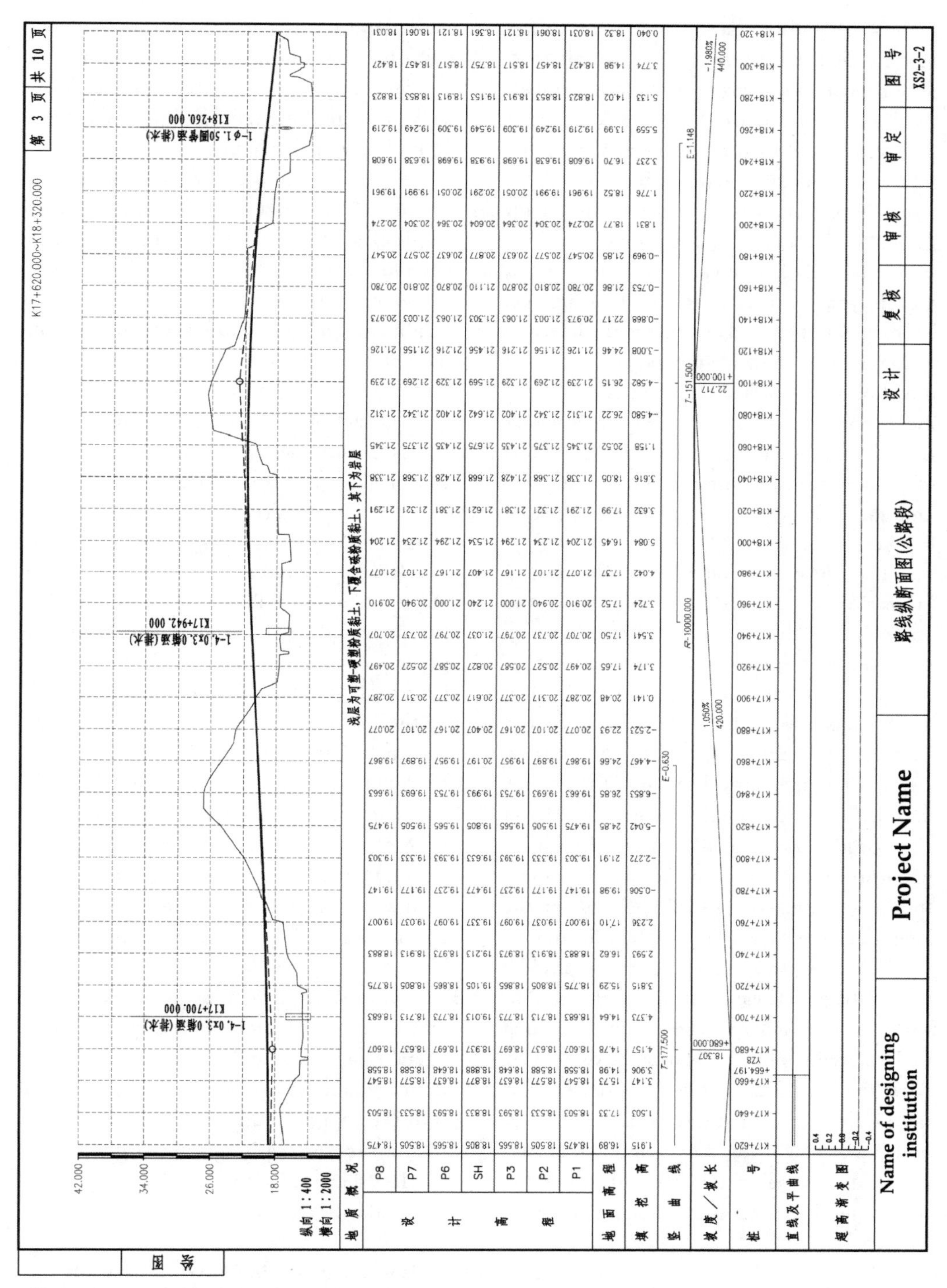

Fig. 5-14 Drawing of longitudinal section of a highway

highway should be drawn. In addition to the content of the route plan, the overall road design plan should also provide the roadbed sideline, slope toe or slope top line, route intersection and its plane form, and show the service area, parking lot, toll station, etc.

The plan drawing of urban roads should generally indicate the route, planned red line, traffic lane line, sidewalk line, parking lot, greening, traffic signs, pedestrian crossing line, building entrances and exits along the line, and various ground and underground pipelines. The direction of rainwater inlets, manholes, etc., indicate intersections and mileage piles along the line. Curve elements and turning radius of corner stones at intersections should be indicated at bends and intersections. The scale is generally 1 : 1000~1 : 500.

The vertical section refers to the vertical projection of the vertical cutting along the center line of the road after unfolding. Since the road route is formed by a combination of straight lines and curves, the cutting surface has both a flat surface and a curved surface (cylindrical surface). In order to clearly show the vertical section of the route, unfold the section plane (without changing the longitudinal gradient of the route when unfolded) into an elevation, namely the longitudinal section. The length of the longitudinal section is the length of the route.

There are two main lines in the middle area of the profile view: one is the ground line, which is an irregular polyline drawn according to the elevation of each stake on the center line. After the horizontal alignment is determined, the ground line is naturally uniquely determined. For new roads, the ground line reflects the undulations of the ground along the center line. The other is the design line, which is determined after technical, economic and aesthetic comparisons. The geometric line has a regular shape, reflecting the ups and downs as well as the longitudinal design slope and vertical curve of the route.

In addition, in order to show the coordination of the horizontal and vertical planes, the design indicators of the vertical section, the fill and excavation, the geological conditions of the road passing area, etc., columns such as height and geological profile, straight lines and horizontal curves, slope/grade length, fill and excavation are set in the data table below the vertical section drawing. In different design stages, requirements for the content of the vertical alignment are different.

The longitudinal design line is composed of straight lines and vertical curves. A straight line (that is, a uniform gradient line) has an uphill and a downhill, which are expressed by the slope and the horizontal length, and the grade length is not counted. The slope and length of the straight line affect the speed of the car, the economy of transportation, and the safety of the car. The determination of some of their critical values and the necessary restrictions are determined by the type of car and the driving performance. For straight lines, the maximum longitudinal gradient of highways is stipulated in China by the *Technical Standard*, which is shown in Table 5-9.

In excavated sections, low-fill sections with side trenches, or other sections with poor lateral drainage conditions, in order to ensure drainage and prevent water from penetrating into the subgrade and affecting the stability of the subgrade, a longitudinal gradient of not less than 0.3% is required. In this case, a longitudinal gradient of not less than 0.5% is

Maximum allowable longitudinal gradient of highways in China **Table 5-9**

Design speed (km/h)	120	100	80	60	40	30	20
Maximum longitudinal gradient (%)	3	4	5	6	7	8	9

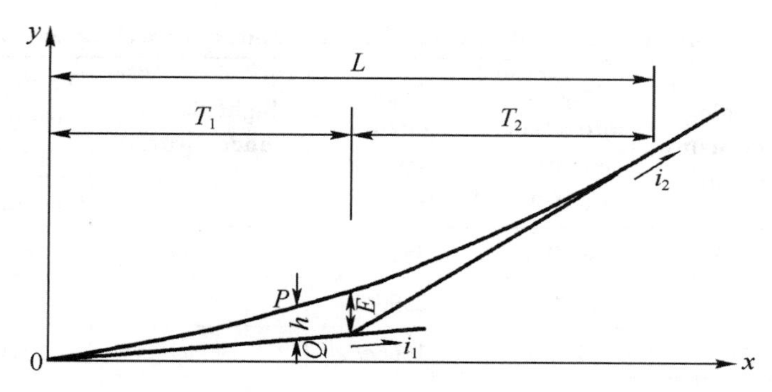

Fig. 5-15 Vertical curve

advisable. In dry area or on highway sections with good lateral drainage and no surface water, this minimum longitudinal gradient is not restricted.

At the turning point of two grades on the profile, a curve is often used to ease the turning for driving safety, comfort and sight distance requirement. This curve is called a vertical curve, as shown in Fig. 5-15. The shape of the vertical curve can be circular or parabolic, which makes no big difference for highway alignment. Since a parabola is much more convenient to align and calculate than a circular curve, so parabolas are generally used as vertical curves in road profile design.

For the purpose of mitigating the impact and meeting the requirement of sight distance, there are requirements regarding the minimum radius and length for vertical curves, as shown in Table 5-10 and Table 5-11.

Minimum radius and minimum length of convex vertical curves Table 5-10

Design speed (km/h)	Stopping sight distance S_T (m)	Impact mitigation required L_{min} (m)	Sight distance required L_{min} (m)	Values specified by the design specification (m)			
				Vertical curve radius		Vertical curve length	
				General value	Limits	General value	Limits
120	210	4000ω	11025ω	17000	11000	250	100
100	160	2778ω	6400ω	10000	6500	210	85
80	110	1778ω	3025ω	4500	3000	170	70
60	75	1000ω	1406ω	2000	1400	120	50
40	40	444ω	400ω	700	450	90	35
30	30	250ω	225ω	400	250	60	25
20	20	111ω	100ω	200	100	50	20

Note: ω means the absolute difference between i_1 and i_2 ($\omega=|i_1-i_2|$).

Minimum radius and minimum length of concave vertical curves **Table 5-11**

Design speed (km/h)	Stopping sight distance S_T (m)	Impact mitigation required L_{min} (m)	Night lightening required L_{min} (m)	Sight distance under flyover L_{min} (m)	Vertical curve radius specified by the design specification (m)	
					General value	Limits
120	210	4000ω	3527ω	1683ω	6000	4000
100	160	2778ω	2590ω	951ω	4500	3000
80	110	1778ω	1666ω	449ω	3000	2000
60	75	1000ω	1036ω	209ω	1500	1000
40	40	444ω	445ω	59ω	700	450
30	30	250ω	293ω	33ω	400	250
20	20	111ω	157ω	15ω	200	100

Note: ω means the absolute difference between i_1 and $i_2(\omega=|i_1-i_2|)$.

5.4 Pavement Structure and Material

A highway pavement is composed of a system of overlaid strata of chosen processed materials that is positioned on the in-situ soil, termed the subgrade. Its basic requirement is the provision of a uniform skid-resistant running surface with adequate life and requiring minimum maintenance. The chief structural purpose of the pavement is the support of vehicle wheel loads applied to the carriage way and the distribution of them to the subgrade immediately underneath. If the road is in cut, the subgrade will consist of the in-situ soil. If it is constructed on fill, the top layers of the embankment structure are collectively termed the subgrade.

The pavement designer must develop the most economical combination of layers that will guarantee adequate dispersion of the incident wheel stresses so that each layer in the pavement does not become overstressed during the design life of the highway. The major variables in the design of a highway pavement are the thickness of each layer in the pavement, the material contained within each layer of the pavement, the type of vehicles in the traffic flow, the volume of traffic predicted to use the highway over its design life and the strength of the underlying subgrade soil. There are three basic components of the highway pavement, namely the foundation, road base and surfacing.

(1) The foundation consists of the native subgrade soil, the layer of subbase and possible capping immediately overlaying it. The function of the subbase and capping is to provide a platform on which to place the road base material as well as to insulate the subgrade below it against the effects of inclement weather. These layers may form the temporary road surface used during the construction phase of the highway.

(2) The road base is the main structural layer whose main function is to withstand the applied wheel stresses and strains incident on it and distribute them in such a manner that the materials beneath it do not become overloaded.

(3) The surfacing combines good riding quality with adequate skidding resistance, while also minimizing the probability of water

infiltrating the pavement with consequent surface cracks. Texture and durability are vital requirements of a good pavement surface as are surface regularity and flexibility.

For flexible pavements, the surfacing is normally applied in two layers—base course and wearing course—with the base course as an extension of the road base but providing a regulating course on which the final layer is applied. In the case of rigid pavements, the structural function of both the road base and surfacing layers are integrated within the concrete slab.

5.4.1 Traffic Load and Material Parameters

Car load is not only the service object of subgrade and pavement, but also the main cause of damage to subgrade and pavement structure. It is a dynamic load that is constantly moving and has the effects of vibration and shock. Characteristics of vehicle load include: the size and characteristics of vehicle wheel load and axle load, the layout of axles, the time distribution characteristics of vehicle axle load and the characteristics of vehicle static and dynamic load.

(1) Effect of Wheel Load on Pavement

Automobiles and vehicles on the road are divided into passenger cars and commercial vehicles. Passenger cars are further divided into small passenger cars, medium buses and coaches; commercial vehicles are further divided into complete vehicles, towed trailers and towed semi-trailers. The total weight of the car and its passenger and cargo is transmitted to the axle through the body, then to the wheels, and finally to the road by the tires. Therefore, the design of the road structure is mainly controlled by axle load or wheel pressure. In the pavement design, the axle type of the vehicle can be divided into 7 categories according to the wheel set and axle group type, and the vehicle type can be divided into 11 categories according to the combination of the axle type. Passenger and truck axles in the form of complete vehicles are divided into front and rear axles. The front axle of the vast majority of vehicles is a single axle composed of two single wheels, and most of the rear axles of vehicles are composed of two-wheel sets, including three types of single axle, double axle and two axles. In order to ensure safety, the axle load limit is specified.

When the car is parked, the force on the road is static pressure, which is mainly the vertical pressure P transmitted by the tires to the road. Its magnitude is affected by the internal pressure p_i of the car tire, the stiffness of the tire and the shape of the tire in contact with the road surface, and the size of the wheel load. A typical shape of the contact surface between the tire and the road surface is shown in Fig. 5-16. Its outline is similar to an ellipse. Since the difference between its long axis and short axis is small, it is represented by a circular contact area in engineering design. The wheel load is simplified into an equivalent circular uniform load, and the tire internal pressure is used as the tire contact pressure p. The equivalent circle radius of the contact surface can be easily calculated.

In China, the standard wheel axle load weighs P=100/4=25kN and tire pressure is p=700kPa for pavement design. The corresponding single and double circle equivalent diameters are d_1=0. 302m and d_2=0.213m.

(2) Traffic Load Survey and Conversion

The traffic volume survey method of pavement design should not only care about the traffic volume of a certain section or cross section, but also attach great importance to the axle load quality of various models. There are many methods for weighing the axle load of vehicles, such as static weighing of floor scales, manual jack weighing, and bridge and culvert induction vehicle weighing.

The annual average daily traffic volume (*AADT*) aside from those vehicles with fewer than 2 axles and 4 wheels is deemed as the

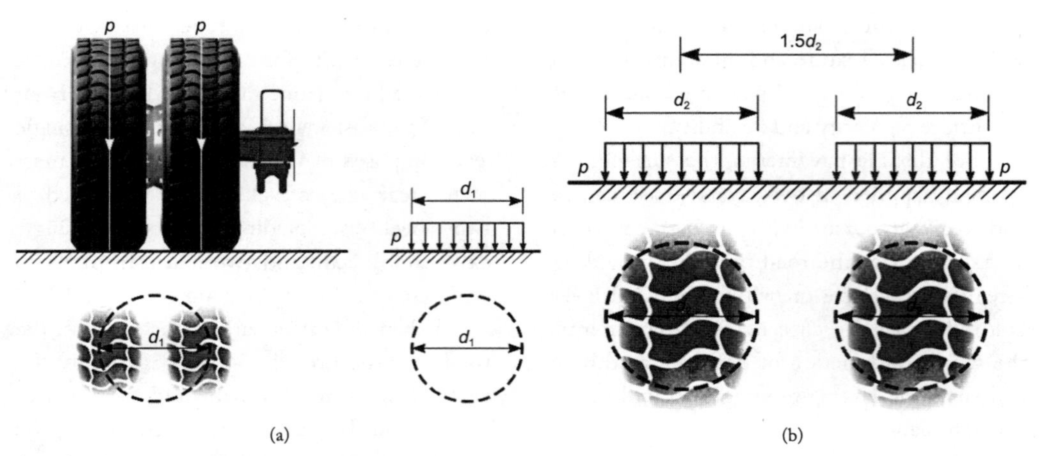

Fig. 5-16 Calculation diagram of load on pavement
(a) Single circle equivalent; (b) Double circle equivalent

annual average daily truck traffic (*AADTT*), which includes exclusively vehicles with 6 wheels on 2 axles and above. The bidirectional *AADTT* in the initial period multiplied by the direction factor (*DDF*) and the lane factor (*LDF*), makes the annual average design traffic volume of the designed lane:

$$Q_1 = AADTT \times DDF \times LDF \qquad (5\text{-}3)$$

where *DDF*—direction factor;

LDF—lane factor.

The direction factor should be determined based on the measured traffic data in different directions, and can be selected within the range of 0.5 to 0.6 when there is no measured data. The lane factor of the asphalt pavement can be determined according to the following three levels: level 1, calculate the number of vehicles on different lanes in the design direction based on the on-site traffic volume observation data, determine the lane factor; level 2, use local experience values; level 3, use the value recommended in Table 5-12. Level 1 should be adopted in the design of reconstructed pavement, and level 2 or level 3 can be adopted in the design of new pavement. The lane factor of cement concrete pavement can be determined according to Table 5-12.

For the design of pavement structure, in addition to the cumulative traffic volume within the design period, another important traffic factor is the ratio of the number of times of axle load action at all levels to the total number of actions, that is, axle load composition or axle load spectrum. According to the actual measured number of axle loads and the corresponding axle load, the histogram shown in Fig. 5-17 is organized as the typical axle load of each level of axle load on the road.

For asphalt pavements, the vehicle type distribution coefficient can be determined according to three levels:

1) Level 1, analyze the percentage of vehicles from different categories of wheel axle composition (totally 11 categories) based on traffic observation data to obtain the vehicle category distribution coefficient;

2) Level 2, based on historical traffic data or empirical data, determine the highway

Lane factor **Table 5-12**

Lane number in one way	1	2	3	≥4
Motorway	—	0.70~0.85	0.45~0.60	0.40~0.50
Highways of other grades	1.00	0.50~0.75	0.50~0.75	—

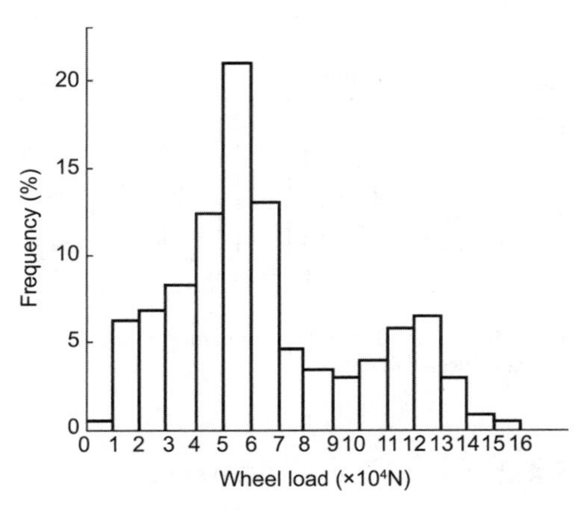

Fig. 5-17 Wheel load spectrum

truck traffic composition (TTC) classification according to Table 5-13, and use the local empirical value of the distribution coefficient of the TTC classification vehicle type;

3) Level 3, determine the highway TTC classification based on traffic history data or empirical data, and use the vehicles type distribution coefficient.

For cement concrete pavement, the focus is to obtain the single-axle axle load spectrum, which can be obtained by two methods: the axle type-based and vehicle type-based methods.

When a vehicle is driving on the road, the trajectory of the wheel always swings around within a certain range near the center line of the cross section. The total number of axle load passes is distributed on the cross section of the lane according to a certain rule. Therefore, the number of passes at a certain point is compared with the total number of passes. The ratio of the number of passes is called the lateral distribution of wheel tracks. Fig. 5-18 shows the frequency curve of the lateral distribution of wheel tracks in one lane when driving in one direction, and Fig. 5-19 shows the frequency curve of the lateral distribution of wheel tracks in two lanes in mixed driving.

In order to quantify and consider the comprehensive cumulative damage effect of traffic volume and different vehicle types on the pavement structure, one kind of axle load is generally selected as the standard axle load for pavement structure design, and while other axle loads are converted into the standard axle loads according to certain principles, so that the traffic volume will be converted into the equivalent cumulative axle load times for structural design.

Two principles of axle load conversion are as follows: firstly, the conversion is based on reaching the same critical state; secondly, for a certain type of traffic composition, no matter which axle load standard is used for conversion, thickness of the road surface calculated by the number of axle loads obtained by the conversion should be the same.

(3) Design Parameters of Pavement Materials

Pavement materials have different design parameter requirements due to their own attributes and different pavement structure levels. They should be based on the road class, traffic load level, climatic conditions, functional requirements of each structural layer, local ma-

Highway TTC classification criteria **Table 5-13**

TTC Category	Ratio of integral truck (%)	Ratio of semi-trailer truck (%)
TTC1	<40	>50
TTC2	<40	<50
TTC3	40~70	>20
TTC4	40~70	<20
TTC5	>70	—

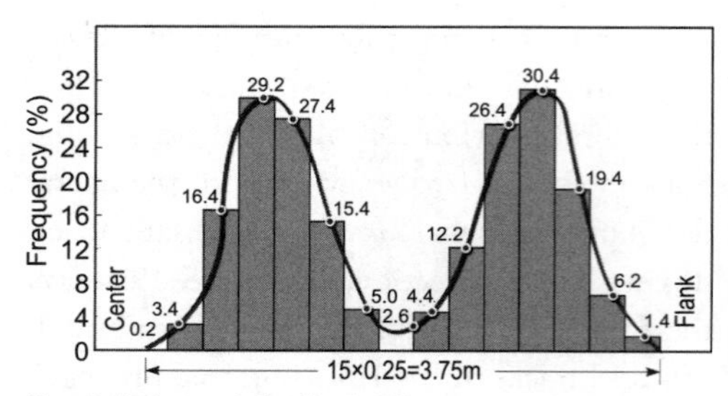

Fig. 5-18 Lateral distribution frequency of wheel path (single lane, one-way traffic)

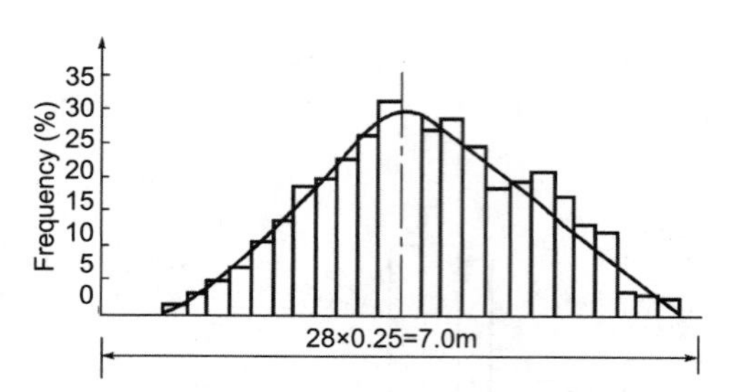

Fig. 5-19 Lateral distribution frequency of wheel path (two-lane, mixed traffic)

terial characteristics, etc.

The design parameters of inorganic binder stabilized materials mainly include: unconfined compressive strength of inorganic binder stabilized materials, unconfined compressive elastic modulus, flexural tensile strength, flexural elastic modulus, uniaxial compression modulus, dynamic compression modulus, fatigue life and other design parameters.

1) The main design parameters of asphalt mixture are compressive modulus, penetration strength, etc., because the modulus of asphalt mixture is related to loading speed, loading time and temperature during loading. Therefore, the modulus and strength of asphalt mixture are condition parameters.

2) The design parameters of cement concrete mainly include: the flexural tensile strength of cement concrete and the flexural elastic modulus of cement concrete, the drilling and splitting test of cement concrete samples, and the value of cement concrete pavement design parameters.

3) The resilient modulus of granular materials is one of the important parameters in the analysis of the mechanical response of asphalt pavement structures. It is a function of the properties, state (moisture content and density) and stress conditions of the granular materials. For all kinds of granular materials and subgrade soils in a specific state (a certain moisture content and compactness value), the stress state is the main factor affecting the value of the modulus.

5.4.2 Base Course Material

The base course of the pavement plays a role of linking the surface and the road base in the pavement structure. According to the difference in stiffness, the material is divided into three types: flexible base, semi-rigid base and rigid base; according to the type of binder, it is divided into non-binding gravel materials, inorganic binding semi-rigid materials, and asphalt (organic) binding material of asphalt stabilized crushed stone.

(1) Unbonded Granular Base

Crushed stone refers to the stone material that meets the requirements of the project, which is mined and processed according to a certain size. Gravel refers to the non-angular granular material that is transported by water for a long time. Graded crushed stone refers to a material composed of crushed stone designed according to a certain grading requirement. Graded gravel refers to a material composed of gravel designed according to certain gradation.

The mixed crushed stone with a certain particle size composition (gradation) can be directly used as a pavement base course; by adding inorganic binder or asphalt stabilization, it can form cement stabilized crushed (gravel) stone, asphalt stabilized crushed stone, etc. High-quality gravel is one of the main raw materials for asphalt concrete and cement concrete.

Graded crushed (gravel) stone pavement types include: crushed stone (gravel) material pavement and soil-crushed (gravel) stone mixture pavement. Loose crushed gravel materials can be used for pavement base course, and can also be used for surface layer to form crushed gravel pavement through proper treatment.

The bonding strength between particles, which is determined by the internal friction force characterized by the bonding force of the material and the internal friction angle, constitutes the structural strength of broken and gravel pavement materials.

(2) Inorganic Binder Stabilized Base Course

The physical and mechanical properties of inorganic binder stabilized materials include stress-strain relationship, fatigue characteristics and shrinkage (temperature shrinkage and dry shrinkage) characteristics. Therefore, inorganic binder stabilized materials have the characteristics of good stability, strong frost resistance and self-contained slabs of the structure itself, but their wear resistance is poor, so they are widely used in the construction of pavement structure base and subbase. The pavement base course built by this is the base course of inorganic binder stabilized material, or called semi-rigid base course, including lime stabilized base, cement stabilized base and industrial waste residue stabilized base.

(3) Asphaltic base

Asphalt binder mixture refers to a mixture composed of asphalt, coarse and fine aggregates and mineral powder, and designed according to a certain mix ratio design method. It is mixed, paved, rolled and formed into a pavement structure as a base course, which is called an asphalt binder base course. According to different design air void and usage, asphalt binder mixtures can be divided into: asphalt treated base (ATB, design air void of 3%~6%, used as base course), semi-open asphalt stabilized macadam (asphalt macadam, AM, with a design air void of 6%~12%, used as a low-grade road surface), and open-graded asphalt stabilized gravel (used for pavement drainage of design air void of 18%~22%, including asphalt treated permeable base, referred to as ATPB, used for base drainage). The mix design and application process of the asphalt binder base is basically the same as that of asphalt concrete, and physical and mechanical properties of the material are also very similar.

(4) Cement Concrete Base

Lean concrete is a kind of concrete made by mixing coarse and fine aggregates with a certain amount of cement and water. This kind of concrete has a lower cement content than ordinary concrete. Compared with commonly used semi-rigid materials such as cement stabilized crushed stone and lime fly ash crushed stone, it has higher strength, rigidity and integrity, and is resistant to erosion and frost. It has good fatigue performance, yields to rigid base materials, and is close to cement concrete pavement in nature. The material composition design and construction mainly refer to cement concrete.

Rolled concrete base refers to the use of ultra-dry hard cement concrete mixture materials, which is constructed with slipform pavers and compacted with road rollers. From the point of material properties, the cement content of the rolled concrete as the base course is basically the same as that of lean concrete, so it can be regarded as a special lean concrete. Its physical and mechanical properties are similar to those of lean concrete base.

5.4.3 Asphalt Pavement

Asphalt pavement is a pavement that uses asphalt materials as binders to bond mineral materials to build a surface layer and various base courses. Asphalt pavement has certain elastic and plastic deformation capacity and is able to withstand strain without breaking. It has good adhesion with car tires, which can ensure driving safety. It has good damping performance, which can make the car drive fast and stable with low noise. The maintenance of asphalt pavement is relatively simpler and is easy to clean and wash Compared with ordinary cement concrete pavement, asphalt pavement has some advantages and disadvantages. The surface of asphalt pavement is even without seams, and the structure is flexible, therefore it provides with more driving comfort and stability. The construction period is short and faster to be open to traffic. And asphalt is easy to repair and is reusable. However, its strength and stability are greatly affected by the base course and soil foundation. The mechanical properties of asphalt mixture are susceptible to temperature and ultraviolet. Asphalt is prone to aging.

(1) Material of Asphalt Pavement

Bitumen is produced artificially from crude oil within the petroleum refining process. It is a basic constituent of the upper layers in pavement construction. It can resist both deformation and changes in temperature. Its binding effect eliminates the loss of material from the surface of the pavement and prevents water penetrating the structure.

With the decreased availability of tar, bitumen is the most commonly used binding/water resisting material for highway pavements. The oil refining process involves petroleum crude being distilled with various hydrocarbons being driven off. The first stage, carried out at atmospheric pressure, involves the crude being heated to approximately 250°C. Petrol is the most volatile of these and is driven off firstly, followed by materials such as kerosene and gas oil. The remaining material is then heated at reduced pressure to collect the diesel and lubricating oils contained within it. At the conclusion of this stage of the process, a residue remains which can be treated to produce bitumen of varying penetration grades. This is the material used to bind and stabilize the graded stone used in the top layers of a highway pavement. A number of tests exist to ensure that a binder has the correct properties for use in the upper layers of a pavement. Two of the most prominent are the penetration test and the softening point test, both of which indirectly measure the viscosity of a sample of bitumen. Viscosity of a fluid refers to its resistance to flow which is of great importance for the bitumen to be sprayed onto a base course or mixed with mineral aggregates to form asphalt mixture at high temperatures. The penetration test is in no way indicative of the quality of the bitumen but it does allow the material to be classified.

The penetration test involves a standard steel needle applying a vertical load of 100g to the top of a standard sample of bitumen at the temperature of 25°C. The depth to which the needle penetrates into the sample within a period of 5 seconds is measured. The answer is recorded in units of 0.1mm. Thus, if the needle penetrates 10mm within the 5-second period, the result is 100 and the sample is designated as 100 pen. The lower the penetration is, the more viscous and therefore the harder the sample will be.

The softening point test involves taking a sample of bitumen which has been cast inside a 15mm diameter metal ring and placing it inside a water bath with an initial temperature of 5°C. A 25mm clear space exists below the sample. A 10mm steel ball is placed on the sample and the temperature of the bath and the sample within it is increased by 5°C per minute. As the temperature is raised, the sample softens and therefore sags under the weight of the steel ball. The temperature at which the weakening binder reaches the bottom of the 25mm vertical gap below its initial position is known as its softening point.

Bitumen should never reach its softening point while under traffic loading. The results from these two tests enable the designer to predict the temperatures necessary to obtain the fluidity required in the mixture for effective use within the pavement.

Hot mix asphalt is a dense material with low air void content, consisting of a mixture of aggregate, fines, binder and a filler material, but in this case the grading is far less continuous (gap-graded) with a higher proportion of both fines and binder present in the mix. The material is practically impervious to water, with the fines, filler and bitumen forming a mortar in which coarse aggregate is scattered in order to increase its overall bulk.

Hot mix asphalt wearing courses typically have from 0% to 55% coarse aggregate content, with base courses having either 50% or 60% and road bases normally at 60%. There are two recipe mixes for gap-graded rolled asphalt wearing course: Type F, characterized by the use of sand fines, and Type C, characterized by the use of crushed rock or slag fines. F denotes a finer grading of the fine aggregate with C denoting a coarser grading of the fine aggregate. Each mix has a designation composed of two numbers, with the first relating to the percentage coarse aggregate content in the mix and the second to the nominal coarse aggregate size.

The compacted asphalt mixture is a multi-phase system with a spatial network composed of stone aggregates, asphalt binders and voids. Its mechanical strength mainly depends on the friction and interlocking effect between aggregate particles and the adhesion effect between asphalt binder and aggregate. According to the particle characteristics of asphalt mixture, the strength of asphalt mixture comes from two aspects: cohesive force due to the presence of asphalt binder and internal friction due to the presence of aggregate.

Asphalt mixture is a typical elastic, viscous and plastic composite. It is close to linear elastic body in the low temperature and small deformation range, behaves as viscoplastic body in the high temperature and large deformation range, and is generally viscous in the transition range of normal temperature.

The mechanical properties of asphalt mixture are examined in a relatively wide temperature and time range, and its changes are extremely regular. This kind of regularity can be described by viscoelastic theory and analyzed as a function of temperature and time.

Under the action of vehicle load, the asphalt pavement surface is in a three-dimensional stress state. The normal stress can change from positive (tensile stress) to negative (compressive stress). The stress state of each point not only changes with the coordinates, but also changes with the movement of the wheel load. There are three main indicators of asphalt mixture to resist damage, namely, shear strength, fracture strength and critical strain.

(2) Construction of Asphalt Pavement

The production of a successful bituminous road surfacing depends not just on the design of the individual constituent layers but also on the correctness of the construction procedure employed to put them in place. In essence, the construction of a bituminous pavement consists of the flowing steps: transporting and placing the bituminous material, compaction of the mixture, and if required, the spreading and rolling of coated chippings into the surface of the material.

The bituminous material is manufactured at a central batching plant where, after the mixing of its constituents, the material is discharged into a truck or trailer for transportation to its final destination. The transporters must have metallic beds sprayed with an appropriate material to prevent the mixture sticking to it. The vehicle should be designed to avoid heat loss which may result in a decrease in temperature of the material, leading to difficulties in its subsequent placement—if it is too cold it may prove impossible to compact properly.

It is very important that the receiving surface is clean and free of any foreign materials. It must, therefore, be swept clean of all loose dirt. If the receiving layer is unbound, it is usual to apply a prime coat, in most cases cutback bitumen, before placing the new bituminous layer. A minimum ambient temperature of 4°C is generally required. If it is irregular, it will not be possible to attain a sufficiently regular finished surface. A typical surface tolerance for a bituminous base course or wearing course would be ±6mm.

A paver is used for the actual placing of the bituminous material. It ensures a uniform rate of spread of correctly mixed material. The truck/trailer tips the mixture into a hopper located at the front of the paver. The mix is then fed towards the far end of the machine where it is spread and agitated in order to provide an even spread of the material over the entire width being paved. The oscillating/vibrating screed and vibro-tamper delivers the mix at the required elevation and cross-section and uses a tamping mechanism to initiate the compaction process.

When the initial placing of the mix is complete, it must be rolled while still hot. Minimum temperatures vary from 75°C to 90°C depending on the stiffness of the binder. This process is completed using either pneumatic tyre or steel wheel rollers. The tyre pressures for pneumatic rollers vary from 276kPa to 620kPa, while the steel wheel rollers vary from 8t to 18t. If the latter are vibratory rather than static, 50 vibrations per second will be imparted. The rolling is carried out in a longitudinal direction, generally commencing at the edge of the new surface and progressing towards the center. If the road is super-elevated, rolling commences on the low side and progresses towards the highest point.

It is important that, on completion of the compacting process, the surface of the pavement is sufficiently regular. Regularity in the transverse direction is measured using a simple 3-metre long straight edge. Deviations measured under the straight edge should in no circumstances exceed 3mm.

5.4.4 Cement Concrete Pavement

A rigid pavement consists of a subgrade/subbase foundation covered by a slab constructed of pavement quality concrete. The concrete must be of sufficient depth so as to prevent the traffic load causing premature failure. Appropriate measures should also be taken to prevent damage due to other causes. The proportions within the concrete mix will determine both its strength and its resistance to climate changes and general wear. Joints in the concrete may be formed in order to aid the resistance to tensile and compressive forces set up in the slab due to shrinkage effects.

(1) Concrete Slab and Joint Details

As the strength of concrete develops with time, its 28-day value is taken for specification purposes, though its 7-day strength is often used as an initial guideline of the mix's ultimate strength. Pavement quality concrete generally has a 28-day characteristic compressive strength of 40MPa, termed C40 concrete. Ordinary Portland cement is commonly used. The cement content for C40 concrete should be a minimum of 320kg/m^3. Air content of up to 5% may be acceptable with a typical maximum water cement ratio of 0.5 for C40 concrete.

The effects of temperature are such that a continuous concrete slab is likely to fail prematurely due to induced internal stresses rather than from excessive traffic loading. If the slab is reinforced, the effect of these induced stresses can be lessened by the addition of further reinforcement that increases the slab's ability to withstand them. This slab type is termed continuous reinforced concrete (CRC). Alternatively, dividing the pavement into a series of slabs and providing movement joints between these can permit the release and dissipation of induced stresses. This slab type is termed jointed reinforced concrete (JRC). If the slab is jointed and not reinforced, the slab type is termed unreinforced concrete (URC). If joints are employed, their type and location are important factors.

Joints are provided in a pavement slab in order to allow for movement caused by changes in moisture content and slab temperature. Transverse joints across the pavement at right angles to its centerline permit the release of shrinkage and temperature stresses. The greatest effect of these stresses is in the longitudinal direction. Longitudinal joints, on the other hand, deal with induced stresses most evident across the width of the pavement. There are four main types of transverse joints: contraction joints, expansion joints, warping joints and construction joints.

Contraction occurs when water is lost or temperatures drop. Expansion occurs when water is absorbed or the temperature rises. The insertion of contraction and expansion joints permit movement to happen. Contraction joints allow induced stresses to be released by permitting the adjacent slab to contract, thereby causing a reduction in tensile stresses within the slab. The joint, therefore, must open in order to permit this movement while at the same time prohibiting vertical movement between adjacent concrete slabs. Furthermore, water should not be allowed to penetrate into the foundation of the pavement. The joint reduces the thickness of the concrete slab, inducing a concentration of stress and subsequent cracking at the chosen appropriate location. The reduction in thickness is usually achieved by cutting a groove in the surface of the slab, causing a reduction in depth of approximately 30%. A dowel bar placed in the middle of the joint delivers the requisite vertical shear strength across it and provides load-transfer capabilities. It also keeps adjacent concrete surfaces level during temperature induced movements. In order to ensure full longitudinal movement, the bar is debonded on one side of the contraction joint.

Expansion joints differ in that a full discontinuity exists between the two sides, with a compressible filler material included to permit the adjacent concrete to expand. These can also function as contraction or warping joints.

Warping joints are required in plain unreinforced concrete slabs only. They permit small angular movements to occur between adjacent concrete slabs. Warping stresses are very likely to occur in long narrow slabs. They are required in unreinforced slabs only, as in reinforced slabs the warping is kept in check by the reinforcing bars. They are simply a sealed break or discontinuity in the concrete slab itself, with tie-bars used to restrict any widening and hold the sides together.

Construction is normally organized so that work on any given day ends at the location of an intended contraction or expansion joint. Where this proves not to be possible, a construction joint can be used. No relative movement is permitted across the joint.

Longitudinal joints may also be required to counteract the effects of warping along the length of the slab. They are broadly similar in layout to transverse warping joints.

(2) Reinforcement

Reinforcement can be in the form of a prefabricated mesh or a bar-mat. The function of the reinforcement is to limit the extent of surface cracking in order to maintain the particle interlock within the aggregate.

In order to maximize its bond with the concrete within the slab, care must be taken to ensure that the steel is cleaned thoroughly before use. Because the purpose of the reinforcement is to minimize cracking, it should be placed near the upper surface of the pavement slab. A cover of approximately 60mm is usually required, though this may be reduced slightly for thinner slabs. It is normally stopped approximately 125mm from the edge of a slab, 100mm from a longitudinal joint and 300mm from any transverse joint. Transverse lapping of reinforcement within a pavement slab will normally be in the order of 300mm.

(3) Construction of Concrete Pavement

There are a number of key issues that must be addressed in order to properly construct a concrete pavement. These include the positioning of the reinforcement in the concrete, the correct forming of both joints and slabs and the chosen method of construction, be it mechanized or manual.

To prepare for paving, the subgrade

must be graded and compacted. Preparation of the subgrade is often followed by the placing of a subbase, a layer of material that lies immediately below the concrete. The essential function of the subbase is to prevent the displacement of soil from underneath the pavement. Subbases may be constructed of granular materials, cement-treated materials, lean concrete, or open-graded, highly-permeable materials, stabilized or unstabilized. Once the subbase has hardened sufficiently to resist marring or distortion by construction traffic, dowels, tie bars or reinforcing steel are placed and properly aligned in preparation for paving.

There are two methods for paving with concrete, slipform and fixed form. In slipform paving, a machine rides on treads over the area to be paved—similar to a train moving on a set of tracks. Fresh concrete is deposited in front of the paving machine which then spreads, shapes, consolidates, screeds, and float finishes the concrete in one continuous operation. This operation requires close coordination between the concrete placement and the forward speed of the paver.

In fixed-form paving, stationary metal forms are set and aligned on a solid foundation and staked rigidly. Final preparation and shaping of the subgrade or subbase is completed after the forms are set. Forms are cleaned and oiled firstly to ensure that they release from the concrete after the concrete hardens. Once concrete is deposited near its final position on the subgrade, spreading is completed by a mechanical spreader riding on top of the preset forms and the concrete. The spreading machine is followed by one or more machines that shape, consolidate, and float finish the concrete. After the concrete has reached a required strength, the forms are removed and curing of the edges begins immediately.

After placing and finishing concrete pavement, joints are created to control cracking and to provide relief for concrete expansion caused by temperature and moisture changes. Joints are normally created by sawing.

(4) Curing and Surface Treatment

Concrete curing is an essential step in achieving a good quality finished product. It requires that both the temperature and moisture content of the mix be maintained so that it can continue to gain strength with time. If moisture is lost due to exposure to sunlight and wind, shrinkage cracks will develop. Such problems due to moisture loss can be avoided if the surface of the concrete is kept moist for at least seven days. This is usually achieved by mechanically spraying the finished surface.

Once joints have been inserted, the surface must be textured. To obtain the desired amount of skid resistance, texturing should be done just after the water sheen has disappeared and just before the concrete becomes non-plastic. Texturing is done using burlap drag, artificial-turf drag, wire brooming, grooving the plastic concrete with a roller or comb equipped with steel tines, or a combination of these methods.

It is extremely important to get the texture to the correct level of quality at the time of construction as potential difficulties may arise with subsequent surface maintenance during its design life. Good skid resistance requires sufficient macrotexture and microtexture. Macrotexture permits most of the rainwater caught between the tires and the surface of the highway to drain rapidly and depends on grooves being developed on the surface of the mix in order to "texture" it. Microtexture, on the other hand, depends on the use of fine aggregate within the mix. It must have abrasion-resistance properties such that the particles of sand stand out of the matrix of the hardened cement paste while subject to traffic loading, therefore allowing it to penetrate the remaining film of water and maintain tire contact with the surface.

The required macrotexture of the surface is achieved by wire brushing or grooving. Wire brushing is done either manually or mechanically from a skewed traveling bridge moving along the line of the pavement. The wire brush is usually a minimum of 450mm long. Grooving is done using a vibrating plate moving across the width of the finished pavement slab forming random grooves. They have a nominal size of 6mm by 6mm, providing excellent surface water drainage properties. A high level of wet-road skid resistance is obtained by deep grooving, but problems may arise with higher tire noise.

The chosen method of texturing depends on the environment, and the speed and density of expected traffic. Curing begins immediately after finishing operations and as soon as the surface will not be marred by the curing medium. Common curing methods include using white pigmented liquid membrane curing compounds. Occasionally, curing is accomplished by waterproof paper or plastic covers such as polyethylene sheets, wet cotton mats or burlaps.

As the concrete pavement hardens, it contracts and cracks. If the contraction joints have been correctly designed and constructed, the cracks will occur below the joints. As the concrete continues to contract, the joints will open, providing room for the concrete to expand in hot weather and in moist conditions. Once the pavement hardens, the joints are cleaned and sealed to exclude foreign material that would be damaging to the concrete when it expands. The pavement is opened to traffic after the specified curing period and when tests indicate that the concrete has reached the required strength. Immediately before the pavement is opened to public traffic, the shoulders are finished and the pavement is cleaned.

5.5 Conclusions

Road engineering is an intersection of civil engineering and transportation engineering. This chapter gives a peek to the vivid world of transportation and traffic engineering. The content focuses on the standards and specifications of highway planning, alignment and pavement structure and material in the background of China. And this is only a very miniate excerpt of the panoramic of road and transportation engineering. As part of the *Introduction to Civil Engineering*, this chapter is dedicated to cover the most important points from the view of highway engineer. The person of interest to highway engineering is advised to study further the courses such as transportation planning, traffic planning, fundaments of transportation engineering, road construction materials, road survey and design, subgrade and pavement engineering, and so on.

Exercises

5-1 What are the differences and connections between the disciplines: transportation engineering, traffic engineering and highway engineering?

5-2 What is the meaning of classification of the various roads in the transportation network and what are the criteria of highway classification?

5-3 What are the key tasks of transportation planning?

5-4 What are the sight distances and how will they be ensured in the process of horizontal and vertical alignment?

5-5 What are the deliveries of geometric

design of highway?

5-6 What are the typical pavement structures of a highway? What is the function of each layer?

5-7 Why do we need to classify the vehicle types and axle types? What is the difference between the traffic volume used in pavement design and the traffic volume used for determining the technical grade of the highway?

5-8 What are the differences between asphalt pavement and cement concrete pavement?

References

[1] National Development and Reform Commission of the People's Republic of China. National Highway Network Planning (2013-2030)[R]. 2013. (in Chinese)

[2] SUN L R. Simplification of the calculation formula of the design hourly traffic volume coefficient[J]. Highways, 1991,9: 14-15. (in Chinese)

[3] American Association of State Highway and Transportation Officials. A Policy on Geometric Design of Highways and Streets (Green Book)[R]. 1994.

[4] Transportation Research Board. Highway Capacity Manual: A Guide for Multimodal Mobility Analysis (6th Edition) [M]. 2016.

[5] Ministry of Transport of the People's Republic of China. Technical Standard of Highway Engineering JTG B01—2014[S]. Beijing: China Communications Publishing & Media Management Co., Ltd., 2014. (in Chinese)

[6] Ministry of Transport of the People's Republic of China. Design Specification for Highway Alignment JTG D20—2017[S]. Beijing: China Communications Publishing & Media Management Co., Ltd., 2017. (in Chinese)

[7] Ministry of Housing and Urban-Rural Development of the People's Republic of China. Code for Design of Urban Road Engineering CJJ 37—2012[S]. Beijing: China Architecture & Building Press. 2012. (in Chinese)

[8] ROGERS M. Highway Engineering [M]. London: Blackwell Publishing, 2003.

[9] TEODOROVIĆ D, JANIĆ M. Transportation Engineering—Theory, Practice, and Modeling[M]. Oxford: Butterworth-Heinemann, 2017

[10] SHAO C F, Traffic Planning (2nd Edition)[M]. Beijing: China Railway Press, 2014. (in Chinese)

[11] ZHAO Y C, LI Y J. Road Traffic Engineering and Control[M]. Lanzhou: Gansu People's Fine Arts Publishing House, 2012. (in Chinese)

[12] XU J J, et al. Road Survey and Design (4th Edition)[M]. Beijing: China Communications Publishing & Media Management Co., Ltd., 2016. (in Chinese)

[13] HUANG X M. Subgrade and Pavement Engineering (6th Edition)[M]. Beijing: China Communications Publishing & Media Management Co., Ltd., 2019. (in Chinese)

Chapter 6
Urban Underground Space

6.1 Overview

The populations grow in dense urban city centers, so does the demand for space and natural resources. To solve this problem, all too often, it has been to build denser and taller buildings in addition to transporting an ever-increasing abundance of resources (e.g. raw materials, water, energy and food) into the city whilst moving waste back out. This has major implications for liveable cities (LC), which in future policy terms might be considered to include aspects of (1) wellbeing, (2) resource security (i.e. "one planet" living) and (3) carbon reduction (now enshrined in international law). An option that has been overlooked, and one which could add significance to this LC agenda, is wider adoption of urban underground space (UUS).

Urban underground space encompasses structures with various functions: storage (e.g. food, water, oil, industrial goods, waste); industry (e.g. power plants); transport (e.g. railways, roads, pedestrian tunnels); utilities and communications (e.g. water, sewerage, gas, electric cables); public use (e.g. shopping centers, hospitals, civil defense structures); and private and personal use (e.g. car garages). A study of UUS in three cities—Paris, Tokyo, Stockholm—provides an example of how UUS is used and classified according to function (Fig. 6-1). Utilities and transport are the most common functions of urban underground infrastructure (UUI) (Fig. 6-1). The cities studied use more than 32% of UUS for transportation (rail and motor tunnels, stations), and more than 8% for utilities (pipelines, cable collectors, sewerage). Other functions vary significantly depending on each city's characteristics and also on how the data were classified. Group "Other" includes public spaces, shopping areas, car garages, storages and industrial use.

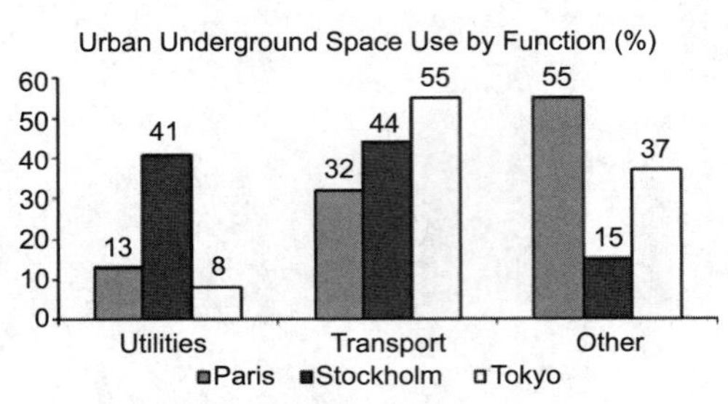

Fig. 6-1 UUS use by function [13]

6.2 Characteristics and Evolution of UUS

6.2.1 Physical Characteristics

Underground space has certain natural peculiarities, or characteristics which are summarized in Table 6-1.

6.2.2 Historical Evolution

Underground space utilization started with caverns that were used as dwellings and for food storage by primitive man. A number of underground dwellings have been thoroughly studied, including examples in China (Banposite, 4000 B.C., Fig. 6-2), and Turkey (Byzantine Cappadocia, 400 B.C., Fig. 6-3). Underground passages for emergency evacuation and covert movements were an inherent part of many medieval cities.

Significant advancements in construction technologies during the 20th century resulted in a boom in UUS development. These technologies include reinforced concrete, tunneling in soft ground and the creation of open underground excavations with minimum subsidence of adjacent ground (facilitated by sheet piling, bored piles and diaphragm walls).

Major characteristics of underground space **Table 6-1**

Characteristic	Description	Examples of underground structures
Isolation	Underground spaces are less susceptible to external influences, and their impact on the external environment is less than aboveground facilities (e.g. noise)	Energy supply facilities
Temperature stability	Less need for heating or cooling: in many cases, underground facilities do not require any temperature adjustment at all	Storage, civil defense shelters
Protection	Underground spaces have limited areas of connection with the outside, and flow or movements through these connection areas are easy to control	Civil defense, valuable or hazardous good storage
Vulnerability to floods	Inundation can cause severe and unpredictable damage to underground structures, e.g. the floor structure of an upper level can collapse under the weight of water	Measures to manage flooding: waterproof and semi-waterproof doors, emergency drainage tunnels and reservoirs, pumping stations
Resilience during earthquakes	Deep underground structures suffer significantly less damage during earthquakes than aboveground structures	Storage facilities, civil defense shelters, emergency response centers
High cost of construction	Construction cost is highly dependent on ground conditions	Underground structures are planned for many years of operation (usually 100~300 years for public infrastructure)
Low cost of operation	No maintenance of outer walls is needed, low cost of heating/cooling. Some extra costs for ventilation, emergency prevention and response systems	Vast range of public structures, particularly in regions with severe climate
Opportunity to locate close to existing facilities	In urban areas, surface space has already been occupied by valuable developments, and underground space is the only available location for new facilities in the required area	Transport and emergency response infrastructure

Fig. 6-2 Banposite, 4000 B.C., China
Source: http://blog.sina.com.cn/s/blog_a2d2cf1e01019h85.html.

Late 20th century was especially beneficial for advancements in underground construction and geotechnical soil improvement technologies, which enabled progressive UUS development in densely build cities' areas, including excavation of large caverns in shallow subsurface.

The highest densities of underground structures can be found in city centers. The city center is a place where public facilities are located, as well as the main public transport transit areas. The reasons for UUS congestion in downtown areas are: (1) high demand for

Fig. 6-3 Byzantine Cappadocia, 400 B.C., Turkey
Source: http://blog.sina.com.cn/s/blog_6f7585300102wevz.html.

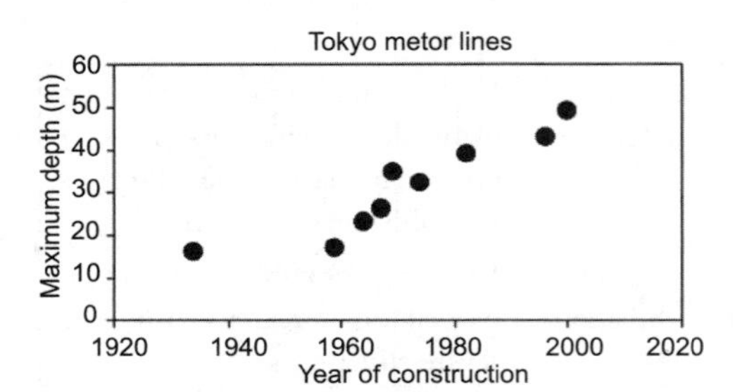

Fig. 6-4 Evolution of the depth of Tokyo metro lines[13]

infrastructure; and (2) the historical evolution of UUI, which includes accumulation of old and unused structures within the UUS.

Through its historical development, UUI has progressed from shallow to deep soil layers. This has created the problem of UUS congestion beneath city centers, where development of new UUI in deeper ground layers is hampered by old UUI in shallow layers. Fig. 6-4 illustrates the historical development of the underground rail system in Tokyo, Japan. In 1915, Japan's first underground railway opened under Tokyo Station. It was only for the railway post office, not for passengers. Tokyo Underground Railway Co., Ltd. opened Japan's first underground line of the subway Ginza Line on December 30,1927, and publicized as "the first underground railway in the Orient." The distance of the line was only 2.2km between Ueno and Asakusa. In 1938, Tokyo Rapid Transit Railway Co., Ltd. opened its subway system between Aoyama 6-chome (present-day Omotesando) and Toranomon. During World War II, the two subway companies merged under the name Teito Rapid Transit Authority by the local government. In 1954, the Marunouchi Line, the first subway line after World War II, opened between Ikebukuro and Ochanomizu. On March 20, 1995, the Tokyo subway sarin attack occurred on the Marunouchi, Hibiya and Chiyoda Lines during the morning rush hour. Over 5000 people were injured and 13 people were killed. All three lines ceased operation for the whole day.

London has long relied on tunnels for security, logistics, transport and utilities. The oldest building in the city, the Tower of London was first built for William, Duke of Normandy following his conquest and subsequent coronation in 1066. The Tower continued to expand and fortify under successive monarchs, many of whom relied on its network of secret tunnels to ferry goods, prisoners and even forbidden lovers.

However, it was 1843 before an underwater tunnel was first attempted. Designed and constructed by Sir Marc Brunel, it opened initially as a pedestrian tunnel, but in 1869 was converted to a railway tunnel that is now part of the London Overground Network. Briefly described as the 8th Wonder of the World, the Thames Tunnel was a tourist attraction with visitors promenading through its below ground arches (Fig. 6-5).

The technology used to construct the Thames Tunnel undoubtedly influenced the design and construction of the London Sewer System. Designed by engineer Sir Joseph Bazalgette, the combined sewer and storm water system was vital to address the public health and amenity crisis caused by sewerage in the River Thames. Constructed between 1859 and 1865, the sewer also incorporated several of London's "lost" underground rivers

Fig. 6-5 Thames Tunnel, 1843[14]

and still serves modern day London. The 1860s were an exciting time for big thinking engineering projects, with London's first underground rail line opened in 1863; the Metropolitan Line was constructed using cut and cover techniques and operated initially with steam trains. The London Underground Train System quickly grew to become a critical part of London's infrastructure and identity, beginning with 40,000 passengers there are now more than 1107 million passenger journeys per year. The expanding underground rail network was used as a shelter for London residents both in World War I and World War Ⅱ (Fig. 6-6). To protect citizens from air bombing raids, between 1940 and 1942, eight deep level shelters were constructed below existing underground rail station tunnels.

London's transport and utilities infrastructure has continued to expand below ground level, most notably are the creation of the Channel Tunnel; the Jubilee Line Extension (1999); Crossrail which is currently under construction (estimated opening in late 2022); and the proposed Thames Tideway Sewer Tunnel (estimated operation in 2025).

Underground space has been utilized in Hong Kong for many decades. It has been developed through various phases of infrastructure development and improvement. The early forms of underground space construction were associated with war time protection and mining operations and this has

Fig. 6-6 London underground station as air raid shelter in World War Ⅱ[14]

Fig. 6-7 Lion Rock Tunnel portal [16]

Fig. 6-8 Tai Koo MTR Station of Hong Kong, China [16]

(a)

(b)

Fig. 6-9 Temporary explosives magazine[16]
(a) Explosive storage niches; (b) Section of the access adit at the West Island Line (WIL) explosives magazine

extended with increased urban densification to numerous examples of underground basements that have incorporated car parks, retail and commercial underground spaces. Notable underground space also includes numerous roads (Fig. 6-7), rail and utility/service tunnels that comprise a network of over 500km of tunnels in the city to accommodate essential services and transport. With the Mass Transit Railway (MTR) construction in the 1970s, it kicked off a spate of excavation of underground space for metro stations (Fig. 6-8) and linking tunnel networks that has developed one of the most efficient and reliable metro systems in the world. The private sector has increasingly, over the last few decades, linked their commercial and retail properties to the MTR station network, providing hubs of interest and commerce within the underground network.

In the late 1980s, the Geotechnical Engineering Office (GEO) of the Civil Engineering and Development Department (CEDD) had been tasked by the government of Hong Kong Special Administrative Region to undertake various cavern development studies that explored the potential for cavern development at the time. It reviewed the overseas approach to cavern development as well as the technologies and the approach to cavern design. As a result of those studies, several notable facilities in Hong Kong, such as temporary explosives magazine (Fig. 6-9) and western salt water service reservoir (Fig. 6-10), were placed in rock caverns to solve the problems of lack of space or to protect nearby communities from the nuisance

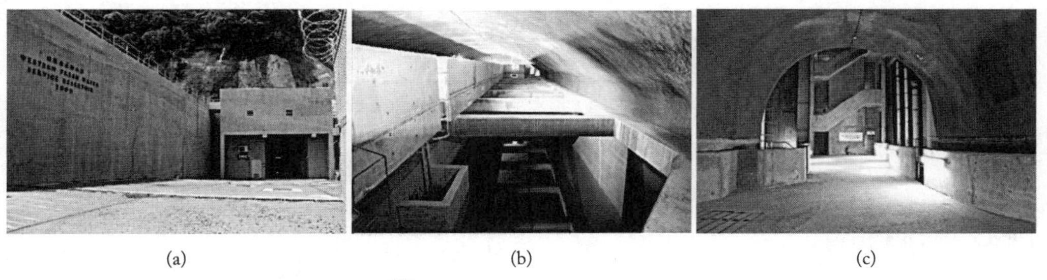

(a) (b) (c)

Fig. 6-10 Western salt water service reservoir[16]
(a) Portal entrance to underground service reservoir; (b) Main tank cavern; (c) Access adit and pipe gallery

(a) (b)

Fig. 6-11 Henriksdal Wastewater Treatment Plant, Stockholm
Source: https://www.water-technology.net/projects/henriksdal-wastewater-treatment-plant-stockholm/.

aspects of the facilities.

Since the 1980s, cavern construction has become a more established solution in various districts for a range of different types of facilities. The technology and equipment used to construct these underground spaces has shown continual improvement in its application and efficiency. Other countries have been using purpose-built rock caverns as their preferred choice to house a variety of facilities, including water and sewage treatment works (Fig. 6-11), data center, oil and gas storage, warehousing, freight transfer, sports hall, etc.

6.2.3 Provided Services

UUS can be considered as a resource that provides certain services. Understanding the services that UUS provides is necessary for allocating a value to UUS. Considering the value of UUS is an important part of the land use planning process and for addressing UUS use in a master plan. Listing and considering UUS services, especially those that might be very important for future urban development and sustainability, can be helpful during planning processes. Often, UUS is not considered in a master plan, or it is assumed that UUS has less value than surface land. The consequence of such undervaluation of UUS is delayed planning of the resource: Often surface land use plans are established first, and UUS use plans are established later. Such lack of proper consideration of UUS can jeopardize its services.

UUS services are listed with examples in Table 6-2. Fig. 6-12 provides a classification

UUS services Table 6-2

Service	Example of use
Physical space	Various underground structures
Space continuum that has certain soil strength properties	Basis for underground and aboveground structures
Excavated materials	Sand and granite are used in the construction industry
Drinking water supply	Wells for drinking water extraction
Groundwater supply to surface vegetation	Urban parks
Surface water reserves supplied by groundwater	Ponds, rivers
Geothermal energy	Geothermal energy bore holes
Cultural heritage	Archaeological sites, natural sites

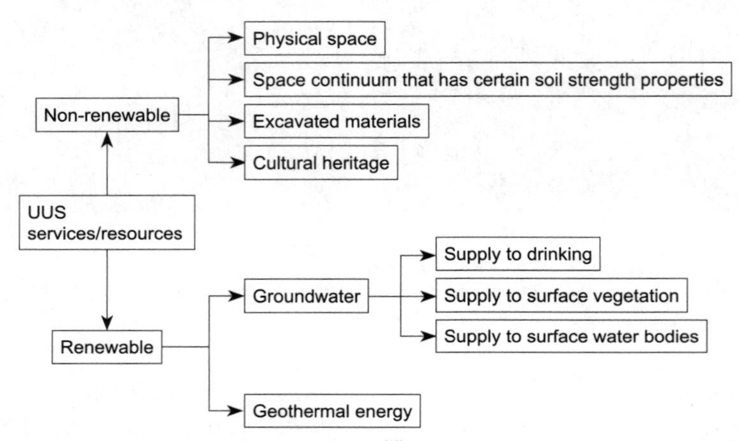

Fig. 6-12 Classification of the UUS services [13]

of UUS services based on whether or not UUS resources are renewable.

Table 6-2 focuses on services inherent to urban areas, and does not include services for resource provision, such as industrial scale excavation of minerals, oil and gas.

UUS services can be prioritized according to the city's needs.

The issue of service prioritization becomes especially difficult if those services are not yet used, e.g. drinking water provision. This question should be addressed in a master plan, along with considerations for sustainable development.

UUS services can be prioritized by considering:

(1) Critical and most important needs (drinking water supply, civil defense infrastructure);

(2) Needs for sustaining the environment (nature as well as the man-made environment);

(3) New and changing community requirements in the future.

Future needs and an increase or decrease in the need for UUS services should take into account:

(1) Urbanization trends;

(2) Population trends;

(3) The type of infrastructure that could provide sustainable urban services;

(4) Global and local environmental change trends.

Both of the above issues, UUS service prioritization and predicting future needs for UUS, need to be addressed during urban development planning. Although this is a difficult task, the worst case would be just to ignore these aspects of planning for future UUS use. Given the complexity of prioritization and planning, and also the importance of UUS services, one possible solution is to reserve the most valuable areas of UUS. An example of this approach would be to avoid tunneling through soil layers with groundwater that might be used as drinking water in the future. Another example would be avoiding installing many conduits under a street in a downtown area that might be a good location for an underground train station.

6.3 Construction Methods

6.3.1 Cut and Cover Method

The cut and cover method is also known as the open cut method. It is a traditional form of tunneling that involves opening up the ground surface and excavating to the required depth. Shallow tunnels are built using this method where a section is excavated and then covered with a support system.

Unlike trenchless construction techniques such as horizontal directional drilling (HDD) and microtunneling, the cut and cover method requires traffic diversion for the duration of the project. The surface of the excavated site is restored to its original condition once the project is complete.

The cut and cover method can be of two types, namely, bottom-up method and top-down method (Fig. 6-13).

In the bottom-up method, the site is first excavated and proper ground support is provided. The tunnel is then constructed of precast concrete, in-situ concrete, steel rings, etc. Once the tunnel is completed, the trench is backfilled over the tunnel roof and the ground surface restoration is carried out. Fig. 6-14 shows the procedure of bottom-up method.

The bottom-up method offers several ad-

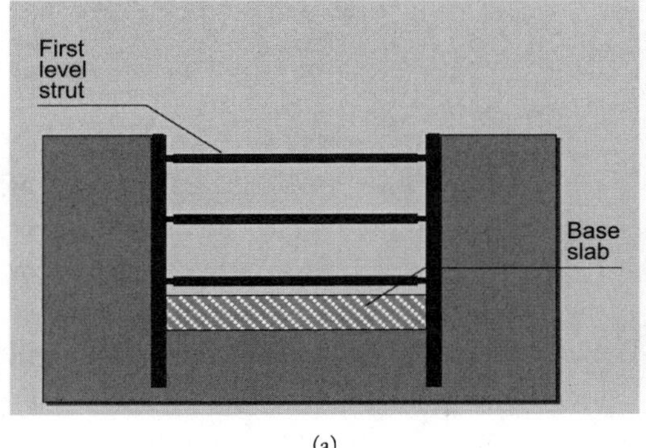

(a)

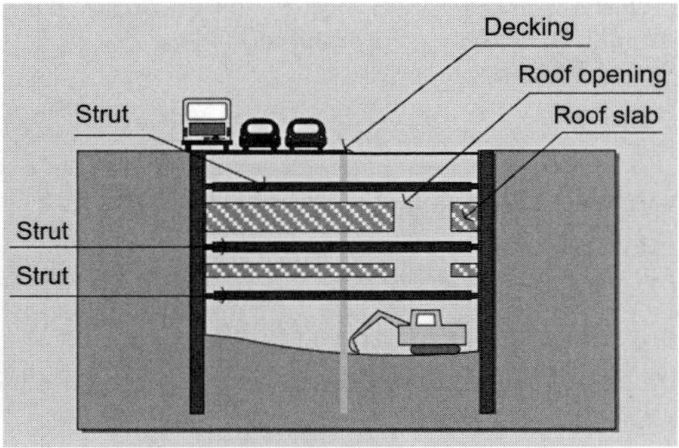

(b)

Fig. 6-13 Two types of the cut and cover method
(a) The bottom-up method; (b) The top-down method
Source: http://www.railsystem.net.

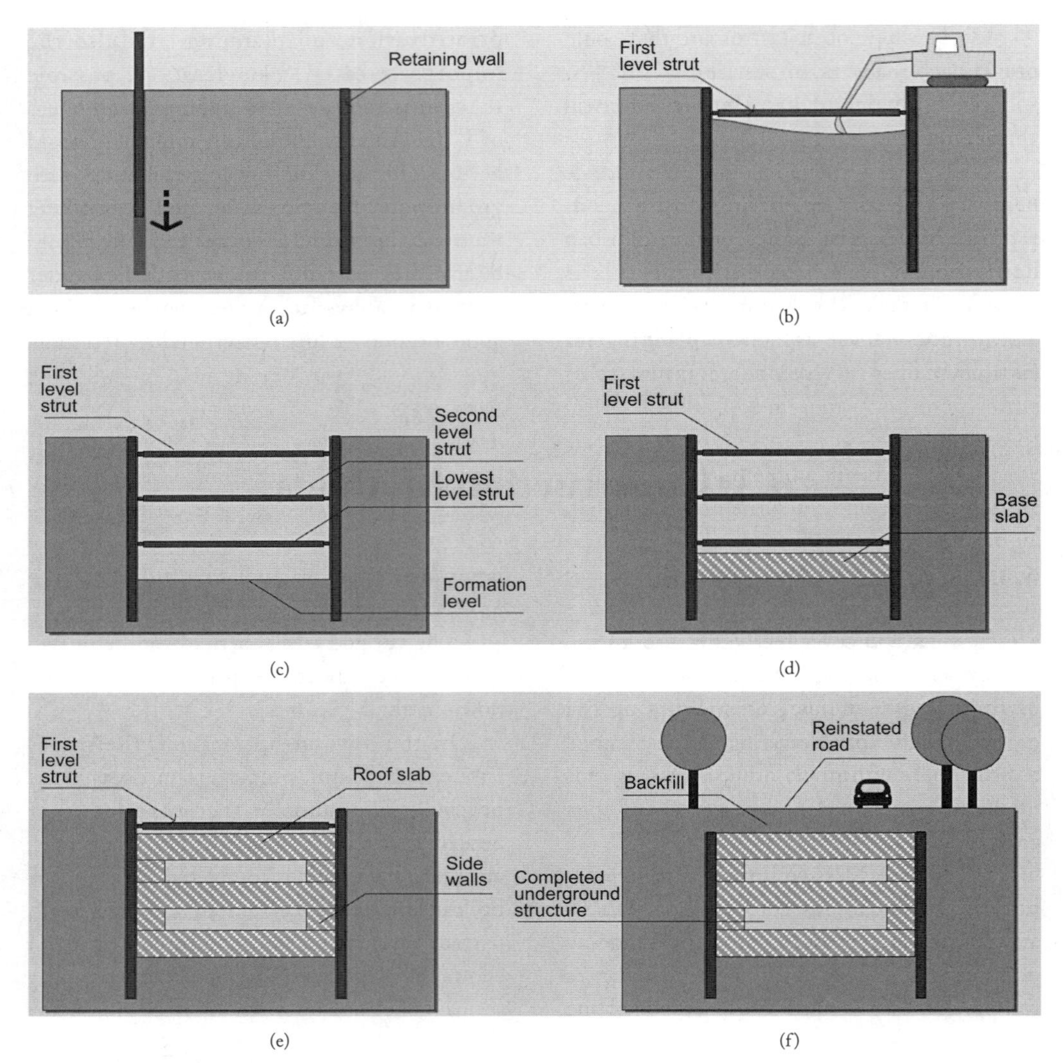

Fig. 6-14 The bottom-up method
(a) Installation of retaining wall; (b)、(c) Excavation & installation of steel strut; (d)、(e) Construction of underground structure; (f) Backfilling & reinstatement
Source: http://www.railsystem.net.

vantages: (1) It is a conventional construction method well understood by contractors; (2) Waterproofing can be applied to the outside surface of the structure; (3) The inside of the excavation is easily accessible for the construction equipment and the delivery, storage and placement of materials; (4) Drainage systems can be installed outside the structure to water or divert it away from the structure.

However, there are some disadvantages of the bottom-up method:(1) Somewhat larger footprint is required for construction than for the top-down method; (2) The ground surface cannot be restored to its final condition until construction is complete; (3) It requires temporary support or relocation of utilities; (4) It may require dewatering that could have adverse effects on surrounding infrastructure.

In the top-down method, cap beams and

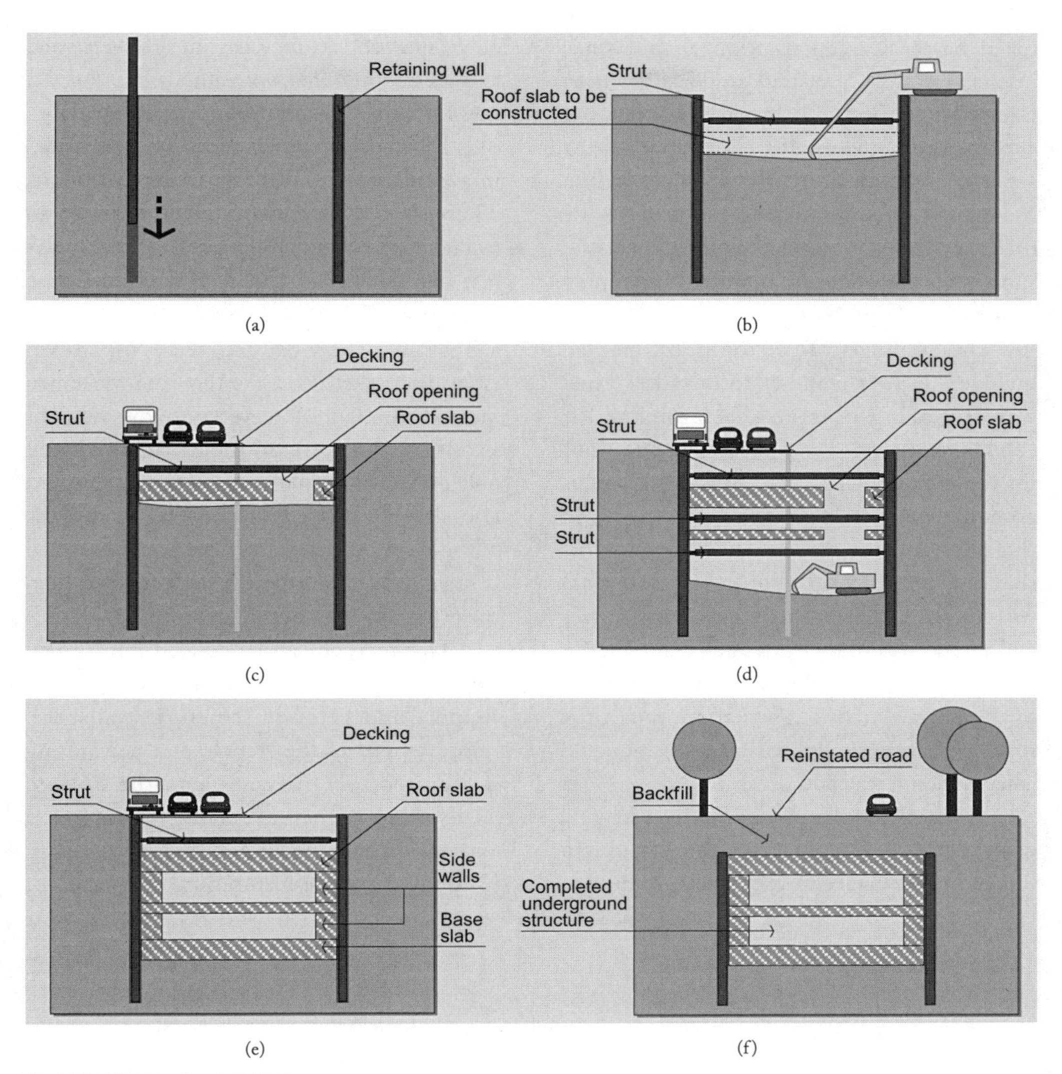

Fig. 6-15 The top-down method
(a) Installation of retaining wall; (b) Excavation & installation of steel strut; (c) Construction of underground structure; (d) Construction of underground structure; (e) Construction of underground structure; (f) Backfilling & reinstatement
Source: http://www.railsystem.net.

side support walls are constructed starting from the ground surface down with slurry walls or secant piling. The tunnel roof is then made of precast beams or in-situ concrete from a shallow excavation. Leaving opening for access, the ground surface is restored allowing restoration of traffic services. The remaining construction is then carried out under the tunnel roof from the access openings. Fig. 6-15 shows the procedure of the top-down method.

The top-down method offers several advantages in comparison to the bottom-up method: (1) It allows early restoration of the ground surface above the tunnel and the temporary support of excavation walls are used as the permanent structural walls; (2) The structural slabs will act as internal bracing for the support of excavation thus reducing the amount of tie backs required, and it requires somewhat less width for the construction area;

(3) It has easier construction of roof since it can be cast on prepared grade rather than using bottom forms; (4) It may result in lower cost for the tunnel by the elimination of the separate, cast-in-situ concrete walls within the excavation and reducing the need for tie backs and internal bracing; (5) It may result in shorter construction duration by overlapping construction activities.

Disadvantages of the top-down method include: (1) It is inability to install external waterproofing outside the tunnel walls; (2) There are more complicated connections for the roof, floor and base slabs; (3) There is potential water leakage at the joints between the slabs and the walls; (4) There are risks that the exterior walls (or center columns) will exceed specified installation tolerances and extend within the neat line of the interior space; (5) Access to the excavation is limited to the portals or through shafts through the roof; (6) Spaces are limited for excavation and construction of the bottom slab.

Since the bottom-up method affects ground traffic heavily and the top-down method takes longer to complete, a semi-top-down approach is often used to achieve a good balance between the two methods.

During a semi-top-down excavation, soils above and near the surface are excavated in accordance with the top-down procedure. Thereafter, a top floor slab is installed to facilitate road traffic. Deeper soils are excavated just like in the bottom-up method. Because excavation of the lower part of the pit follows a bottom-up procedure, semi-top-down excavation takes less time than the top-down method required. Moreover, since the top floor slab is installed early in the excavation, it not only provides a stronger supporting system, but also allows traffic above to continue. Because of these advantages, semi-top-down excavation is assumed to have better performance than the other two methods in urban areas.

This method can be divided into several steps (Fig. 6-16): (1) installation of excavation support/underground structural walls, soil improvement in the trench, and dewatering if required; (2) excavation to the bottom level of top structural slabs, construction and waterproofing of top structural slabs; (3) ground surface restoration, excavation of

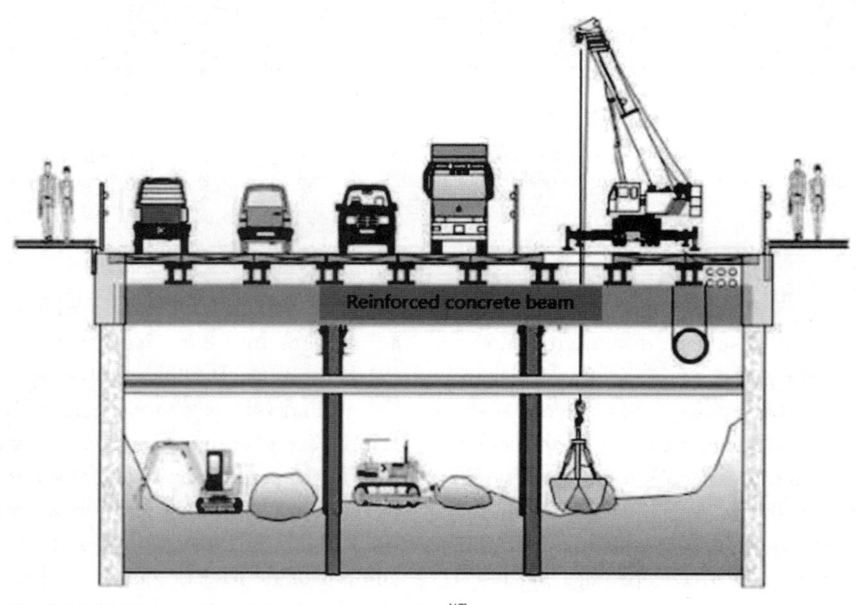

Fig. 6-16 Sketch map of semi-top-down construction [17]

underground interior, installation of bracing of supporting walls; and (4) construction of underground structures.

6.3.2 Conventional Tunneling

The conventional tunneling method is a cyclical process of tunnel construction that involves excavation by drilling and blasting or by mechanical excavators [except the full-face tunnel boring machine (TBM)]. This is followed by application of an appropriate primary support.

Mainly used in rock formations, several variants exist including the mining method, drill-and-blast method, new Austrian tunneling method, etc. Conventional tunneling usually includes the following steps (Fig. 6-17): (1) drilling of blast holes, (2) charging, (3) ignition, (4) ventilation, (5) mucking: loading and hauling, (6) scaling, (7) temporary support, and (8) surveying.

Because there is access to the excavation face, conventional tunneling is considered a very flexible process and is preferred in situations where there is a variation in ground conditions or tunnel shape. This flexibility includes the ability to vary the support, the blasting technique and size, ring closure time and the excavation technique as well as other factors.

6.3.3 Mechanical Tunneling

TBMs are extensively used throughout the world to bore various tunnel diameters in various types of rock and soil strata. A TBM consists of a main body and supporting elements that enable it to perform cutting, shoving, steering, gripping, shielding, exploratory drilling, ground control and support, lining erection, muck removal and ventilation.

(1) Gripper Tunnel Boring Machine

A gripper TBM (Fig. 6-18) is suitable for driving in hard rock conditions when there is no need for a final lining. The rock supports (rock anchors wire-mesh, shotcrete, and/or steel arches) can be installed directly behind the cutter head shield and enable controlled relief of stress and deformations. The existence of mobile partial shields enables gripper TBMs

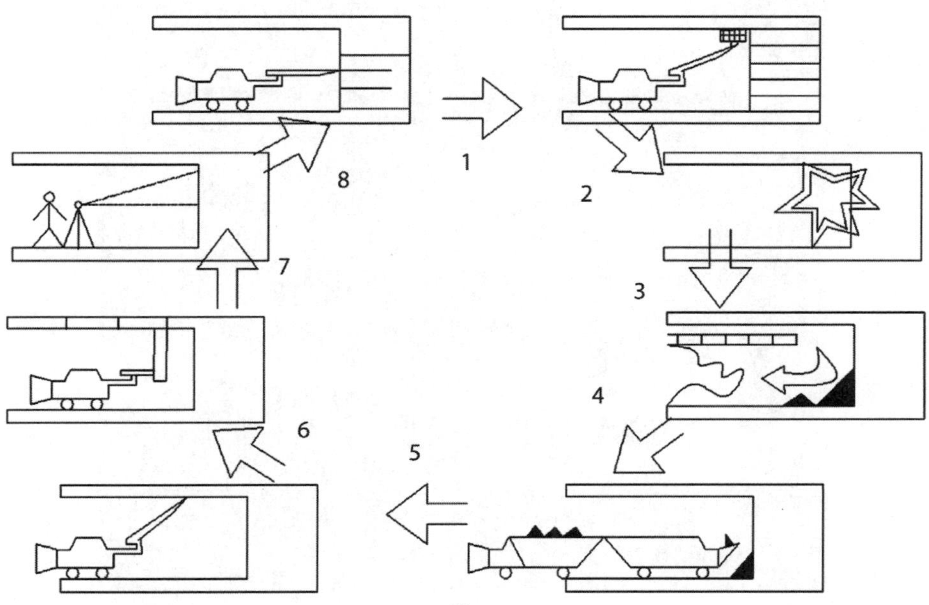

Fig. 6-17 General procedure of drill-and-blast method [17]

to be flexible even in high-pressure rock. This is useful when excavating in expanding rock to prevent the machine from jamming.

(2) Double-shield Tunnel Boring Machine

A double-shield TBM is generally considered to be the fastest machine for hard rock tunnels under favorable geological conditions with installation of the segment lining. It is possible to drive 100m in 1 day. This type of TBM consists of a rotating cutter head and double shields, a telescopic shield (an inner shield that slides within the larger outer shield), and a gripper shield together with a tail shield (Fig. 6-19).

While boring, gripper shoes radially press against the surrounding rock to hold the

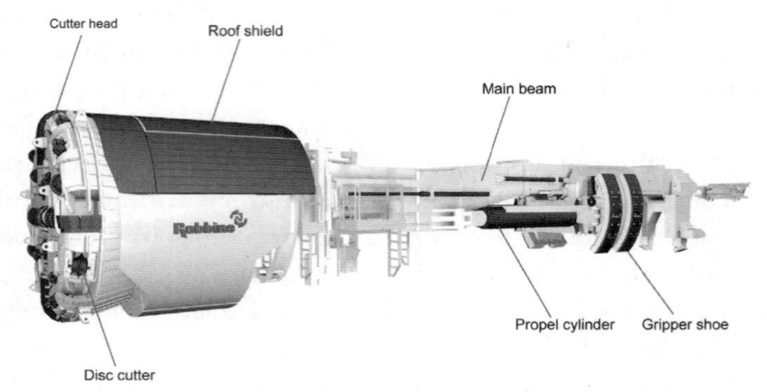

Fig. 6-18 Typical diagram of an open gripper main beam of TBM [17]

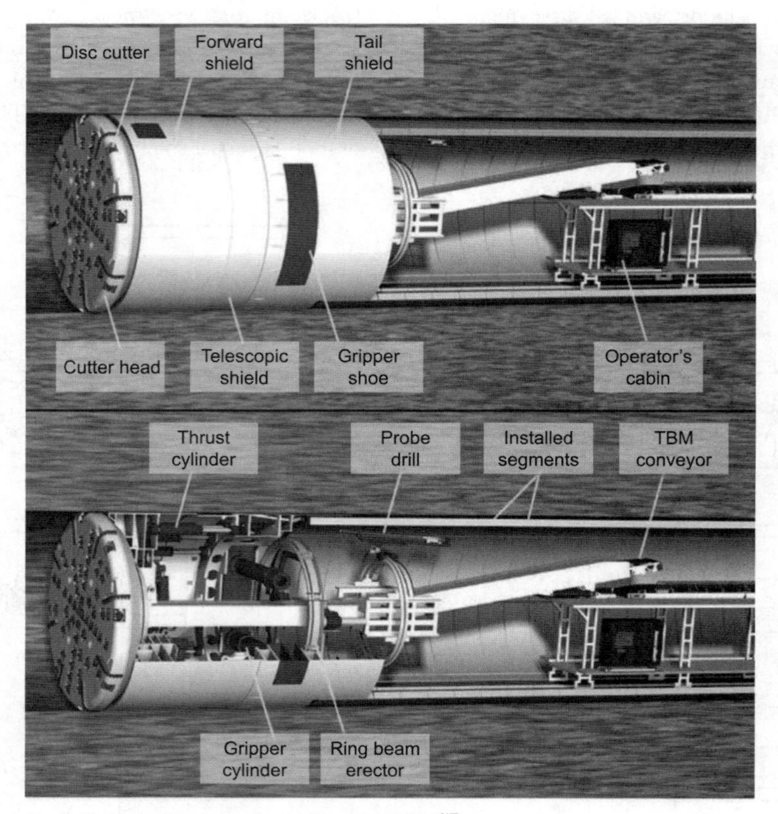

Fig. 6-19 Typical diagram of a double-shield TBM [17]

machine in place and take some of the load from the thrust cylinders. For the motion of the front shield, the gripper shoes are loosened, before the front shield is pushed forward by thrust cylinders protected by the extension of the telescopic shield. Because regripping is a fast process, double-shield TBMs can almost continuously drill. As for the tail shield, it is used to provide protection for workers while erecting, installing the segment lining and pea gravel grouting.

(3) Single-shield Tunnel Boring Machine

Single-shield TBMs are used in soils that do not bear groundwater and where rock conditions are less favorable than for double shields, such as in weak fault zones. The shield is usually short so that a small radius of curvature can be achieved (Fig. 6-20).

(4) Earth Pressure Balance Machine

Earth pressure balance (EPB) technology is suitable for digging tunnels in unstable ground such as clay, silt, sand or gravel. An earth paste face formed by the excavated soil and other additives supports the tunnel face. Injections containing additives improve the soil consistency, reduce soil stick, and thus its workability. The components of an EPB machine is shown in Fig. 6-21.

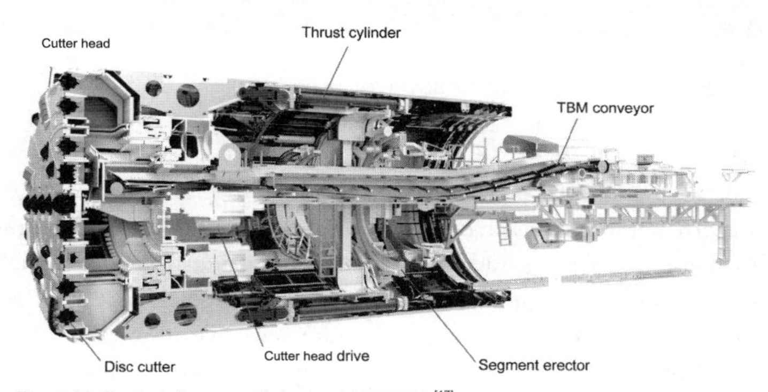

Fig. 6-20 Typical diagram of single-shield TBM [17]

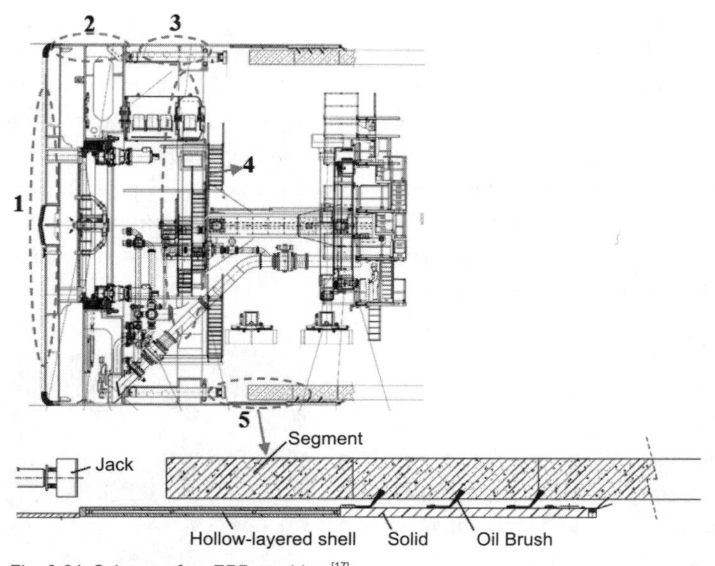

Fig. 6-21 Scheme of an EPB machine [17]

1-Cutter wheel; 2-Cutter ring; 3-Support ring; 4-Internal steel support; 5-Shield talkskin

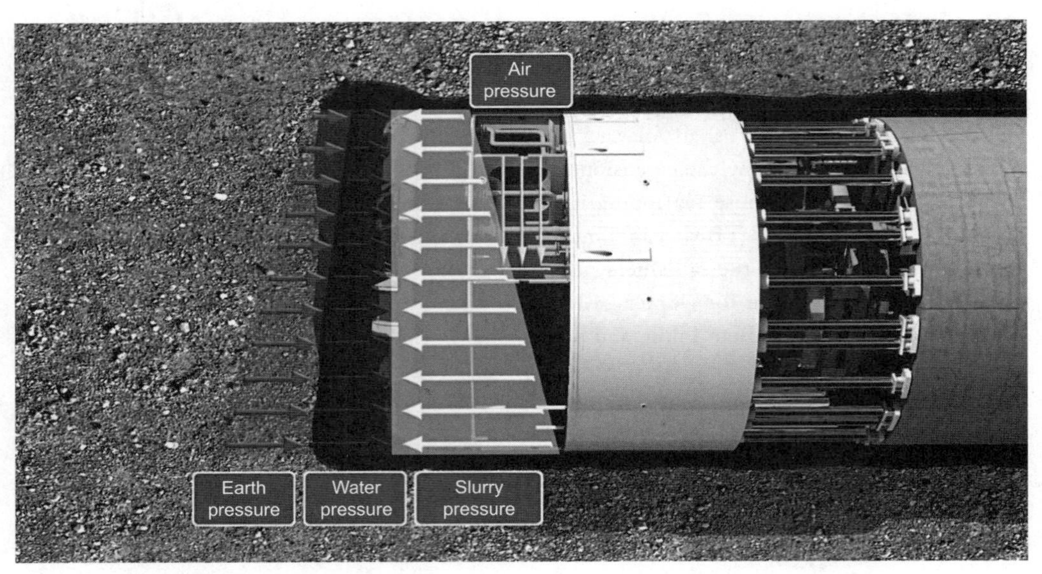

Fig. 6-22 Slurry tunnel boring machine
Source: http://www.creg-germany.com/products_tunnel-boring-machines_slurry-tbm.php.

To ensure support pressure transmission to the soil, the earth paste is pressurized through the thrust force transferring into the bulkhead. The TBM advance rate (inflow of excavated soil) and the soil outflow from the screw conveyor regulate the support pressure at the tunnel face. This is monitored at the bulkhead by the readings of pressure sensors.

(5) Slurry Tunnel Boring Machine

Slurry TBMs (Fig. 6-22) are used for highly unstable and sandy soil and when the tunnel passes beneath structures that are sensitive to ground disturbances. Pressurized slurry (mostly bentonite) supports the tunnel face. The support pressure is regulated by the suspension inflow and outflow. The slurry's rheology must be chosen in accordance with the soil parameters and should be carefully and regularly monitored.

(6) Mixshield Technology

Mixshield technology (Fig. 6-23) is a variant of conventional slurry technology for heterogeneous geologies and high-water pressure. In mixshield technology, an automatically controlled air cushion controls the support pressure, with a submerged wall that divides the excavation chamber. This wall seals off the machine against the excess pressure from the tunnel's face. As air is compressible in nature, the mixshield is more sensitive in pressure control and thus will provide more accurate control of ground settlement.

(7) Pipe Jacking

Pipe jacking (Fig. 6-24) is a small-scale tunneling method for installing underground pipelines with minimum surface disruption. It is used for sewage and drainage construction, sewer replacement and lining, gas and water mains, oil pipelines, electricity and telecommunication cables, and culverts.

A fully automated mechanized tunneling

Fig. 6-23 Overview of a mixshield TBM [17]

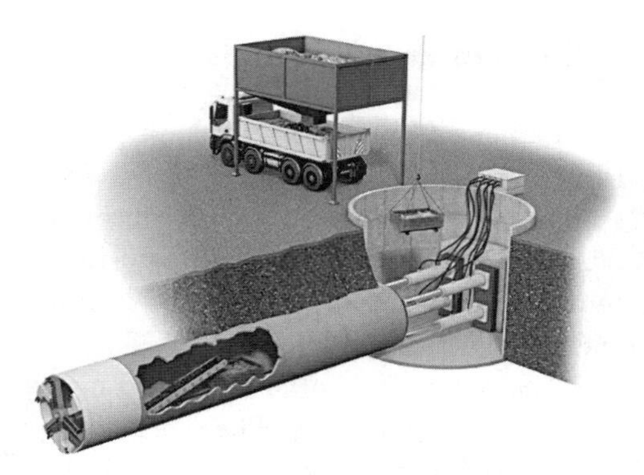

Fig. 6-24 Typical pipe jacking arrangement [20]

Fig. 6-26 The cross section of DOT shield [17]

(a)

(b)

Fig. 6-25 Two kinds of partial-face excavation machines [17]

shield is usually jacked forward from a launch shaft toward a reception shaft. Jacking pipes are then progressively inserted into the working shaft. Another significant difference between the pipe jacking method and shield method is that the lining of the pipe jacking is made of tubes and the lining of the shield method is made up of segments.

In order to significantly reduce the resistance of the pipes, a thixotropic slurry is injected into the outside perimeter of the pipes. The thixotropic slurries can also reduce disturbance to the ground while pipe jacking slurry thickness increases. The thickness should be six to seven times the void between the machine and pipes.

(8) Partial-face Excavation Machine

Partial-face excavation machines (Fig. 6-25) have an open-face shield and can sometimes be more economical in homogeneous and semistable ground with little or no groundwater. In boulder layer, the open-face can deal with boulders much easier than closed shield machines. In cavity ground, the open-face can avoid the risk of falling down into the bottom of the cavity. Due to their simple design and that the operator workplace is close to the open tunnel face, these machines can easily be adapted to changing geological conditions. Good excavation monitoring can also be carried out.

(9) Noncircular Shields

Only approximately two-thirds of a circular tunnel section can effectively be used. Consequently, the TBMs of the future are expected to have noncircular cross-sectional machines and be so-called noncircular shields. Different types already exist, such as double-O-tube shield tunneling (DOT) shields (Fig. 6-26) for which a middle column is installed in order to form a stable tunnel lining.

Various tunnel cross sections can also be

made using a shield machine with a primary circular disk cutter in the center and multiple secondary planetary cutters on the peripheries (Fig. 6-27).

(10) Deep Drilling Technology

Today, different companies have been working on developing technology that can cope with specific demands. Oil, gas and geothermal energy sources can be explored using deep drilling rings. The Terra Invader deep drilling rig is an efficient drilling technology used to explore deep energy deposits and can be used for onshore and offshore drills to depths of down to 8000m (Fig. 6-28).

Fig. 6-27 A rectangular shield [17]

Fig. 6-28 Terra Invader rig used for deep drilling [17]

6.3.4 Immersed Tunnels

Immersed tunnels are made of prefabricated sections and are widely used to cross shallow water (Fig. 6-29). Tunnel elements are fabricated elsewhere and then towed to the tunnel location before being joined together and buried permanently.

The construction of an immersed tunnel consists of several steps. These are namely trench excavation, foundation preparation, tunnel element fabrication, transportation and handling of tunnel elements, element lowering, element placing and backfilling (Fig. 6-30).

Immersed tunnels can be divided into two types depending on the method of fabrication used: 1) Steel-concrete composite tunnels: stiffened steel plates are used that act together with an inner concrete filling. Tunnel elements are usually fabricated in yards and then the concrete is poured. 2) Reinforced concrete tunnels: concrete is reinforced by steel bars or prestressing cables. Tunnel elements are cast in a dry dock and transported to the site before being flooded.

(1) Single-shell Steel Immersed Tunnel

For single-shell steel immersed tunnels, the external structural shell plate works compositely with the inner reinforced concrete. As there is no external concrete, the steel plates are protected against corrosion. The Hong Kong cross-harbor tunnel (Fig. 6-31) is an example of such a structure.

For single-shell steel immersed tunnels, leaks in the steel shell may be difficult to find and to be sealed. Although subdividing the surface into smaller panels by using ribs will improve the chances of sealing a leak, great care and considerable testing is required to ensure that the welds are defect free. The risks of permanent leakage can therefore be higher in single-shell steel immersed tunnels than in other types.

(2) Double-shell Steel Immersed Tunnel

For double-shell steel immersed tunnels, an internal structural shell acts together with

Fig. 6-29 An immersed tunnel [17]

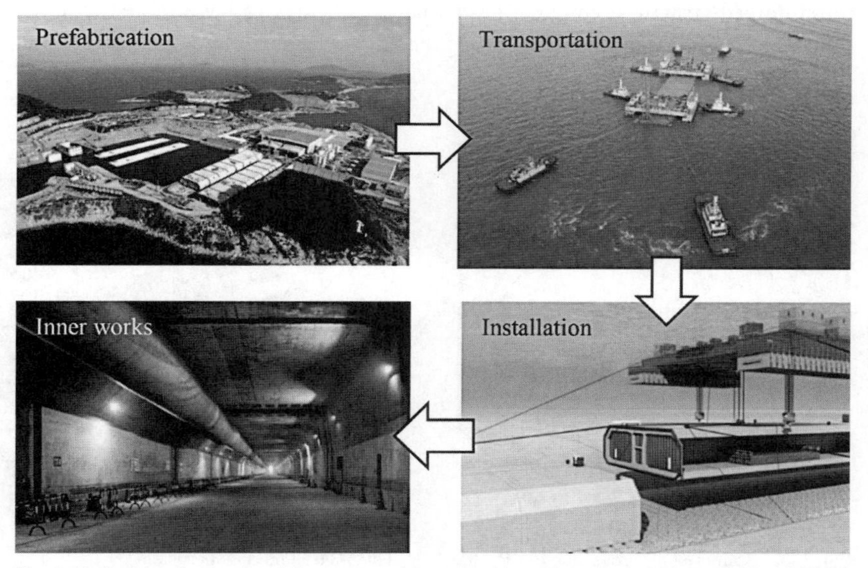

Fig. 6-30 Construction works of the immersed tunnel in Hong Kong-Zhuhai-Macao Bridge (HZMB) Island-tunnel project [19]

Fig. 6-31 Hong Kong cross-harbor tunnel elements nearly ready for side launching [17]

concrete placed within it. Another structural element is the top and invert concrete which are placed outside the structural shell plate.

A second steel shell is constructed to act as a frame for ballast concrete at the sides. The interior structural shell is therefore protected by external concrete that is placed with nonstructural steel plates. Fig. 6-32 shows the cross section of the Second Hampton Roads Tunnel in Virginia. The thick black lines correspond to steel plates.

(3) Sandwich Construction

Sandwich construction consists of an inner and outer steel shell that act together with an unreinforced structural concrete layer that is sandwiched between them (Fig. 6-33). For a successful composite action, the inner surfaces of the steel shells must have connectors. The result is that steel shells carry tension loads while the concrete bears compression loads.

(4) Concrete Immersed Tunnels

The advantage of concrete is that it is durable and can be molded to any shape. Being heavy, concrete tunnel elements are usually constructed in dry basins near the project-site fabrication factory (Fig. 6-34 and Fig. 6-35) before being flooded and towed into position.

In concrete structures, uneven shrinkage is the major cause of cracking. It is therefore important to implement a control crack method when using concrete immersed tunnels.

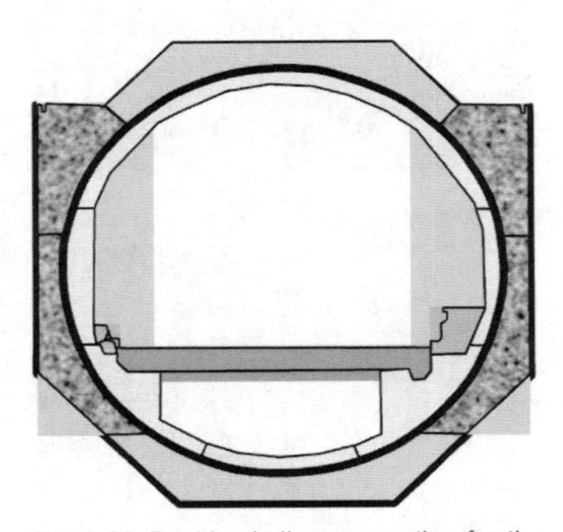
Fig. 6-32 Double-shell cross section for the Second Hampton Roads Tunnel, Virginia [17]

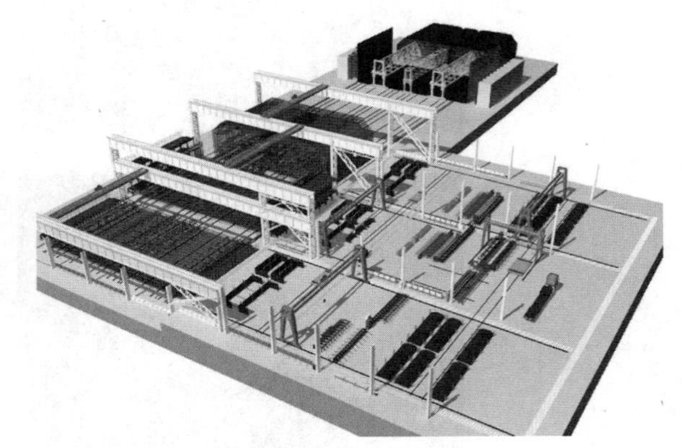
Fig. 6-34 HMZB tunnel system [17]

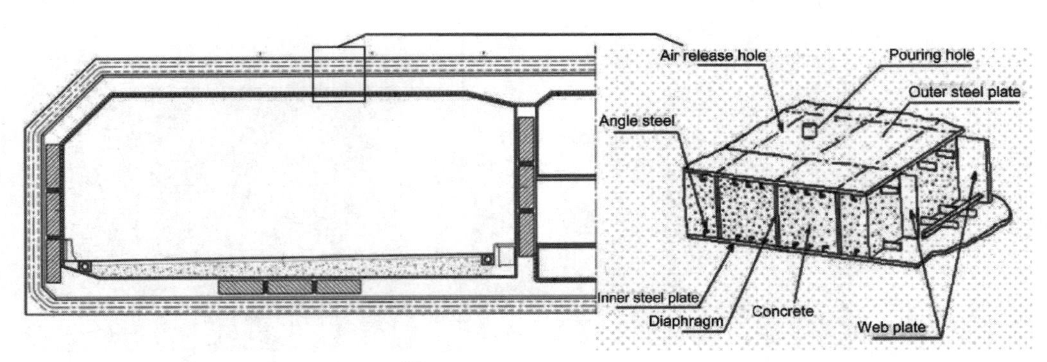

Fig. 6-33 Scheme of a sandwich construction [17]

Fig. 6-35 Full section of the hydraulic automated formwork system of HMZB tunnel [17]

6.4 Case Studies

6.4.1 Underground Space in Helsinki

There are 10,000,000m^3 underground spaces in Helsinki (parking, sports, oil and coal storages, the metro, etc.), more than 400 premises, 220km of technical tunnels, 24km of raw water tunnels and 60km of all-in-one utility tunnels (district heating and cooling, electrical and telecommunication cables, and water). Some unique examples of the use of underground spaces are shown in Fig. 6-36 and Fig. 6-37.

Buildings in Helsinki are mainly quite low with skyscrapers only being built in some special areas. The historic inner city (Fig. 6-38) is therefore remarkably different from the center of Singapore, for instance. Helsinki can be classified by the term "down-rise city" (using underground resources effectively) while Singapore, in turn, is a "high-rise city", which was fashionable in the 1900s. The deepest underground space in Helsinki is situated only about 100m below sea level. Nevertheless, underground resources may also be found in the inner city in the future, if needed.

Helsinki is among the smallest by area and clearly the biggest by population in Finland. It needs to have open spaces even in the city center. A good example of land property

Fig. 6-36 Interior of the Temppeliaukio Church [22]

Fig. 6-37 Underground Swimming Pool in "Itäkeskus " [22]

(a)

(b)

Fig. 6-38 Helsinki and downtown in Singapore
(a) Helsinki market square; (b) Downtown in Singapore

resources made use of several times is the Katri Vala Park situated in the city center (Fig. 6-39). Nowadays, there are four underground activities under the park totally independent from each other. The possibility to build one more space between the existing underground "floors" is currently being investigated. The Katri Vala Park is also an example of the concept called "0-land use".

As many deep cellars, underground spaces and tunnels already exist in the center of Helsinki, the new underground cold-water reservoir for district cooling was excavated between 50~90 meters from ground level (Fig. 6-40). Although all underground space below the surface of real estate owners' land belongs to them, they may only restrict its use or get compensation if the space to be used is harmful or it causes some loss to the owner. This is mainly the case in (local) government underground projects. In non-government projects, such as private car parks, a (servitude) agreement is drawn up between the construction company and the land owner even when the company is not paying for the use of the underground space.

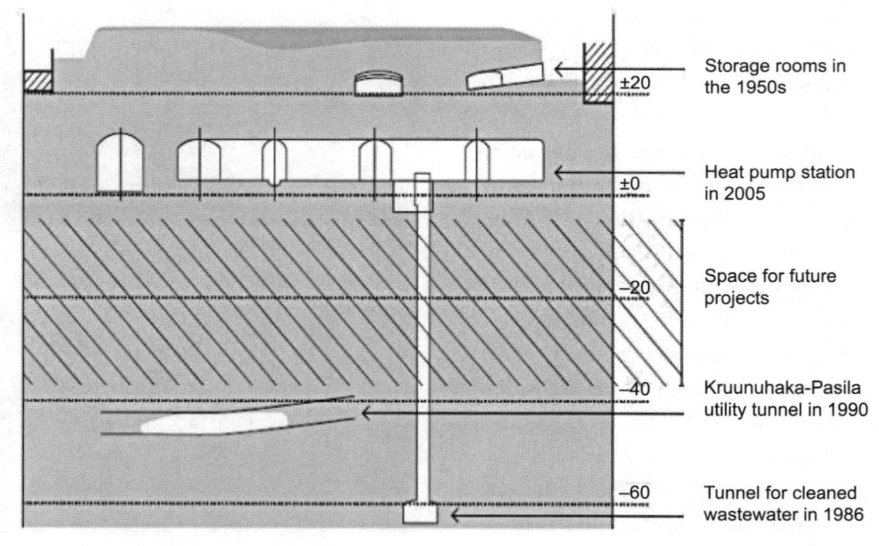

Fig. 6-39 Example of "0-land use": Katri Vala Park in Helsinki [22]

Fig. 6-40 Cold-water reservoir for district cooling in Helsinki city center [22]

6.4.2 Underground Space in Moscow

The Moscow Metro is considered to be one of the most beautiful transit systems in the world. Architecturally, it is a true underground museum which can be accessed for the price of one ride. Each of its 222 stations is unique, with many of those built in the Soviet Era being recognized as masterpieces of architecture. You will find elements of the Empire style, Art Deco, gothic architecture and, naturally, Russian national motifs. The Moscow Metro is an integral part of any guided tour of Moscow intended for those interested in architecture and history. Not only do they illustrate the evolution of architectural styles but also they refer to specific ideological aspects or historical events, glorifying the life of the country and its individual nations and heroes.

Kiyevskaya Station (Fig. 6-41) opened in 1954 to celebrate 300 years of Russo-Ukrainian unity. Designed by Yevgeny Katonin, Vadim Skugarev and Georgy Golubev, the station was constructed after the death of Joseph Stalin, when Nikita Khrushchev had come to power in the Soviet Union. The new leader wanted to eternalize his home republic in bright and festive motifs, and the project designers did their best to fulfill his desire. The pylon feet and the vaults are white, making a perfect background to 18 large mosaics with unique plots, mostly representing the friendship between the Russian and Ukrainian peoples and their unity within the Soviet Union. There is an interesting story about this station. One of the mosaics allegedly reveal a man from the future, a guerilla warrior who is talking on the phone and holding a box on his knees. The device in his hands thought to be a cell phone and the box to be a laptop, which

Fig. 6-41 Kiyevskaya Station
Source: https://www.moscovery.com/moscow-underground/.

caused a little stir. However, a closer analysis has shown that the telephone is a famous field model used in wartime and the box is just the telephone system cover.

There are also some remarkable stations among those which have only recently been built. Dostoyevskaya Station on the Lyublinsko-Dmitrovskaya Line (Fig. 6-42), opened in 2010, is decorated with panels depicting scenes from various works by Fyodor Dostoyevsky. His memorial apartment is located nearby, in the house where the great writer was born and raised. The station walls are decorated with black-and-white Florentine mosaics.

The Soviet underground is undoubtedly one of the most beautiful underground systems in the world. Limited budgets do not allow modern architects to compete with the stations of the 1930s and 1950s, when every new station was built with the intent to make it a majestic underground palace. However, today's architects carry on the tradition established by their Soviet predecessors by designing every new station in a unique way. Rumyantsevo Station on the red Sokolnicheskaya Line (Fig. 6-43) is one of the recently built stations that deserve special mention. The decoration of Rumyantsevo, opened in 2016, revolves around fantasies inspired by the works of Piet Mondrian, a famous Dutch abstract painter.

6.4.3 Underground Space in New York

The iconic Rockefeller Center in New York commenced construction in 1930 and is a complex of 19 commercial buildings covering 89,000m^2 and is declared a National Historic Landmark.

In 1999, the basement level below the Rockefeller Center used by commuters for shortcutting across the block was refurbished to create a "concourse" with retail shops. As shown in Fig. 6-44, shops and restaurants line the pedestrian concourse that links six key buildings, a public ice-skating rink, and concourse entrance is also accessible through a subway station below Sixth Avenue. Interestingly, the Rockefeller Center management explains on their website that "although the concourse has been replicated in hundreds of arcades around the world, it was never commercially successful for the owners until a thorough renovation brightened its spreading network of corridors and passageways".

The Second Avenue Subway Project was first mooted in 1929, but has been delayed by recession and war, along with the usual challenges associated with financing and constructing major projects. In 2005, New York State passed the *Transportation Bond Act* to facilitate the project. Ground officially broke

Fig. 6-42 Dostoyevskaya station
Source: https://www.moscovery.com/moscow-underground/.

Fig. 6-43 Rumyantsevo Station
Source: https://www.moscovery.com/moscow-underground/.

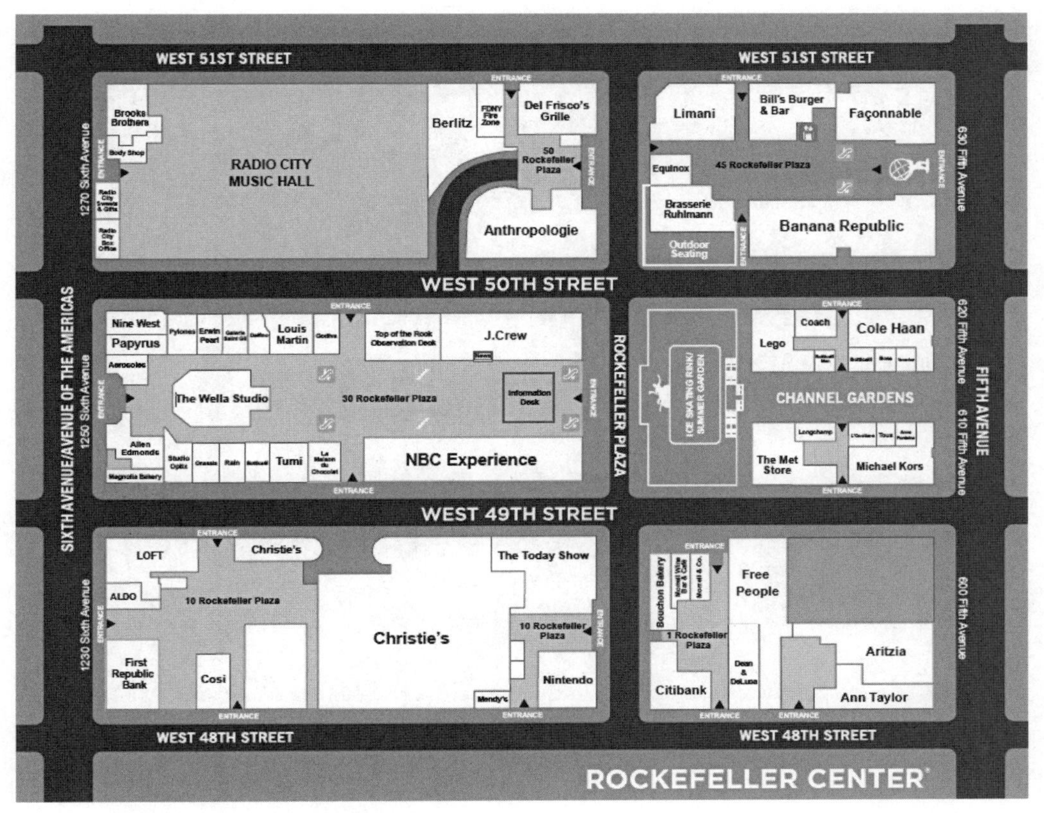

Fig. 6-44 Rockefeller Center Plaza (basement) [14]

in April 2007, followed several months later by a $1.3 billion federal funding commitment for phase one of the project. The Second Avenue Subway is a monumental construction project involving underground explosive excavation as well as use of a tunnel boring machine (Fig. 6-45).

The first phase opened for service on New Year's Day in 2017, decreasing crowding on the Lexington Avenue Line by up to 13%, and reduced travel time by 10 minutes or more for many passengers traveling from the Upper East Side.

However, in addition to the transit connectivity and capacity benefits, this underground infrastructure will provide a stimulus for transit oriented development, and potentially create a modal shift from private cars to public transport.

The Lowline project proposes to use solar technology to illuminate an historic trolley terminal on the Lower East Side of New York City, in order to create an underground park (Fig. 6-46). The site was opened in 1908 for trolley car passengers, but has been unused since 1948 when the service was discontinued. Despite such a long period of neglect, the space is said to still retain beautiful original features such as cobblestones and vaulted ceilings. It is also adjacent to an existing subway station to encourage people from all across the city to visit and enjoy the park, even in winter. Although funding and planning approval are yet to be finalized, the project appears to have strong public backing in principle. If it is opened as scheduled, the Lowline could act as a game changer in people's perceptions of underground public spaces.

Fig. 6-45 Construction works on Second Avenue Subway [14]

Fig. 6-46 Lowline underground park [14]

6.4.4 Historical Underground Space in Iran

Kariz underground city (Fig. 6-47) is the remainder of Kish main qanat that was made in the past. This qanat is over 2500 years old. Also this qanat was used to provide fresh potable water for the island inhabitants.

Today this qanat complex with change of use has transformed to an amazing underground city that has more than 10,000m^2 and now it's called Kariz of Kish.

By developing the construction of Kariz, viewers can easily go from ground surface to qanats and spend pleasant moments by passing in halls and corridors. Temperature in Kariz underground city is 10 to 12 degrees colder than the outside weather, leading tourists and inhabitants to stay safely from the heat in warm seasons of a year.

Ouyi underground city (Fig. 6-48) has been architected in three levels and dug to 20m deep. Entrance of this underground city is a narrow hall, in the size of one person. When you enter the city, the smell of moist soil is felt. This underground town was created in three levels in depth of 4m to 21m from the ground surface. Area of this town was mostly for connection between neighborhoods and to protect life and property of people in unsafe situations and has been spread both

(a) (b) (c)

Fig. 6-47 Kariz underground city[29]

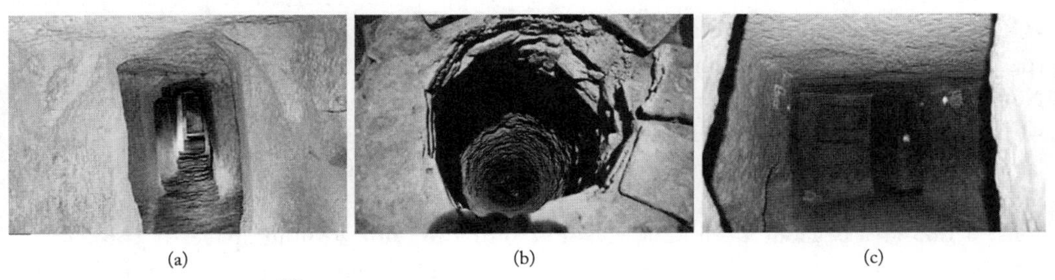
(a) (b) (c)

Fig. 6-48 Ouyi underground city[29]

(a) (b)

Fig. 6-49 Samen underground city[29]

horizontally and vertically.

Along the way of underground city, rooms have been dug for temporary refuge. The height of each room is 180cm. People in Ouyi have done many measures for resting spaces and being safe from enemies so that even in the depth of 18m, enemies couldn't attack them.

The area of Samen underground city (Fig. 6-49) is estimated over 3 acres, and to date almost 25 rooms have been discovered. This underground town was discovered in depth of 3m to 6m of ground level. Canals have been dug inside which had been detached during historical periods. Around the canals, multiple rooms can be seen as well. According to archaeological studies, the town consists of nested canals. This place during historical periods had expanded, and rooms and hallways of this town have different ages.

6.4.5 Underground Space in Jiangbei New Area, Nanjing

On June 27, 2015, the State Council officially approved the establishment of the Nanjing Jiangbei New Area, which is located in Nanjing, Jiangsu Province, China.

Overview of the economic development of Nanjing, Jiangbei region seriously behind the Jiangnan situation has become an obstacle to further development, the construction of state-level "Jiangbei New Area" grand plan thus came into being.

The first phase project of underground space in the central area of Nanjing Jiangbei New Area covers a total area of 51 hectares, with a total of 24 plots in which 21 plots of land are planned for commercial offices and 3 plots for squares and green spaces. The total construction area is about 1.4 million square meters, with functions including commerce, underground loop, municipal pipe gallery, rail transit, parking, equipment room, etc., as shown in Fig. 6-50.

Recently, the "Integrated Development and Construction Model of Underground Space in the CBD of Jiangbei New Area" was selected as the model case by the Provincial Natural Resources Department for land and natural resource savings. The CBD underground space, known as "City Reflection", has a total scale of 4.5~4.8 million square meters, incorporating elements such as urban green space, parks and businesses (Fig. 6-51). The project officially started construction on September 20, 2017 and is scheduled to be completed in 2023. It is the largest urban underground space project in China at this stage.

A total length of about 7.1km underground loop is designed in the upper section of the underground space. It connects to an underground parking which fits 13,200 vehicles. The underground space adopts

(a) (b) (c) (d)

Fig. 6-50 Planning of underground space of Nanjing Jiangbei New Area

(a)

(b)

Fig. 6-51 The CBD underground space of Nanjing Jiangbei New Area

the "three level traffic system" which are underground roads, underground connecting roads and the underground parking space. The vehicle evacuation distance is no more than one kilometer, and the distances for all safety exits to the ground level are also no more than one kilometer. This has ensured the maximum number of people and vehicles to be evacuated in time if there were ever any emergencies.

The underground pipe gallery project is now underway in Nanjing Jiangbei New Area. To achieve the highest quality, authorities of Nanjing Jiangbei New Area have been cooperating with Southeast University, Nanjing Tech University and Hohai University. The underground pipe gallery is 53.4 kilometers long. This project is predicted to be completed in February, 2021. Considering the soil quality and terrain of Nanjing Jiangbei New Area, researches have been inspecting the site to determine the best construction plan. Southeast University worked in the research of the assembly pipe material. Nanjing Tech University did the research regarding the project's influence to the environment of the surrounding area. Hohai University did the research on the soil-cement of the construction. The project also provides field data for the research teams of the universities.

6.5 Conclusions

(1) In parallel with rapid urbanization, UUS is becoming more actively utilized, particularly in dense aggregations such as central business districts (CBDs) or downtowns, due to the increasing demand for urban space while ensuring ecological modernization.

(2) Underground space structures can be classified in several ways. Underground space structures are always classified in terms of the concept of space of structures. Thus, underground space structures can be classified according to structural shape, geological condition, construction method, relationship with ground buildings and buried depth.

(3) Several construction methods of underground space engineering can be selected. Each construction method has its characteristics. According to geological conditions, environment and so on, specific construction methods are used for specific projects.

Exercises

6-1 Describe the functions of UUS.

6-2 Describe the characteristics and scope of application of the main construction methods of UUS.

6-3 Cut and cover method can be of several types. Describe the advantages and disadvantages of these types.

6-4 What is the general procedure of conventional tunneling method?

6-5 Describe the characteristics of various kinds of mechanical tunneling methods.

6-6 Describe the construction steps of an immersed tunnel.

References

[1] QIAO Y K, PENG F L, SABRI S, et al. Low carbon effects of urban underground space [J]. Sustainable Cities and Society, 2019, 45: 451-459.

[2] LI W, TAN Y. Collection and management of urban underground space information [J]. Journal of Geomatics, 2018, 43(3): 70-72.

[3] FAN Y Q. How to construct the structural system of urban underground space engineering construction standard [J]. Procedia Engineering, 2016, 165: 19-28.

[4] LABBÉ M. Architecture of underground spaces: From isolated innovations

to connected urbanism [J]. Tunnelling and Underground Space Technology, 2016, 55: 153-175.

[5] FROLOV Y S, KONKOV A N, LARIONOV A A. Scientific substantiation of constructive-technological parameters of St. Petersburg subway underground structures [J]. Procedia Engineering, 2017, 189: 673-680.

[6] MIRALIMOV M, ADILOV F, ABIROV R, et al. To numerical approach for calculation of underground structures [J]. IOP Conference Series: Materials Science and Engineering, 2020, 883: 012204.

[7] CHEN Z, GAO X, ZHAO X, et al. Survey on the application of prefabricated structure in underground engineering [J]. IOP Conference Series: Materials Science and Engineering, 2020, 741: 012075.

[8] LEBEDEV M O, ROMANEVICH K V. Engineering and geophysical research in reconstruction of underground structures [J]. Mining Informational and Analytical Bulletin, 2019, 5: 97-110.

[9] VON DER TANN L, STERLING R, ZHOU Y X, et al. Systems approaches to urban underground space planning and management–A review [J]. Underground Space, 2020, 5(2): 144-166.

[10] ZHAO J. Analysis on design of underground space in Tianjin: An example of Langxiang underground street [J]. IOP Conference Series: Earth and Environmental Science, 2019, 233: 022030.

[11] GOEL R K. Use of underground space for the development of cities in India [C]. Tunnelling Asia 2015 on Underground Space for Development for Better Environment and Safety: Issues and Challenges, 15 & 16 April 2015, TAI & CBIP, New Delhi.

[12] HUNT D V L, MAKANA L O, JEFFERSON I, et al. Liveable cities and urban underground space [J]. Tunnelling and Underground Space Technology, 2016, 55: 8-20.

[13] BOBYLEV N. Mainstreaming sustainable development into a city's master plan: A case of urban underground space use [J]. Land Use Policy, 2009, 26: 1128-1137.

[14] REYNOLDS E, REYNOLDS P. Planning for underground spaces "NY-Lon Underground" [M]. //ADMIRAAL H, SURI S N. Think Deep: Planning, Development and Use of Underground Space in Cities. Naples: International Society of City and Regional Planners and International Tunnelling and Underground Space Association, 2015: 6-33.

[15] BOWNES D, GREEN O, MULLINS S. Underground: How the tube shaped London [M]. London: Allen Lane, 2012.

[16] WALLACE M I, NG K C. Development and application of underground space use in Hong Kong [J]. Tunnelling and Underground Space Technology, 2016, 55: 257-279.

[17] YUN B. Underground Engineering: Planning, Design, Construction and Operation of the Underground Space[M]. London: Academic Press, 2019.

[18] US Department of Transportation Federal Highway Administration. Technical Manual for Design and Construction of Road Tunnels—Civil Elements [M]. Publication No. FHWA-NHI-10-034. 2009.

[19] LIN W, LIN M, YIN H Q, et al. Design of immersed tunnel and how we research submerged floating tunnel [J]. Chapters, 2020.

[20] Pipe Jacking Association. An Introduction to Pipe Jacking and Microtunnelling [M]. Telford: Pipe Jacking Association, 2017.

[21] BICKEL J O, KUESEL T R, KING E H. Tunnel engineering handbook (2nd Edition.) [M]. New York: Chapman & Hall, 1996.

[22] VÄHÄAHO I. Development for urban underground space in Helsinki [J]. Energy Procedia, 2016, 96: 824-832.

[23] STERLING R, ADMIRAAL H, BOBYLEV N, et al. 0-Land use through sustainable use of underground space [C].

Proceedings of Euregional Conference, Sustainable Building, Towards 0-impact Building and Environment, SB 10 Western Europe, Maastricht, Hesden-Zolder, Aachen, Liége, 2010: 112-113.

[24] IL'ICHEV V A, MANGUSHEV R A, NIKIFOROVA N S. Development of underground space in large Russian cities [J]. Soil Mechanics and Foundation Engineering, 2012, 49(2): 63-67.

[25] Guidelines for Complex Development of the Underground Space of Large Cities [M]. Moscow: RAASN. 2004.

[26] IL'ICHEV V A, MANGUSHEV R A. Substructure installation at the State Academic Marinsky Theater building in Saint Petersburg [J]. Soil Mechanics and Foundation Engineering, 2010, 47(4):113-120.

[27] MANGUSHEV R A. Examples of modern constructive and procedural methods used in developing the underground space in Saint Petersburg [C]. Collection of Papers Presented at the Scientific-Technical Conference "Urgent Geotechnical Questions in the Solution of Complex Problems of New Construction and Reconstruction", Saint Petersburg. 2010: 24-32.

[28] COLLINS G. Bringing up the basement; Rockefeller Center is turning its underground concourse into a shiny new shopping zone. Lost in the Bargain, preservationists say, is an Art Deco treasure [N]. New York Times, 21 February 1999.

[29] MONTAZEROLHODJAH M, POURJAFAR M, TAGHVAEE A. Urban underground development: An overview of historical underground cities in Iran [J]. International Journal of Architectural Engineering & Urban Planning, 2015, 25(1): 53-60.

Chapter 7
Municipal Engineering

7.1 Overview

Urban development has followed economic reform in China and a large number of central cities rise and develop, making cities in an increasingly important position. Municipal administration is closely related to modern cities, and the development of cities has driven the formation and improvement of municipal administration. People living in cities require a better living environment to be satisfied from urban life. For all aspects of cities to function normally, a special management system and management mode is needed. The continuous improvement of urban modernization also requires the synchronous development of municipal facilities.

Modern municipal engineering finds its origins in the 19th century in the United Kingdom (UK), following the Industrial Revolution and the growth of large industrial cities. The threat to urban populations from epidemics of waterborne diseases such as cholera and typhus led to the development of a profession devoted to "sanitary science" that later became "municipal engineering".

Municipal engineering is concerned with municipal infrastructure. This involves specifying, designing, constructing and maintaining streets, sidewalks, water supply networks, sewers, street lighting, municipal solid waste management and disposal, storage depots for various bulk materials used for maintenance and public works, public parks and bicycle paths. In the case of underground utility networks, it may also include the civil portion of the local distribution networks of electrical and telecommunication services. It can also include the optimizing of garbage collection and bus service networks. Some of these disciplines overlap with other civil engineering specialties, however, municipal engineering focuses on the coordination of these infrastructure networks and services, as they are often built simultaneously and managed by the same municipal authority.

Today, municipal engineering may be confused with urban design or urban planning. Whereas the urbanist or urban planner may design the general layout of streets and public places, the municipal engineer is concerned with the detailed design. For example, in the case of a new street, the urbanist may specify the general layout of the street, including landscaping, surface finishing and urban accessories, but the municipal engineering will prepare the detailed plans and specifications for the roads, sidewalks, municipal services and street lighting. As practiced a century ago, municipal engineering, however fully embraced the function of urban design and urban planning, even though the terms had yet to be coined.

A major characteristic of Chinese urban construction is that ground utilities/housing are developed rapidly in many cities due to the introduction of modern marketing methods, whereas urban infrastructure develops relatively slowly. This causes problems such as delayed marketization, long construction periods and high construction costs for urban infrastructure. As a result, urban infrastructure construction becomes costly and more difficult. When building a new urban infrastructure, specialist technologies, in terms of constructing in limited spaces, or reducing impact on the surrounding environment, are required. Congested commercial and residential buildings surrounding the construction site have seriously complicated the excavation work.

In addition, during construction, special attention should be paid to existing water and sanitation pipes built underground. In Beijing, the construction of subway Line 10 suffered the Beichen Road collapse accident in 2005, due to leakage from an existing water pipeline underneath the construction site. Therefore, comprehensive investigation and advanced municipal engineering technologies are required to avoid a repeat of these tragedies.

As discussed above, due to construction without sufficient consideration of sustainable development in the early phases of urbanization, new urban infrastructure construction is becoming extremely difficult. Existing utilities and buildings not only increase construction costs but also affect plans for new utilities' construction.

On the other hand, urban infrastructure must adapt to environmental change. Global warming has raised the issue of city flooding, which is becoming a big problem for most Chinese cities due to inadequate storm drainage utilities. Development of new stormwater drainage systems is therefore urgently required. Despite heavy rains in some regions, lack of rainfall in other regions brought a remarkable water shortage problem, which is still problematic for many inland cities and one of their most serious problems. In recent years, several water transfer projects have been undertaken, for instance the South-North Water Transfer Project. In addition to the current solutions, alternative methods should be considered; for instance, stormwater can be reused by constructing stormwater storage infrastructure. Other problems such as environmental protection and energy saving should be taken into consideration in urban development.

Although the Chinese economy continues to grow at about 8% of gross domestic product (GDP), investments in urban construction are unable to meet demands in many cities. The funding shortage for local urban development is still a serious problem. All these problems bring new challenges to governments and municipal engineers.

This chapter is composed of five parts: urban public transportation, urban water supply and drainage engineering, urban gas and heating systems, urban flood control projects, and urban garbage disposal. As students majoring in water supply and drainage engineering, civil engineering, municipal engineering, and roads and bridges, which are closely related to municipal engineering, it is very necessary to learn and understand the basic knowledge of municipal engineering.

7.2 Urban Public Transportion

7.2.1 Urban Road and Bridge System

1. Urban Road System

(1) The Current Status of the Road System

Urban traffic refers to the flow between people and things in urban functional land, and these flows are inseparable from the use of urban land, which is the urban road system. As a hub, the urban road system plays a vital role in the urban environment extending in all directions, and is the most critical part of the urban system. It can not only provide a good support for the urban form and structure, but also effectively divide the use of land. The development of urban road system directly affects the process of urban development. A stable and unblocked urban transportation network can provide convenience to people's daily life and make the city develop more significantly.

The design and improvement of urban road is a never-ending work. With the continuous progress and development of the city, urban road must keep pace with the overall process speed. At the beginning of the road design, the most important thing is to grasp the regional urban environment construction planning. At the present stage, our country's economic strength continues to improve. The scale of urban development has continued to increase. People's living standards are gradually improving. The number of motor vehicles is increasing. Therefore, road capacity overload will occur. The most important problem is congestion and chaos. The transportation department needs to choose reasonable solutions to solve the traffic problems in a timely manner based on the actual local conditions. From this point of view, the urban road system and urban development are closely related.

Compared with the developed countries, there is still a certain gap in the level of urban road planning and design in China. In the case of imperfect urban road planning and design, it is difficult to ensure the smooth progress of urban road planning in order to develop urban road engineering well. Our country should increase the exploration and innovation of urban road planning and design scheme, strengthen the research on time-saving, labor-saving, safe and effective urban road planning and design technology, and encourage universities and enterprises to research and innovate urban road planning and design technology.

(2) Classification Structures of the Road System

In 2012, *Specification for Design of Urban Road Engineering* was released in Shanghai. It adopted the two-dimensional road typology principle of "hierarchy in mobility capacity with function in context character". Urban roads are still divided into four types in terms of "hierarchy", including expressway, trunk road, secondary road and branch road in accordance with *Code for Transport Planning and Urban Road* (Herein after referred to as the *Code TPUR*). And they shall meet the following requirements:

Expressways should be centrally separated, fully controlled access, and the spacing and form of entrances and exits should be controlled. Continuous traffic should be realized. One-way traffic should not be less than two lanes, and supporting traffic safety and management facilities should be provided. On both sides of the expressway, there should be no entrance or exit of public buildings that attract a large number of traffic and people.

The trunk road should be connected to each major district of the city and should be mainly used for traffic function. It is not suitable to set up public buildings on both sides of the main road to attract a large number of traffic and people.

The secondary road should be combined with the main road to form a trunk road network, with the main function of distributing traffic, as well as the service function.

The branch road should be connected with the secondary road, residential area, industrial area, traffic facilities and other internal roads. It should solve the local traffic and give priority to the service function.

The design speed of roads at all levels should meet the requirements of Table 7-1.

In terms of "function", which is defined by the dominate land uses, roads can be summarized as eight categories such as "central business district, urban area, rural area, independent development zone, scenic area, cultural heritage area, commercial area and transit corridor" . This two-dimensional road classification not only keeps consistent with the *Code TPUR*, but also refines road function types and the arrangement of design elements in typical cross sections. It is conducive to ensuring the multiple roles of roads with responding to both local conditions and travel demands in road planning and design.

The structural form of urban road network refers to the plane projection geometric figure of urban road network, which is formed

Design speed of roads at all levels **Table 7-1**

Highway grade	Expressway			Trunk road			Secondary road			Branch road		
Design speed(km/h)	100	80	60	60	50	40	50	40	30	40	30	20

according to the needs of urban development to meet the requirements of urban scale, form, land use layout, urban traffic and other requirements. The common structural forms of urban road network at home and abroad can be abstractly summarized into three basic types: grid type, radial ring type and free type.

Grid network (also known as checkerboard road network) is suitable for small and medium cities in flat areas. It is divided into neat streets to facilitate the layout of buildings along the street. The road network is decentralized and flexible. The disadvantages lying in the road function include: It is not clear; the number of intersections is large; and the diagonal direction of the traffic is inconvenient. The old urban areas of many large cities in China are of this structure.

Radiation ring road network is derived from the city center around a number of radiation trunk roads, and between each radiation trunk road there are a number of ring roads. The advantage of this road network is that it is beneficial to the traffic connection among the city center, suburbs and the neighboring districts outside the city, and it has its advantages in defining the road function. The disadvantage is that it is easy to lead all directions of traffic to the city center, resulting in over-concentrated downtown, traffic flexibility is not as good as grid network. Therefore, this structure is suitable for large cities and megalopolis.

The free road network is generally formed due to the irregular shape of urban topographic relief and road combined with topographic change. Its main advantages are not sticking to one pattern, fully combining with the natural terrain, less damage to the environment and landscape, and saving project cost. Disadvantages lie in the larger detour distance, more irregular neighborhood, and more scattered building land. Such networks are common in mountainous and hilly cities such as Chongqing and Qingdao.

The above three basic forms are often combined together to form a hybrid structure. This structure is often gradually formed according to the actual needs of urban development, adapting measures to local conditions, maximizing strengths and avoiding weaknesses, and rationally organizing and distributing traffic.

(3) Features of the Road System

The urban road network consists of various levels of urban roads (excluding roads in residential areas). It basically determines the outline of urban land layout and land use, and its impact on urban construction and development will continue.

Compared with other roads, urban roads have the following characteristics:

1) The urban road network (mainly refers to the arterial road network) constitutes the basic skeleton of urban land, and the use and development of various blocks in urban planning will inevitably be affected by it. In the overall urban planning stage, it is necessary to make corresponding considerations and arrangements for the urban trunk road network to meet the needs of urban land use, and the further improvement of the trunk road network will promote the development and construction of urban land.

2) The function of urban road network is diverse, and the road composition is complex. Generally, only motor vehicle traffic is considered in highway network, but the traffic composition of urban road network is more complex. All kinds of motor vehicles, non motor vehicles and pedestrians use the road network space to realize their travel.

3) Landscape art requires high quality. Whether the general layout of the city is beautiful and reasonable, to a large extent, is reflected in the planning and layout of the road network, especially the trunk road network. The urban road network, the buildings on both sides of the road and the landscape constitute a beautiful picture of the city.

2. Urban Bridge System

(1) The Development of Chinese Bridge System

Bridges are artificial structures that cross the barriers of rivers, lakes, seas, deep gullies and canyons. They are key nodes and hinge projects for the connectivity of transportation facilities, and important guarantees for the development of national economy and the safety of social life. The development of the bridge industry can effectively resolve the overcapacity of traditional industries such as steel, cement and energy, and vigorously promote strategic emerging industries such as new materials, intelligent manufacturing, high-end equipment, new generation information technology, energy conservation and environmental protection, and tertiary industries such as logistics and transportation. The vigorous development of China has played an important role in promoting industrial integration and upgrading, and stimulating economic growth.

The four decades since China's Reform and Opening-up were a golden period for the development of bridge construction in China. Following the general laws of technological development and the path of integration–development–innovation, Chinese bridge engineering has undergone three stages: learning and following in the 1980s, tracking and improving in the 1990s, and innovating and transcending since the start of the 21st century. The development of bridge engineering in China has now taken a substantial leap forward, with the construction of many extra-large bridges adopting novel structures, difficult designs and construction, and complicated high-tech materials and procedures; examples include the Su Tong Yangtze River Highway Bridge, Tianxingzhou Bridge and Lupu Bridge. Furthermore, China actively participates in and hosts international competitions, and has played a role in the construction of many well-known international bridges, including the Malaysia Penang Second Bridge, the Panama Canal Third Bridge and the New Oakland Bay Bridge. These projects have won China 34 outstanding international awards such as the International Federation of Consulting Engineers (FIDIC) excellence award for "Major Civil Engineering Project of the Last 100 Years", the "Outstanding Civil Engineering Achievement Award" issued by the American Society of Civil Engineers (ASCE), and the "Outstanding Structural Engineering Award" issued by the International Association for Bridge and Structural Engineering (IABSE). These awards mark the development of the Chinese bridge industry and represent the respect and recognition of the international bridge industry. Chinese bridge engineering has gradually moved into the center of the world stage.

However, compared with developed countries, there are still problems and deficiencies in innovative design concepts, durability, high-performance materials, software, equipment, standards and specifications, professional talent teams and other aspects. Some basic theoretical research and common key technologies still need to be broken through. The degree of construction refinement is not high. The durability problem is prominent. The level of industrialization, information technology and intelligence is not high and the degree of industrialization is low. Those restrict the development of Chinese bridges towards sustainable bridges featuring safety, longevity, green, efficiency and intelligence. In the next decade, China's bridge industry will enter an important period of strategic opportunity driven by innovation, transformation and upgrading, which will promote the rapid and healthy development of Chinese bridge technology, and we will strive to realize the dream of a bridge power as soon as possible.

(2) The Basic Composition of the Bridge

The bridge has three main components: the superstructure, the substructure and the accessory.

The superstructure is composed of bridge span structure and bearing system. The bridge span structure is the part of the load-bearing structure that crosses the bridge hole and above the support. According to different forces, it can be divided into beam type, arch type, steel frame, suspension cable and other basic systems, and these basic systems constitute a variety of combined systems. It includes the main load-bearing structure, arch building, bridge deck building, bridge deck pavement, waterproof drainage system and safety protection measures, etc. The bearing system is a force transmission and connection device installed between the upper and lower structures of the bridge. Its function is to transfer various loads of the superstructure to the pier and adapt to the displacement caused by factors such as live load, temperature changes and concrete shrinkage. Generally it is divided into fixed bearing and movable bearing.

The substructure is composed of bridge piers, abutments and pier foundations. Piers and abutments are structures that support the upper spans of bridges on either side of a river or on a bank. The abutments are at both ends and the piers are between the abutments in order to protect the bridge abutment and embankment filling, some protection and backflow works are often done on both sides of the bridge abutment. Pier foundation is the structural part that ensures the safety of bridge pier and transfers the load to the foundation.

(3) Classifications of the Bridge System

Bridges are classified in various ways, each of which reflects a certain aspect of their characteristics. According to the use of bridges, there are railway bridge, highway bridge, pedestrian bridge, urban bridge and so on. According to the nature of bridges crossing obstacles, there are river bridge, line bridge, deep valley bridge, viaduct, trestle and so on. According to the structural system, there are beam bridge, arch bridge, steel frame bridge, suspension bridge, cable-stayed bridge and all kinds of composite bridges.

Beam bridges are of bending structures and the oldest and simplest type of practical bridges. In ancient China, limited by the physical properties and the practical length of stone or timber, quantity and span length of beam bridges can not be compared with those of arch bridges. After the wide application of steel and concrete when the Industry Revolution achievements were introduced to China, beam bridges were taken as the most common bridge type in China. Since the span length of modern beam bridges is mainly based on material development, a large number of structural branches are derived, and beam bridge structure system as the most ancient and simplest bridges are substantiated and enjoy great popularity.

Cable-stayed bridges, with a self-anchored structure, own many technological advantages, such as reasonable performance of load bearing, material-saving, high capacity in spanning, good adaptation, safety and convenience in construction, and so on. Cable-stayed bridges are constructed along a structural system which is comprised an orthotropic deck and continuous girders which are supported by stays, i.e. inclined cables passing over or attached to towers located at the main piers. Many exiting cable-stayed bridges are masterpieces of steel construction. They are pleasing in outline, clean in their anatomical conception and totally free of meaningless ornamentation.

The suspension bridge is currently the only solution for spans in excess of 1000m. The principal structural elements of a suspension bridge are: flexible main cables, towers, anchorages, hangers, deck and stiffening girder. Suspension bridges, when well designed and proportioned, are clearly the most aesthetically pleasing of all bridges. The simplicity of the structural arrangement, with the function of each part being clearly expressed, combines with the graceful curve of the main cables, the slender suspended deck and vertical towers, to be studied and designed. It clearly indicates that the development of suspension bridges will have a splendid future in China and will face greater challenges as well.

7.2.2 Urban Railway System

1. The Current Status of Urban Railway System

In 2013, China proposed a momentous strategic decision for the Silk Road Economic Belt (SREB) and Maritime Silk Road (MSR) in the twenty-first century known as the Belt and Road Initiative (BRI). Driven by the BRI, development of urban rail transit in China is likely to steer toward a favorable "Going Global" situation.

Since the beginning of the 13th Five-Year Plan period, BRI has become China's paramount plan for the globalization and internationalization of its economic development, and it hopes to push economic globalization toward a more inclusive and balanced development. This trend is conducive to the internationalization of the China urban rail transit, as it has a more comprehensive industrial chain and has formed industrial clusters. In the future, driven by the BRI, the China urban rail transit will take great strides toward the global market and become a major highlight of the 13th Five-Year Plan.

Over 2016~2020, with the development of urbanization, approximately 100 million people will migrate to the cities and nearly 100 million new vehicles will be added on the roads. Urban development also requires a substantial increase in the proportion of travels by public transport, while 40 cities have ongoing urban rail transit projects and 80 cities are planning to construct urban rail transit systems. The joint effect of the five forces (rise in urban population, non-public vehicle ownership increasing, green modes of urbain transport, ongoing urban rail transit projects, and future urban rail transit plans) could facilitate a larger-scale development of urban rail transit. The relaxation of policies related to urban household registration, development of urban agglomerations and metropolitan areas, and construction of state-level new areas will generate an immense demand for the development of coordinated, multimodal urban rail transit systems. Internet Plus technology revolution will lead urban rail transit into a new era of intelligent and Fully Automated Operating (FAO) development. The BRI will lead the development of urban rail transit industry in China toward the global market. However, it is only by cultivating qualified talent, expanding funding channels, and implementing preliminary preparation that it is more likely to control the development progress of the urban rail transit industry in the 13th Five-Year Plan period.

More than 100 cities have formulated 13th Five-Year Plans for 2016~2020 urban rail transit construction, which will generate a new phenomenon of simultaneous construction of urban rail transit in 100 cities. New lines are simultaneously rising sharply. In 2016, the length of the newly added rail lines exceeded 500km for the first time (525km). A hundred cities working on and annually opening 1000km urban rail transit will be soon realized with such an explosive growth. Based on this speed of development, by the end of 2020, more than 60 cities would have established urban rail transit systems, with the total length of operating rail lines reaching 8000km. China would be one of the few countries in the world with such a large-scale urban rail transit network in place and covering so many cities as well as with such a high growth in the operating lines.

2. Types and Features of Urban Railway System

Urban rail transit mainly consists of subway, light rail, tram, intercity rail, maglev train and new transportation system.

The full name of the subway is a large-capacity rapid rail transit system that is operated in a fully enclosed underground tunnel by electric traction, wheel-rail guidance, and vehicle formation, or runs on the ground or on elevated lines according to the specific conditions of the city. Light rail is a type of electric traction rail transit system used for urban passenger transportation between standard trams and express transit systems. Tram is a low-volume rail transit system that uses electric traction, wheel-rail guidance, single or two groups to run on urban road lines. Intercity rail is a high-volume rapid rail transit system that is driven by electric or diesel locomotives, wheel-rail-oriented, and vehicle formation runs between the city and satellite cities, and is based on ground dedicated lines. Maglev train is a rail transportation method that uses the electromagnetic principle of "Like poles repel each other, unlike poles attract" and rely on electromagnetic force to make the train levitate and travel. It is a new type of rail transit system without wheels and using contactless travel. There is no uniform and strict definition of the new transportation system. In a broad sense, it refers to the general term for various short-distance new transportation modes that are different from the existing transportation modes.

Compared with other public transportation, urban rail transit has the following characteristics: small footprint and large transportation volume. The transportation capacity of rail lines is about 10 times that of highway transportation. Energy consumption per unit of transportation is low, thus saving energy. Electric traction is adopted, which has little environmental pollution. The noise is concentrated, with low per capita noise and easy to manage. Passengers ride safely, comfortably, conveniently and quickly.

3. Opportunities and Challenges of Urban Railway System

The China Association of Metros (CAMET) has studied the "Going Global" of urban rail transit and believes that, compared to those of other countries, the urban rail transit systems in China have strengths and weaknesses. The unique strengths are as follows: (1) The implementation of the BRI has guaranteed the intensified support of government policies for the development of urban rail transit. (2) The comprehensive industrial chain, a number of leading enterprises, and the echelons of their supporting businesses have formed a pattern of development characterized by industrial clusters. (3) The immense domestic demand for development has become a powerful impetus for the internalization of urban rail transit. (4) Urban rail transit is more cost-effective in China; in particular, the domestic market remains overall cost-effective. The relative weaknesses are as follows: (1) the overall underdeveloped strength of enterprises, with inherent deficiencies in their core competitiveness and innovation levels; (2) lack of certain core technologies; (3) lack of a sophisticated standardization and incomplete standard system; and (4) weak brand-building ability and insufficient brand promotion.

Urban rail transit construction in the 13th Five-Year Plan period is unprecedented in terms of the number of cities covered, the scale of projects, the level of standards, the modes of rail transit included, the amount of investment involved, and the likelihood of China's urban rail transit entering the global market. Given this large-scale and rapid development, issues caused by lack of talent resource, limited funding and inadequate preliminary preparation will become more prominent. These three factors may affect the development of the urban rail transit industry in this new era.

The differentiation in the development of urban rail transit between cities become increasingly apparent over 2016~2020. Different cities are at varied stages of development, thus requiring different modes of urban rail transit, and with different priorities. Therefore, it is necessary to create respective differentiated development goals and set priorities in accordance with the different development stages of each city, in order to promote the systematic construction of urban rail transit based on local conditions. During the 13th Five-Year Plan period, China's urban rail transit system was superior to that in other countries in terms of the number of networked cities, the scale of networks, and the speed of establishing networks. Hence, it is important to develop a systematic and network-oriented thinking, and to formulate network-based guidelines that provide standardized guidance for operation management and resource sharing. The development of the urban rail transit industry in 2016~2020 placed higher requirement for industrial standardization. Hence, developing and expanding collective standards, establishing and promoting standard systems, and vigorously improving project quality are top priorities in following the trend of standardization reforms initiated by the government. In 2016~2020, the exploration and application of a series of information technologies, such as Internet Plus, has led the urban rail transit industry into a new era of automated and intelligent development. The application of such technologies provides users with higher quality, safer and more convenient services and offers technical support for more efficient and safer planning, construction and operations. With increasing market demand, strong government support and rise of independent enterprises, the development of the urban rail transit industry will shift from "localized" to "self-developed" technologies, thus achieving breakthroughs in independent innovation. Driven by the BRI, China's urban rail industry will further extend its participation in the global market and enhance its competitiveness. The urban rail transit together with high-speed rail systems will jointly change the global market scenario.

7.3 Urban Water Supply and Drainage Engineering

7.3.1 Water Supply Engineering

1. Classification of Water Supply System

(1) Classification by Water Source

Water supply can be divided into surface water (rivers, lakes, reservoirs, oceans, etc.), groundwater (shallow groundwater, deep groundwater, springs, etc.) and urban water supply network system according to the type of water source.

(2) Classification According to Water Supply Mode

Water supply can be divided into artesian system (gravity water supply), pump water supply system (pressure water supply) and mixed water supply system.

(3) Classification According to the Purpose of Use

Water supply can be divided into domestic water supply, production water supply and fire water supply system according to the use purpose.

(4) Classification by Service Objects

Water supply can be divided into urban water supply system and industrial water supply system.

The vast majority of cities use a unified water supply system, namely the same system to supply all kinds of water for living, production and fire fighting. But the water quality and water pressure requirement of industrial water have its particularity. In the industrial water quality, water pressure requirements and domestic water under different circumstances, sometimes according to specific conditions, in addition to considering the unified water supply system, extra considerations include the use of water supply systems such as quality, pressure.

The layout of water supply system is mainly affected by urban planning, water source, topography and other factors. The water supply system shall be arranged in a variety of ways according to the specific situation of the city planning, water source conditions, topography, and users' requirements for water quantity, water quality and water pressure.

2. Components of Water Supply Engineering

The main sources of water supply treatment are surface water and groundwater. The conventional surface water treatment process is mainly to remove turbidity in water and pollutants such as bacteria and viruses. Water treatment system is mainly composed of clarification and disinfection process.

(1) Conventional Water Treatment Process (First-generation Urban Drinking Water Purification Process)

The security of urban drinking water is one of the basic conditions of urban development. Before the 19th century, the safety of urban drinking water could not be guaranteed, resulting in the spread of water-borne infectious diseases (such as cholera, typhoid fever and dysentery), which caused great harm to people's lives and health. This was the first major drinking water safety problem faced by mankind. Under this social background, water treatment technology developed in the early 20th century for the purpose of removing suspended substances and killing pathogenic bacteria. The first generation of urban drinking water purification technology is shown in Fig. 7-1.

Coagulation and flocculation processes are defined as follows: "Coagulation" means a reduction in the forces which tend to keep suspended particles apart. The joining together of small particles into larger, settleable and filterable particles is "flocculation". Thus, coagulation precedes flocculation and the two processes must be considered conjunctively. Selection of the type and dose of coagulant depends on the characteristics of the coagulant, the concentration and type of particles, the concentration and characteristics of natural organic matter (NOM), water temperature and water quality. Presently, the interdependence of these five parameters is only understood qualitatively, and prediction of the optimum coagulant combination from characteristics of the particles and the water quality is not yet possible.

Sedimentation follows flocculation. As shown, surface waters contain naturally suspended materials that can be observed in the water as cloudiness or turbidity. If turbid waters are placed into a large quiescent basin and left over time, the suspended material can settle to the bottom of the basin. Particles settle out of solution because they are large enough to settle out by gravitational forces. This process is called sedimentation. Most raw surface waters contain mineral and organic particles. Depending on their density, suspended particles larger than 1μm can be removed by sedimentation. Sedimentation may be employed at the beginning of a water plant to remove mineral particles from highly turbid waters, called pre-sedimentation, or after coagulation and flocculation processes, which is referred to as conventional sedimentation.

Filtration of water is defined as the separation of colloidal and larger particles from water by passage through a porous medium, usually sand, granular coal, or granular activated carbon. The suspended particles can be removed during filtration range in diameter from about 0.001μm to 50μm and larger. Several different types of medium arrangements and rates of flow through filters can be used. The filtration process most commonly used is gravity filtration, but

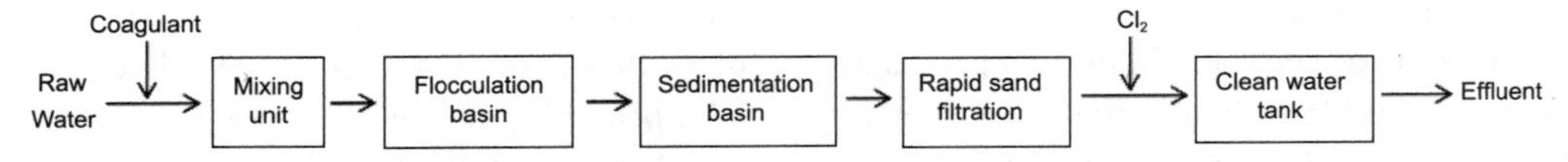

Fig. 7-1 Flowchart of the conventional water treatment process

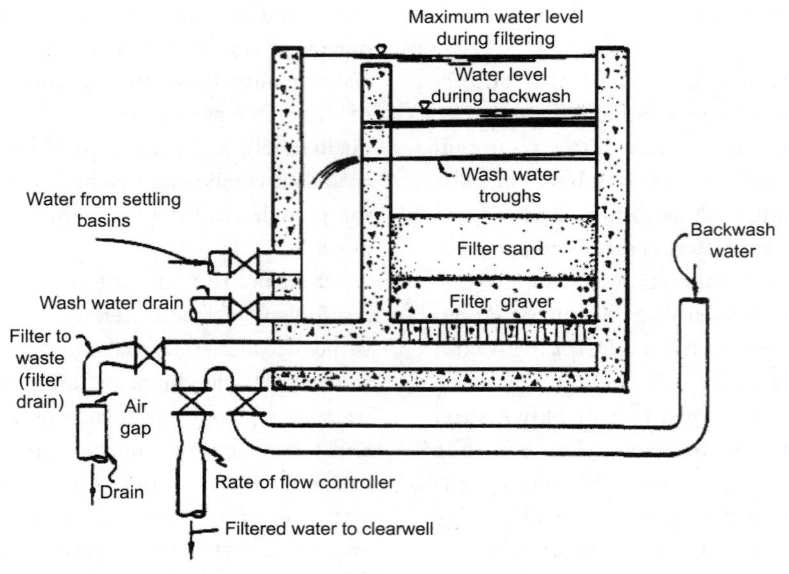

Fig. 7-2 Schematic of rapid sand filtration[10]

pressure filters and diatomite filters are used at smaller installations. Recently high-rate filters have been developed which require less space and have higher solid-loading capacity than conventional filters. Rapid sand filtration is the most commonly used in domestic waterworks (Fig. 7-2).

Disinfection involves destruction or inactivation of organisms which may be objectionable from the standpoint of either health or esthetics. Inasmuch as the health of water consumers is of principal concern to those responsible for supplying water, design of facilities for disinfection must necessarily be carefully executed.

(2) Deep Water Treatment Process (Second-generation Urban Drinking Water Purification Technology)

With the aggravation of water environmental pollution and the development of water quality analysis technology, in the 1970s, it was found that urban drinking water contained a variety of toxic and harmful organic pollutants and chlorination disinfection by-products, another major drinking water safety problem facing mankind chemical safety problem.The conventional treatment process mentioned above has a certain effect on the removal of organic matter in water, but has a poor effect on the removal of trace organic pollutants, with a removal rate of only a few percent in general. So new urban drinking water treatment processes need to be developed.

The advanced treatment process mainly includes activated carbon adsorption process, chemical oxidation process, and the combination of the two processes.

Activated carbon (AC) is formed by carbonization and activation under high temperatures; it has a massive surface area, a well-developed pore structure, and excellent adsorption ability. Its adsorption effectiveness is determined by the volume of well-developed pores and the pore-size distribution relative to the molecular weight (MW) of organic matter. Natural organic matter (NOM) adsorption primarily occurs in mesopores and large micropores. The MW fraction (<5kDa) of natural organic matter (NOM) is readily adsorbed by granular activated carbon (GAC). A GAC filter eventually turns into a biological activated carbon (BAC) filter after several

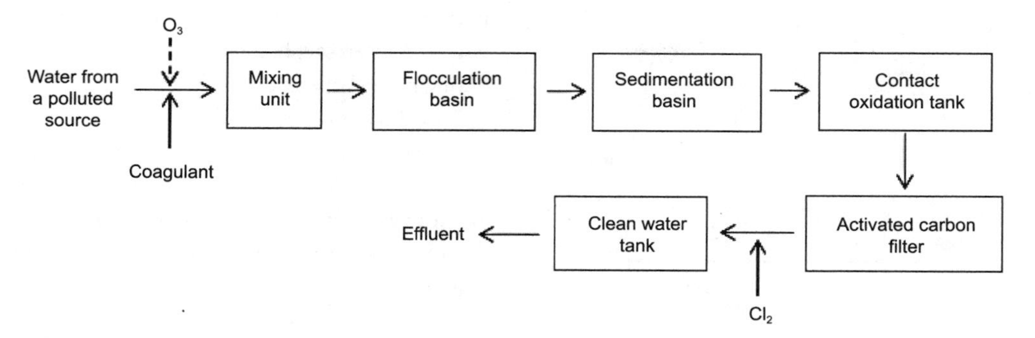

Fig. 7-3 Flowchart of the deep water treatment process

months' operation. Biological activated carbon can improve the removal rate of organic matter in water, and the regeneration period of activated carbon can be prolonged under lower pollutant load. Some ozone-biological activated carbon processes in the activated carbon regeneration cycle can be extended to more than two years. Ozone-activated carbon process is generally set after the conventional treatment process, as shown in Fig. 7-3.

3. The Layout Principle and Forms of Water Supply System

(1) The Layout Principles of Water Supply System

The main content of water supply pipeline network planning and layout is to determine the location of water supply pipes and the overall layout of the pipeline network. The layout principles of water supply system are as follows:

Technical and economic comparison of multiple schemes should be conducted according to the overall planning of the city, combined with the local actual situation.

Determine the water mains and main pipes first, and then determine other pipelines and facilities.

In order to save project investment, and operation and management costs, the length of pipeline should be shortened as far as possible.

It is necessary to coordinate the location relationship with other pipelines (drainage pipelines, rainwater pipelines and gas pipelines), cables and roads.

Ensure the reliability of water supply.

Avoid buildings as much as possible in the planning process to reduce demolition.

Consider the convenience of construction, operation and maintenance of the pipe and canal.

Take into account the recent and long-term development needs.

The comprehensive planning of water supply system includes the routing of water supply pipe and the layout of pipeline network. Delivery pipe is a pipe or canal connecting a water source to a water plant or a water plant to a distant water supply network. In order to ensure the stability of water conveyance, water conveyance pipes are generally not less than two and connection pipes with valves should be set between water conveyance pipes. The minimum slope of the water supply pipe shall be greater than $1:5D$ (D denotes the diameter). The main line should be extended in the same direction as the water coming out of the pumping station, and it is better to arrange several nearly parallel main pipes to make a circular network. The pressure pump station in the pipeline network should be located in the area with high water consumption.

(2) The Forms of Water Supply Systems

Distribution system may be generally classified as screen systems, branching systems, or a combination of these. The

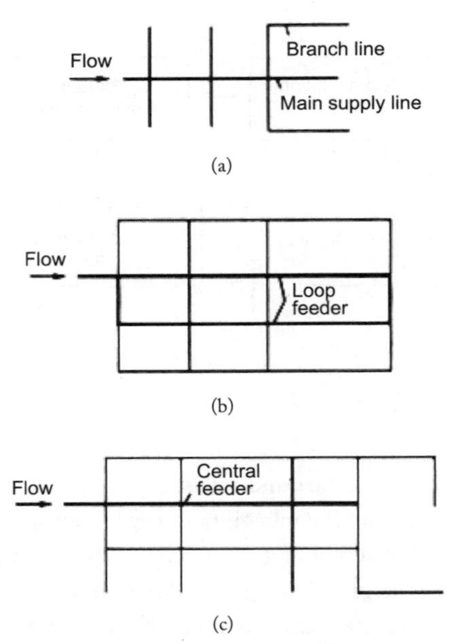

Fig. 7-4 Types of water supply systems[11]
(a) Branching; (b) Screen; (c) Combination

configuration of the system is dictated primarily by street patterns, topography, degree and type of development of the area, and location of treatment and storage works. Fig. 7-4 illustrates the nature of several basic types of systems. The branching system has numerous terminals or dead ends. When any section of the pipeline is damaged, all lines after that section will be cut off, so the water supply reliability of this type of pipeline network is poor. So, a screen system is more reliable and usually preferred to a branching system, since it can furnish a supply to any point from at least two directions. A screen or combination system can also incorporate loop feeders, which act to distribute the flow to an area from several directions.

In location where sharp changes in topography occur (hilly or mountainous regions), it is common practice to divide the distribution system into two or more service areas or zones. This precludes the difficulty of extremely high pressure in low-lying areas in order to maintain reasonable pressures at higher elevations. Usual practice is to interconnect the various systems, with the interconnections closed off by valves during normal operations.

(3) Hydraulic Analysis

The solution methods described for the analysis of series parallel and branched pipes are not very suitable for the more complex case of networks. Apply the continuity equation to a node,

$$\sum_{i=1}^{n} q_i = 0 \tag{7-1}$$

Where n is the number of pipes joined at the node. The sign convention used here sets flows into a junction as positive.

Apply the energy equation to a loop, then

$$\sum_{i=1}^{m} h_{fi} = 0 \tag{7-2}$$

Where m is the number of pipes in a loop. The sign convention sets flow and head loss as positive clockwise.

In addition,

$$h_{fi} = f(q_i) \tag{7-3}$$

Where $f(q_i)$ represents the Darcy-Weisbach/Colebrook-White equation. Equation (7-1)~Equation (7-3) comprise a set of simultaneous non-linear equations, and an iterative solution is generally adopted. Then a standard solution technique, the loop method, is now discussed.

This method, originally proposed by Hardy-Cross in 1936, essentially consists of eliminating the head losses from Equation (7-2) and Equation (7-3) to give a set of equation in discharge only. It may be applied to loops where the external discharges are known and the flows within the loop are required. The basis of the method is as follows:

1) Assume values to satisfy $\sum q_i = 0$;

2) Calculate h_{fi} from q_i;

3) If $\sum h_{fi} = 0$, then the solution is correct;

4) If $\sum h_{fi} \neq 0$, then apply a correction

factor $\delta q\,(\delta q=-\dfrac{\sum h_{f_i}}{2\sum \dfrac{h_{f_i}}{q_i}})$ to all q_i and return to Equation (7-2).

7.3.2 Drainage Engineering

1. Chemical Parameters of Water Quality

Many organic and inorganic chemicals affect water quality. In drinking water, these effects may be related to public health or to senses and economics. In surface waters, chemical quality can affect the aquatic environment. Several chemical parameters are also of concern in wastewater. The most common chemical parameters of water quality are the following.

(1) Dissolved Oxygen

Dissolved oxygen is generally considered to be one of the most important parameters of water quality. The molecular oxygen dissolved in water is called dissolved oxygen and is usually abbreviated simply as DO. The content of dissolved oxygen in water is closely related to the partial pressure of oxygen in the air and the temperature of water. Oxygen is only slightly soluble in water. For example, the saturation concentration at 20℃ is about 9mg/L or 9ppm.

Dissolved oxygen usually comes from two sources: one is that when the dissolved oxygen in water is not saturated, the oxygen in the atmosphere penetrates into the water. Another source is oxygen released by plants in water through photosynthesis. Thus the dissolved oxygen in water is constantly replenished by the dissolution of the oxygen in the air and by the photosynthesis of green aquatic plants. But when the water body is polluted by organic matter, oxygen consumption is serious, dissolved oxygen cannot be added in time, anaerobic bacteria in the water body will reproduce quickly, and organic matter will make the water body black and smelly because of corruption.

The dissolved oxygen value is a basis for studying the self-purification capacity of water. The dissolved oxygen in the water is consumed, and it takes a short time to return to the initial state, indicating that the water body has a strong self-purification capacity and the water body is not seriously polluted. Otherwise, it indicates that water pollution is serious, and self-purification ability is weak, or even lose self-purification ability.

(2) Biochemical Oxygen Demand

Bacteria and other microorganisms use organic substances for food. As they metabolize organic material, they consume oxygen. The organics are broken down into simpler compounds, such as CO_2 and H_2O, and the microbes use the energy released for growth and reproduction. When this process occurs in water, the oxygen consumed is the DO. If oxygen is not continually replaced in the water by artificial or natural means, then the DO level will decrease as the organics are decomposed by the microbes. This need for oxygen is called the biochemical oxygen demand. In effect, the microbes "demand" the oxygen for use in the biochemical reactions that sustain them. The abbreviation for biochemical oxygen demand is BOD; this is one of the most commonly used terms in water quality and pollution control technology.

Organic waste in sewage is one of the major types of water pollutants. It is impractical to isolate and identify each specific organic chemical in these wastes and to determine its concentration. Instead the BOD is used as an indirect measure of the total amount of biodegradable organics in the water. The more organic material there is in the water, the higher the BOD exerted by the microbes will be.

The 20 days or so required for the ultimate BOD to be developed is much too long a time to wait for lab results. This is particularly true when the BOD data are used to monitor the efficiency of a water pollution control plant. It has been found that more than two thirds of the BOD, is usually exerted within the first 5 days of decomposition. For practical purposes, the 5-day BOD, or BOD_5, has been chosen as a representation of the organic content of water or wastewater. For standardization of results, the test must be conducted at a temperature of 20℃.

In summary, the parameter of BOD_5, is the amount of dissolved oxygen used by microbes in 5 days to decompose organic substances in water at 20℃.

(3) Chemical Oxygen Demand

The BOD test provides a measure of the biodegradable organic material in water, that is, of the substances that microbes can readily use for food. But there also might be nonbiodegradable or slowly biodegradable substances that would not be detected by the conventional BOD test. The chemical oxygen demand, or COD, is another parameter of water quality that measures all organics, including the nonbiodegradable substances. Chemical oxygen demand (COD) is the oxidation dose consumed when a certain strong oxidant is used to treat a water sample under certain conditions.

It is a chemical test using a strong oxidizing agent (potassium dichromate), sulfuric acid and heat. The general oxidant used to measure COD is potassium permanganate or potassium dichromate. The values obtained by using different oxidants are also different, so it is necessary to specify the detection method. According to the different strengthened oxidants, the results obtained by the two methods are respectively called potassium dichromate oxygen consumption (commonly known as chemical oxygen demand, COD) and potassium permanganate oxygen consumption (commonly known as oxygen consumption, OC, also known as permanganate index).

Chemical oxygen demand (COD) is often used as a measure of the amount of organic matter in the water. The higher the COD is, the more serious the water pollution by organic matter will be. The result of the COD test can be available in just 2 hours, a definite advantage over the 5 days required for the standard BOD test. COD values are always higher than BOD values for the same sample, but there is generally no consistent correlation between the two tests for different wastewaters. In other words, it is not feasible to simply measure the COD and then predict the BOD. Because most wastewater treatment plants are biological in their mode of operation, the BOD test is more representative of the treatment process and remains a more commonly used parameter than the COD.

(4) Ammonia Nitrogen

In the natural surface and underground water bodies, nitrate nitrogen (NO_3^-–N) is the main form, and in the form of free ammonia (NH_3) and ammonium ion (NH_4^+), ammonia nitrogen in the polluted water body is called hydronium, also known as non-ionic ammonia. Non-ionic ammonia is the main cause of aquatic organism toxicity, while ammonium ion is relatively non-toxic. Ammonia nitrogen is a nutrient in water body, which can lead to water eutrophication. It is the main oxygen-consuming pollutant in water body and is toxic to fish and some aquatic organisms.

The methods of ammonia nitrogen detection usually include colorimetric method of Nessler reagent, phenol-hypochlorite (or salicylic acid-hypochlorite) colorimetric method and electrode method. The colorimetric method of Nessler reagent is characterized by simple operation and sensitivity, but the metal ions, sulfides, aldehydes and ketones, color, turbidity and so on in the water will interfere with the determination results, and corresponding pretreatment is required. The phenol-hypochlorite colorimetry has the advantages of sensitivity and stability. The electrode method usually does not require pretreatment of the water sample, and it has many advantages such as wide measuring range. When ammonia nitrogen content is high, distillation-acid titration method can also be used.

(5) Solids

Solids occur in water either in solution or in suspension. These two types of solids are distinguished by passing the water sample through a glass-fiber filter. By definition, the suspended solids are retained on top of the filter, and the dissolved solids pass through the filter with the water. If the filtered portion of the water sample is placed in a small dish and then evaporated , the solids in the water remain as a residue in the evaporating dish. This material is usually called total dissolved solids, or TDS. The concentration of TDS is expressed in terms of mg/L. It can be calculated as follows:

$$TDS= 1000(A-B)/C \tag{7-4}$$

Where A—weight of dish plus residue (mg);

B—weight of empty dish (mg);

C—volume of sample filtered (mL).

In drinking water, dissolved solids may cause taste problems. Hardness, corrosion, or esthetic problems may also accompany excessive TDS concentrations. In wastewater analysis and water pollution control, the suspended solids retained on the filter are of primary importance and are referred to as total suspended solids, or TSS.

The TSS concentration can be computed using Equation (7-4) , where A represents the weight of the filter plus retained solids, B represents the weight of the clean filter, and C represents the volume of sample filtered.

Another classification of solids that is of particular significance in wastewater treatment is volatile solids. These are organic substances that can be burned off or volatilized at 550℃ in a furnace. The residues remaining after burning at that temperature are the fixed or nonvolatile solids. The concentration of volatile suspended solids gives an indication of the organic loading on biological treatment units. It can be determined by measuring the loss in weight of the glass fiber filter plus solids after burning.

2. Components of Drainage Engineering

Modern sewage treatment technology can be divided into physical treatment, chemical treatment and biochemical treatment according to the treatment principle.

Physical treatment: Using physical action to separate the suspended solid in sewage. Methods include filtration, sedimentation, air flotation and reverse osmosis, etc.

Chemical treatment: Using chemical reaction action for separation and recovery of sewage in various forms of pollutants (including suspended, dissolved, colloidal, etc.). The main methods include neutralization, coagulation, electrolysis, redox reaction, extraction, adsorption, ion exchange and electrodialysis. Chemical treatment is mostly used to treat industrial sewage.

Biochemical treatment: Using microbial metabolism to make the dissolved, colloidal state of organic pollutants into a stable harmless substances. The main methods can be divided into two categories, that is aerobic oxidation and anaerobic reduction. The former is widely used to treat municipal wastewater and organic production wastewater, including activated sludge method and biofilm method. The latter is mostly used to treat high concentration organic wastewater and sludge produced in sewage treatment process.

There are many kinds of pollutants in urban sewage and production sewage, and it is necessary to adopt a combination of several methods to remove pollutants of different properties. In this way, the product water can achieve the purpose of purification and discharge standards. Modern sewage treatment technology, according to the degree of treatment, can be divided into primary, secondary and advanced treatment.

(1) Primary Treatment

The most objectionable aspect of discharging raw sewage into watercourses is the floating material. It is only logical, therefore, that screens were the first form of wastewater treatment and present-day treatment plants still use screens as the first step. A typical screen, shown in Fig. 7-5, consists of a series of steel bars which might be about 16~25mm (Mechanical) or 25~40mm (Manual) apart. The purpose of a screen in modern treatment plants is the removal of materials which might damage equipment or hinder further treatment. In some older treatment plants, screens are cleaned by hand, while mechanical cleaning equipment is used in almost all new plants. The cleaning rakes

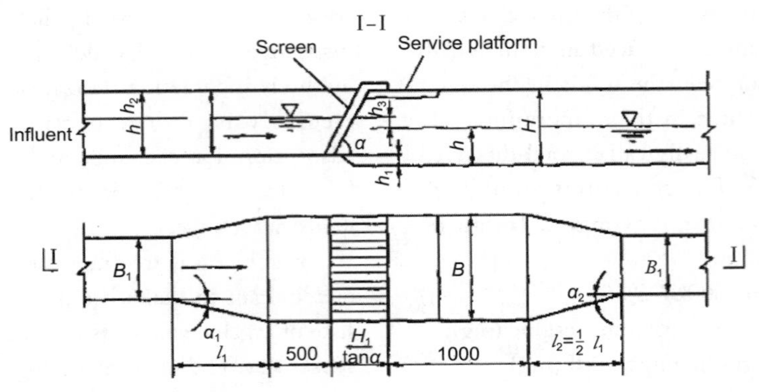

Fig. 7-5 Screen for wastewater treatment

are automatically activated when the screens get sufficiently clogged to raise the water level in front of the bars. The next treatment step involves the removal of grit or sand. This is necessary because grit can wear out and damage such equipment as pumps and flow meters. The most common grit chamber is simply a wide place in the channel where the flow is slowed down sufficiently to allow the heavy grit to settle out. Sand is about 5/2 times as heavy as most organic solids and thus settles much faster than the light solids.

Following the grit chamber, most wastewater treatment plants have a sedimentation basin to settle out as much of the solid matter as possible. Accordingly, the detention time is kept long and turbulence is kept to a minimum. The solids settle to the bottom and are removed through a pipe while the clarified liquid escapes over a V-notch weir, a notched steel plate over which the water flows, promoting equal distribution of liquid discharge all the way around a tank. Settling tanks are also known as "sedimentation tanks" and often as "clarifiers". The settling tank which follows preliminary treatment such as screening and grit removal is known as a "primary clarifier". The solids which drop to the bottom of a primary clarifier are removed as "raw sludge" , a name which doesn't do justice to the undesirable nature of this stuff. Large sewage treatment plants generally use circular sedimentation basins (Fig. 7-6a) or rectangular sedimentation basins (Fig. 7-6b). The function of the sedimentation basin is shown in Table 7-2.

Raw sludge is generally odoriferous and full of water, two characteristics which make its disposal difficult. It must be both stabilized to retard further decomposition and dewatered for ease of disposal. In addition to the solids from the primary clarifier, solids from other processes must similarly be treated and disposed. The treatment and disposal of wastewater solids

The function of the sedimentation basin **Table 7-2**

Index	Removal effect	Operational condition
BOD_5	Removal rate is 10%~30%, the specific removal rate is related to suspended matter	Residence time 1.5~2.5h Surface load 1~2$m^3/(m^2 \cdot h)$
SS	Removal rate is 35%~60%, associated with SS concentrations of sewage; SS≥300mg/L, removal rate≥50%; SS<300mg/L, removal rate<50%	

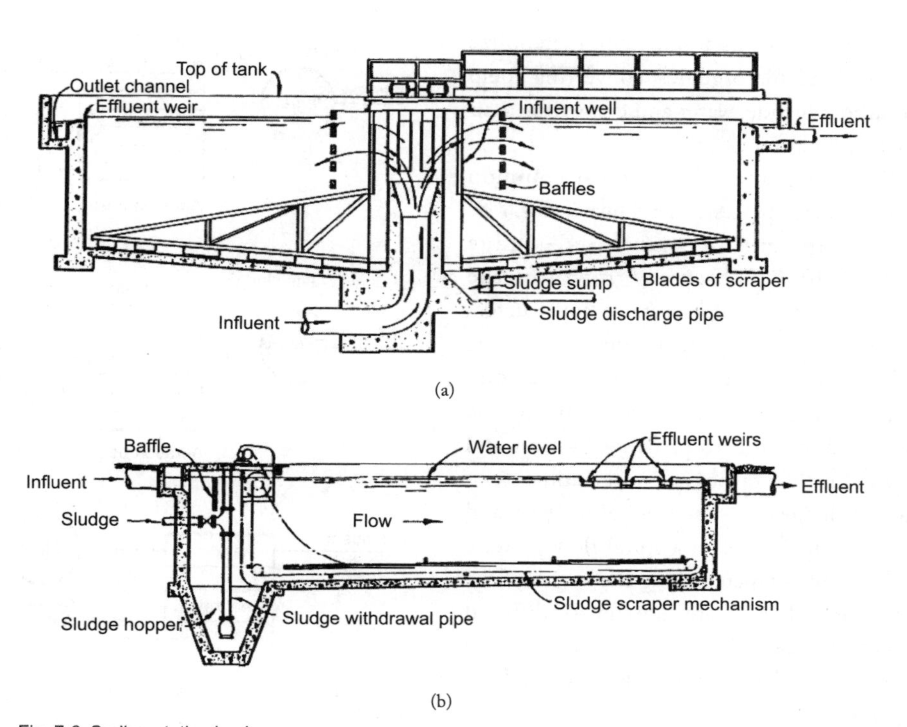

Fig. 7-6 Sedimentation basin
(a) Circular; (b) Rectangular
Source: TM A. Water Supply, Water Treatment. Washington.D.C.: Departments of the Army and the Air Force, 1985.

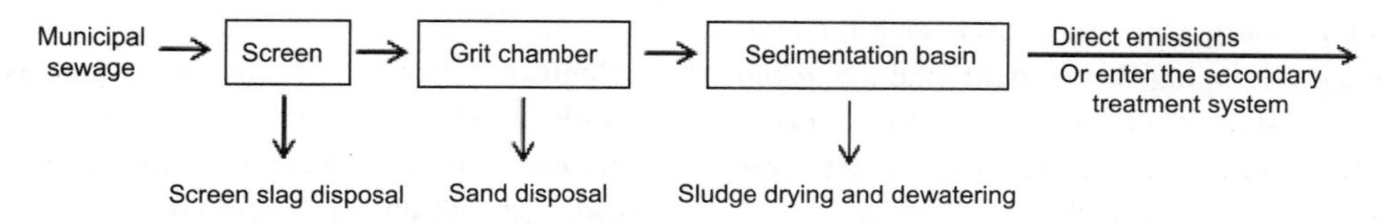

Fig. 7-7 Typical Flow Chart for Primary Treatment

(sludge) is an important part of wastewater treatment and is discussed later in this chapter.

In review, primary treatment, consisting of physical processes designed to remove the solids from wastewater, includes screening, de-gritting and settling. The most typical primary treatment process is shown in Fig. 7-7.

(2) Secondary Treatment

Primary treatment processes remove only those pollutants that will either float or settle out by gravity. But about half of the raw pollutant load still remains in the primary effluent. The purpose of secondary treatment is to remove the suspended solids that did not settle out in the primary tanks and the dissolved BOD that is unaffected by physical treatment. Secondary treatment is generally considered to mean 85 percent BOD and TSS removal efficiency and represents the minimum degree of treatment required in most cases. In the United States, secondary treatment processes are almost biological systems.

Biological treatment of sewage involves the use of microorganisms. The microbes, including bacteria and protozoa, consume the organic pollutants as food. They metabolize the biodegradable organics, converting them into carbon dioxide, water and energy for their growth and reproduction.

Flow diagrams for the most common activated- sludge processes are shown in Fig. 7-8.

The conventional activated-sludge process (Fig. 7-8a), an outgrowth of the earliest activated-sludge systems constructed, is used for secondary treatment of domestic wastewater. The aeration basin is a long rectangular tank with air diffusers on one side of the tank bottom to provide aeration and mixing. Settled raw wastewater and return activated sludge enter the head of the tank and flow down its length in a spiral flow pattern. An air supply is tapered along the length of the tank to provide a greater amount of diffused air near the head where the rate of biological metabolism and resultant oxygen demand are the greatest.

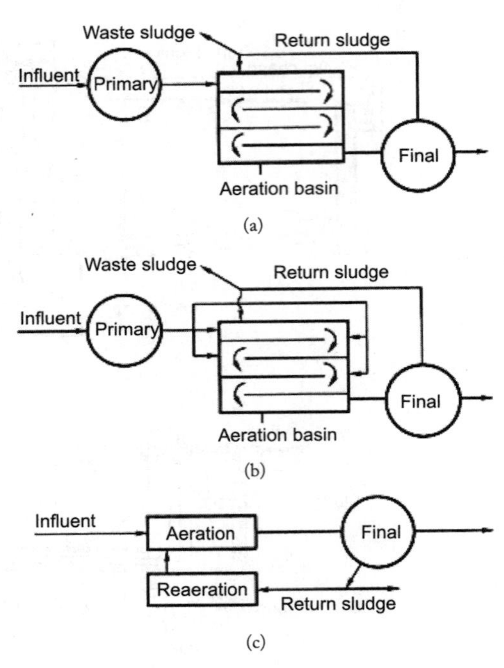

Fig. 7-8 Flow diagrams for common activated-sludge process[11]
(a) Conventional activated-sludge process; (b) Step-aeration activated-sludge process; (c) Contact stabilization without primary sedimentation

The stept aeration activated-sludge process (Fig. 7-8b) is a modification of the conventional process. Instead of introducing all raw wastewater at the tank head, raw flow is introduced at several points along the tank length. Stepping the influent load along the tank produces a more uniform oxygen demand throughout. While tapered aeration attempts to supply air to match oxygen demand along the length of the tank, step loading provides a more uniform oxygen demand for an evenly distributed air supply.

The contact stabilization activated-sludge process (Fig. 7-8c) provides for reaeration of the return activated sludge from the final clarifier, allowing this method to use a smaller aeration tank. The sequence of aeration-sedimentation-reaeration has been used as a secondary treatment process.

A typical flow chart for secondary treatment is shown in Fig. 7-9. Secondary treatment is generally composed of screen, grit chamber, primary sedimentation basin, aeration basin and secondary sedimentation basin. The aeration basin is a key equipment for sewage biological treatment. By changing the way of inflow and return sludge as required, the areation basin can be flexibly operated in accordance with the three methods in Fig. 7-8.

(3) Advanced Treatment

Secondary treatment can remove between 85% and 95% of the BOD and TSS in raw sanitary sewage. Generally, this leaves 30mg/L or less of BOD and TSS in the secondary effluent. But sometimes this level of sewage treatment is not sufficient to protect the aquatic environment. For example, periodic low flow rates in a trout stream may not provide the amount of dilution of the effluent that is needed to maintain the necessary DO levels for trout survival.

Another limitation of secondary treatment is that it does not significantly reduce the fluent concentrations of nitrogen and phosphorus in the sewage. Nitrogen and phosphorus are important plant nutrients. If they are discharged into a

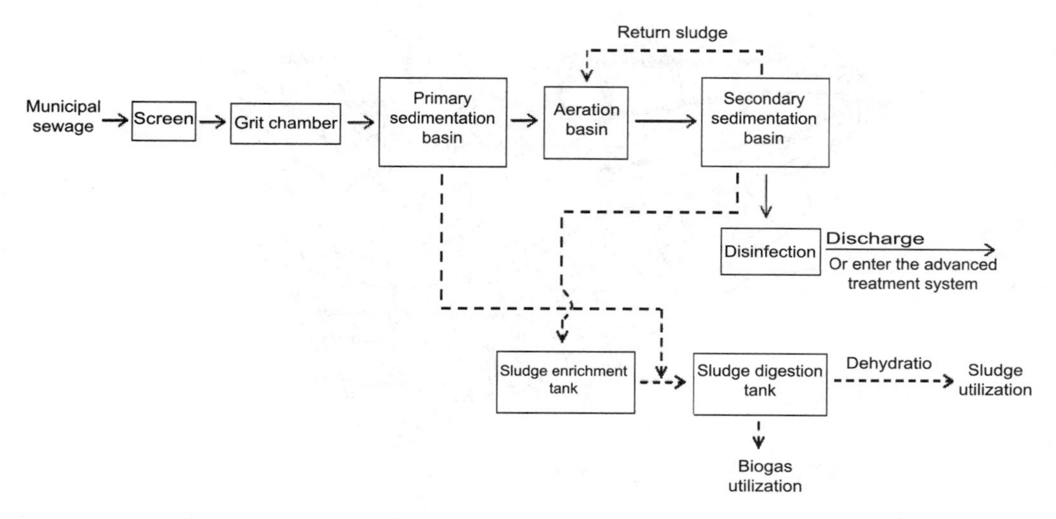

Fig. 7-9 Typical flow chart for secondary treatment

lake, algal blooms and accelerated lake aging or cultural eutrophication may be the result. Also, the nitrogen in the sewage effluent may be present mostly in the form of ammonia compounds. These compounds are toxic to fish if the concentrations are high enough. Yet another problem with the ammonia is that it exerts a nitrogenous oxygen demand in the receiving water, as it is converted to nitrates. This process is called nitrification.

When pollutant removal greater than that provided by secondary treatment is required, either to further reduce the BOD or TSS concentrations in the effluent or to remove plant nutrients, additional or advanced treatment steps are required. This is also called tertiary treatment, because many of the additional processes follow the primary and secondary processes in sequence.

Tertiary treatment of sewage can remove more than 99 percent of the pollutants from raw sewage and can produce an effluent of almost drinking water quality. But the cost of tertiary treatment, for operation and maintenance as well as for construction, is very high, sometimes doubling the cost of secondary treatment. The benefit-to-cost ratio is not always big enough to justify the additional expense. Nevertheless, application of some forms of tertiary treatment is not uncommon.

3. The Layout Principles and Forms of Drainage System

(1) The Layout Principles of Drainage System

A general guideline for sewer system layout and design has been proposed by Thackson. Basic steps include the following:

1) Obtain or develop a topographic map of the area to be served.

2) Locate the drainage outlet. This is usually near the lowest point in the area and is often along a stream or drainage way.

3) Sketch in a preliminary pipe system to serve all of the contributors (Fig. 7-10).

4) Pipes must be located so that all users or future users can readily tap on. They must also be located so as to provide access for maintenance and thus are ordinarily placed under the streets or other places that can be excavated and laid.

Insofar as practical, sewers should follow natural drainage ways so as to minimize excavation and pumping requirements. Large trunk sewers are usually constructed in low-lying areas closely paralleling streams

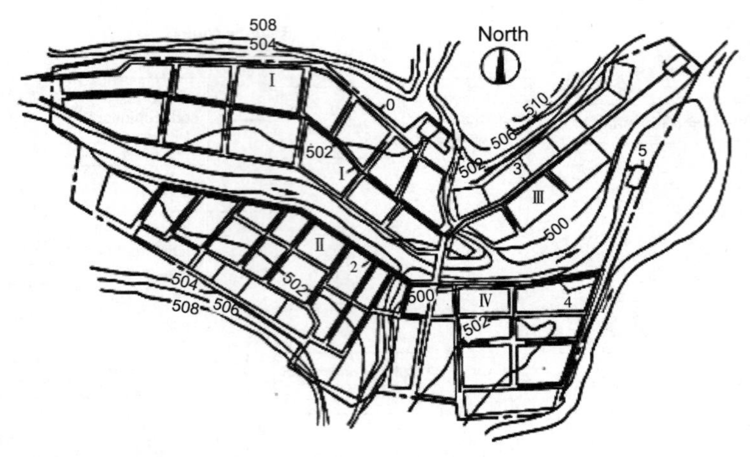

Fig. 7-10 Typical layout for the design of a sewerage system[12]

or channels. In general, pipes should cross contours at right angels.

5) Establish preliminary pipe size.

6) Revise the layout so as to optimize flow-carrying capacity at minimum cost. Pipe lengths and sizes should be kept as small as possible, pipe slopes should be maximized within tolerances for velocity, excavation depth should be minimized, and the number of appurtenances should be kept as small as possible.

7) Try to avoid pumping across drainage boundaries. A balance must be made between excavation and pumping.

The planning and design of municipal sewers shall be carried out in the following order. The first step is to divide drainage basins in areas requiring drainage. The second step is to choose the location of the sewage treatment plant and the outlet. The third step is to draw up the route of the main drainage pipeline, generally determine the main pipe firstly, then determine the sub-main pipe and finally determine the branch pipe. The fourth step is to determine the drainage area to be lifted and the location of the lifting pump station.

The urban storm water system is composed of storm water outlet, storm sewer, inspection well, outlet and other structures. Firstly, it is necessary to determine the drainage area and drainage mode in the layout of urban storm water pipe and drainage system, then plan the route of storm water pipe and drainage, and finally determine the location of storm water pump house, storm water storage tank, storm water outlet and other structures. It is particularly important to note that storm drains should be planned according to the topography so that storm drains can be

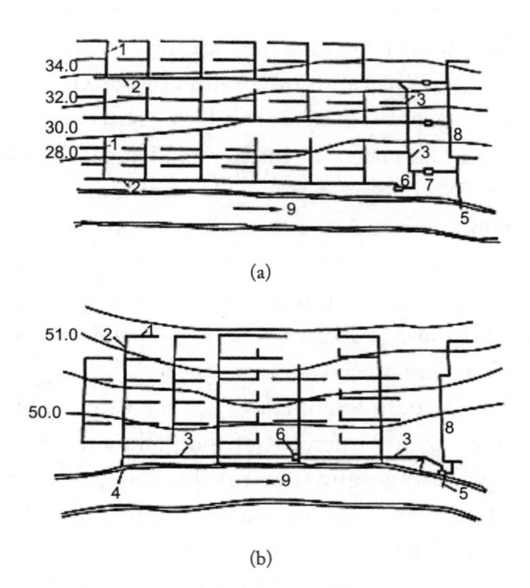

Fig. 7-11 Types of drainage systems[12]
(a) Paralleled; (b) Orthogonal
1-Branch pipe; 2-Sub-main pipe; 3-Main pipe; 4-Overflow pipe; 5-Outlet of the canals; 6-Pumping house; 7-Sewage treatment plant; 8-Irragation canal; 9-River

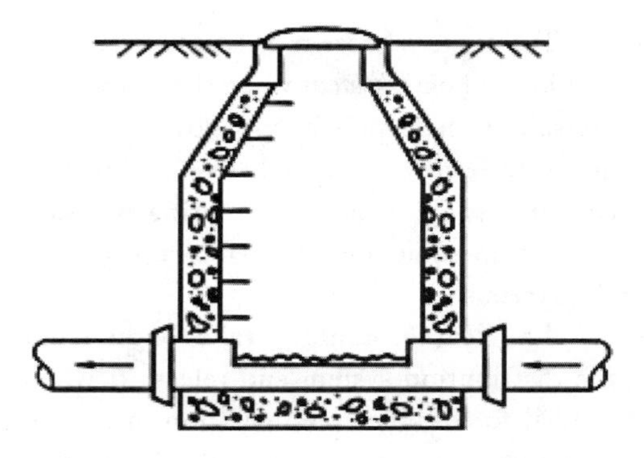

Fig. 7-12 Typical manhole for sanitary sewers[11]

discharged into nearby bodies of water with the shortest possible distance.

(2) The Forms of Drainage System

Drainage system is generally arranged into branching systems. According to the different terrain, two basic layout forms, paralleled form (Fig. 7-11a) and orthogonal form (Fig. 7-11b), are generally adopted. If the drainage trunk is parallel to the topographic contour line, it is paralleled, otherwise it is orthogonal. The paralleled formula is suitable for cities with steep terrain, while the orthogonal formula is suitable for cities with flat terrain and sloping to one direction. Practical projects often combine these two types of layout. The slope of main pipe with large flow is small, while that of branch pipe with small flow is large.

Inspection wells and drop wells are the main means of connecting drainage pipes.

(3) Connection Mode of Drainage Pipes

As population densities increase, the disposal of wastewater in the soil becomes dangerous, especially if wells are used for water supply. It becomes necessary to carry the wastewater to a more remote point of disposal by means of underground pipes called sewers.

Sewers are usually made of concrete, cast iron, clay or asbestos-cement, and are laid so that the wastewater will flow by gravity (downhill). Sewers are not designed to withstand positive pressure, so the full flow should be avoided. On the contrary, sewers are designed as "open channel" devices, and the grade or slope of the pipes is important. Sewerage systems must be carefully designed before construction, since deep cuts and rock must be avoided while always maintaining a downhill slope. Even in the best designed system, however, it is sometimes necessary to pump sewage to a higher elevation. Sewage pumping stations are often placed underground, fenced and landscaped. The pipes from such pumping stations are designed to withstand pressure and full flow.

A manhole, is used whenever a sewer line changes size, grade (slope) or direction. A cross section of a typical manhole is shown in Fig. 7-12.

Spacing of manholes on straight pipes **Table 7-3**

Name	Pipe diameter or canal depth (mm)	The largest spacing (m)	Commonly used spacing (m)
The sewer	≤400	30	20~30
	500~700	50	30~50
	800~1000	70	50~70
	1100~1500	90	65~80
	1600~2000	100	80~100
The storm-water pipe and confluence pipe	≤400	40	30~40
	500~700	60	40~60
	800~1000	80	60~80
	1100~1500	100	80~100
	>1500	120	100~120

Sewers on straight lines are required for the ease of cleaning. The maximum spacing of manholes of drainage pipes with different functions is shown in Table 7-3.

7.3.3 House Water-supply and Drainage Engineering

1. House Water-supply System

House (interior) water-supply systems are intended to draw water from the exterior network and to distribute it among consumers in a building. The boundary between the exterior distribution system and a house system is constituted by the upstream flange of the first gate valve in the communication pipe past the exterior wall of a building or a heat-exchange substation.

In a micro-district, water taken from a distribution system is supplied through a communication pipe to an area substation and thence through a supply piping network to groups of buildings or individual buildings.

Interior water-supply systems may likewise be divided into domestic, fire-protection and industrial. To cut down the cost of construction and service, these are frequently combined with one another. For example, there exist domestic/fire lines, industrial/fire lines, etc.

Depending on the temperature of water, cold and hot (t=50~75℃) water-supply lines maybe distinguished.

A cold water-supply system (Fig. 7-13a) consists essentially of a communication pipe 1, a meter box 2, pressure boosters 3, storage tanks 4, a piping line 5 and 6, valves 7, and water-dispensing fixtures 8.

A hot water-supply system also has water heaters, filters, deaerators, etc.

The communication pipe 1 is a pipe between the supply main and the meter box.

The meter box 2 houses a water meter—a mechanical device used to measure the volume of water passing into the building and valves used to shut off the water meter during inspection.

The pressure boosters 3 serve to raise pressure in the house system when the guaranteed pressure in the supply main is lower than required to feed water to high-level and remote consumers. Systems incorporating pressure boosters are sometimes referred to as high-service systems.

The storage tanks 4 take water from the distribution system and release it when needed for trouble-free operation in case of interruption in service or when the distribution system fails to maintain the pressure at the level required for house water supply. They are available in the form of tanks 4a mounted at the highest point in the building or in the form of air-operated water-dispensing tanks 4b installed at or below ground level in the basement.

The function of the piping system is to distribute water supply among the consumers. It incorporates a house piping system 6 which feeds the individual consumers and a supply piping system 5 which delivers the water from the meter box 2 to the house system 6.

The valves 7 serve to control the flow of water, and the water-dispensing fixtures 8 serves to let the water out for direct use.

The layout of a house water-supply system, that is, the type and number of component parts and the piping arrangement, depends on the pressure available from the main and that required for normal functioning of the house system. Whatever the layout, the house water-supply system must ensure trouble-free operation.

A direct arrangement (communication pipe–water meter–piping–water–dispensing fixtures, shown in Fig. 7-13b) is used where the pressure in the main is always greater than required to supply water to the highest or farthest consumer in the building. It finds particularly wide application in buildings of 5 or 6 storeys high.

A system with a storage tank (Fig. 7-13c) is employed where the pressure in the main falls below normal for several hours every day or where the water demand follows a highly nonuniform pattern. The tank 4 accumulates water when it is available under a high pressure to release it when the pressure falls. This arrangement also serves to store water for continuous service when the supply main fails to deliver water in the required quantity, for fire protection, in bath-houses and laundries, and at factories.

A system incorporating a pressure booster (Fig. 7-13d) is useful in case of permanent or long-term pressure deficit and nonuniform consumption. Sometimes, it is also employed where pressure shortages occur periodically. The pressure booster (booster pump) 3 is turned on either automatically or manually when the pressure in the supply main falls below normal.

A system with a storage tank and a pressure booster (Fig. 7-13e) is used where water consumption varies considerably over the day and also where a large quantity of water has to be stored to make up for a permanent or long-term pressure deficit. The joint use of the pressure booster 3 and storage tank 4 makes it possible to minimize the storage tank capacity even when demand fluctuations are considerable.

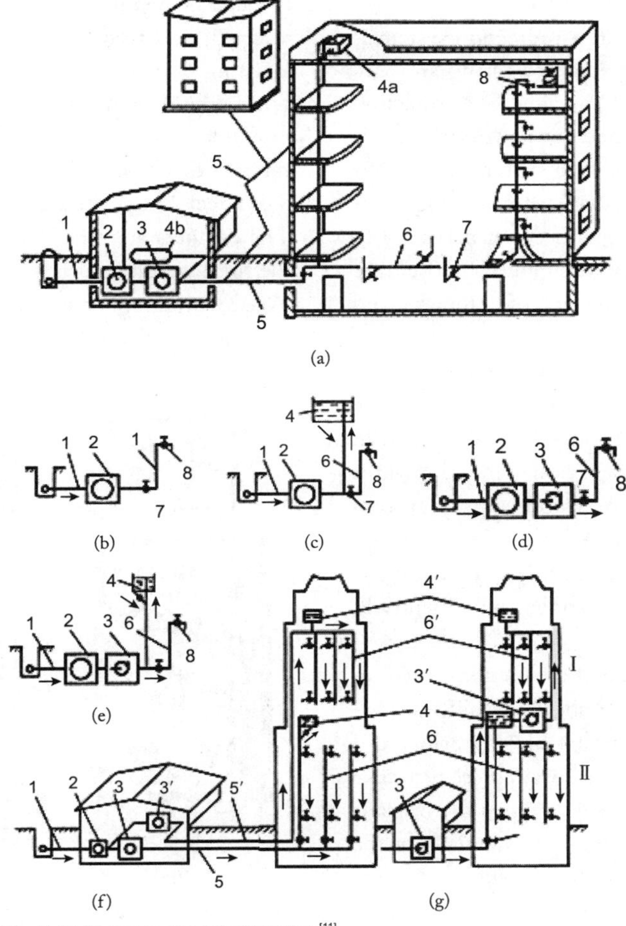

Fig. 7-13 House water–supply system[11]
1-Communication pipe; 2-Meter box; 3-Pressure booster; 4-Storage tank;
5、6-Pipeing line; 7-Valve; 8-Water-dispensing fixture

Split system finds application in buildings more than 50m high (17 storeys upwards) where the pressure in the house system exceeds the maximum safe value (0.6MPa). Their upper and lower circuits may be arranged in parallel (Fig. 7-13f) or in series (Fig. 7-13g), The parallel configuration offers a more reliable performance, but it is longer than the series arrangement. The system of Fig. 7-13(f) is also used in lower buildings to cut down the operating cost of pumping plants. In this case, the lower storeys are supplied with water under the pressure maintained in the supply main and the upper storeys have a piping system of their own to which water is fed by a low-capacity pump.

2. Fire-protection System

The choice of a fire-protection system is governed by the fire-hazard rating and fire resistance of the building. The fire-protection system can be divided into hydrant water-supply system, sprinkler system and other fixed fire extinguishing systems using non-water extinguishing agent. Water necessary to extinguish a fire may be supplied by fire lines with fire hoses or by means of automatic and semi-automatic systems.

(1) Hydrant Water-supply System

Fire-protection system is a fixed fire extinguishing equipment set up to extinguish fire in buildings by pressurizing the water provided by outdoor water-supply system (when the pressure of external pipeline network does not meet the use demand). It is the most basic fire extinguishing facility in buildings.

Fire lines with hoses are used in structures whose component members are built of non-flammable and combustible materials and where the occupants are always present to take care of a fire before the arrival of a fire brigade. Although a fire line consists essentially of the same elements as a domestic supply system, it has a number of specific features largely stemming from the necessity to meet more stringent requirement as regards reliability of service and rate of flow. Fire lines are arranged in a gridiron pattern. They use steel piping and valves designed to handle a high working pressure. Where a fire-protection line is combined with domestic supply, galvanized steel pipe is a more common choice, plastic pipe may never be used. For trouble-free operation, standby pump units must be provided in pressure boosters.

Water is dispensed from fire-line outlets (Fig. 7-14a). A typical fire outlet consists of a fire valve 4, a hose 7, a metal hose barrel or lance 1, and fittings 6 for connection of the hose to the barrel and valve. The nominal diameter of the outlets is 50mm or 65mm.

The fire hose (usually 10m or 20m long) is made of hemp linen lined with rubber for strength and tightness. The hose barrel has a fitting 6 at one end and a spray tip at the other. The tip outlet is 13mm, 16mm, 19mm in diameter. Fire outlets are usually located in cabinets 5. The hose is stored on a rotary shelf 8 or a reel 2 mounted on a swinging bracket 3. The cabinets are attached to the wall so that the fire-hydrant mouth is 1.1m above the floor. Where several water streams are required from a single point, twin outlets are installed in the

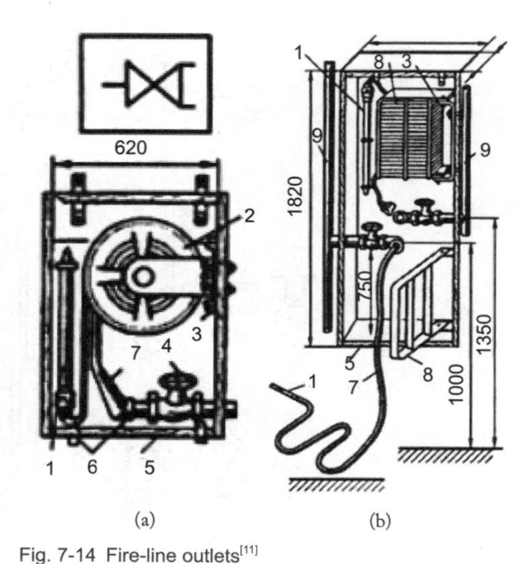

Fig. 7-14 Fire-line outlets[11]
(a) Single; (b) Twin
1-Hose barrel; 2-Reel; 3-Bracket; 4-Valve; 5-Cabinet; 6-Fitting; 7-Hose; 8-Shelf; 9-Standpipe

same cabinet (Fig. 7-14b).

Generally, fire-hydrants should be placed in a well-positioned corridor, entrance or stairwell for easy access.

(2) Fire-protection Sprinkler System

Automatic sprinkler system and drencher system are common in fire-protection sprinkler system. Automatic sprinkler system and drencher system are intended to fight fires and to sound an alarm all on their own. They are provided in areas where the risk of a fire breaking out and spreading rapidly is considerable (for example, in book depositories, libraries and painting shops).

A sprinkler system (Fig. 7-15) consists of a water-supply system (incorporating a service pipe, an air-operated water-dispensing tank 9 and a cistern 15), standpipes 14, a control valve 13, a sprinkler network (incorporating supply pipelines 12 and sprinkler branches 11), and sprinkler heads 10. The control valve and the piping downstream of the valve form a section that can readily be shut off for inspection and maintenance.

The sprinkler heads 10 open up when the temperature rises to a predetermined level and discharge water to put out the fire. When a fire breaks out, the solder melts, and the plug falls apart. Being supplied under pressure, water strikes the baffle 6 and knocks out the glass plate. However, the water in the pipeline network has certain pressure, so when the pipeline network leaks, the decoration inside the building will be damaged and the use of the building will be affected. The system is suitable for buildings with ambient temperatures of 4 to 70 degrees.

An automatic drencher system (Fig. 7-15b) is similar in construction to the sprinkler system just discussed. Here, water is sprayed by drenchers 18 which resemble sprinkler but have no fusible plugs. The drenchers are fed through branches 11 when a sprinkler inserted in a forced-supply pipeline 17 opens a control valve 16. When a fire breaks out, the sprinkler opens up and actuates the control valve which supplies water to all the drenchers. Moreover, semi-automatic drencher system (Fig. 7-15c) are turned on by the occupants when there is a fire that is likely to spread over a large area. This system is turned on by an electrically driven or hand-operated gate valve installed at the control board.

Sprinkles may be "wet", "dry", or "alternate-

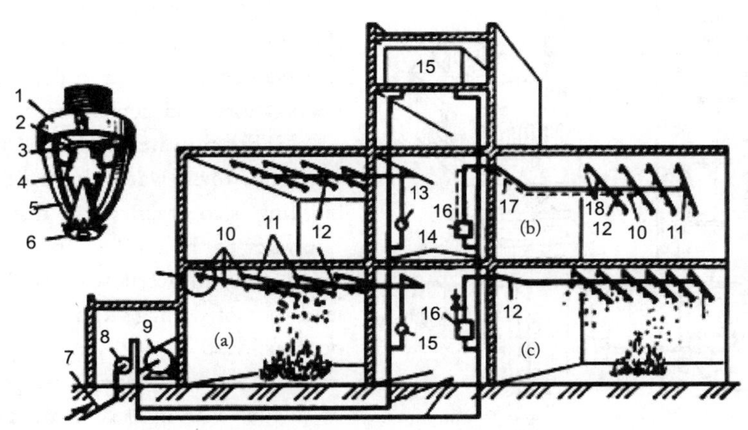

Fig. 7-15 Schematic diagram of fire-protection sprinkler system[11]
(a) Automatic sprinkler system; (b) Automatic drencher system;
(c) Semi-automatic drencher system
1-Shank; 2-Diaphra; 3-Valve plate; 4-Fusible plug; 5-Frame; 6-Baffle; 7-Inlet; 8-Pump; 9-Air-operated water-dispensing tank; 10-Sprinkler head; 11-Sprinkler branches; 12-Supply pipelines; 13-Control valve; 14-Standpipes; 15-Cistern; 16-Control valve; 17-Forced-supply pipeline; 18-Drencher

wet-and-dry". The wet sprinkler system is permanently filled with water, but it is not proof against frost. The dry sprinkler system is filled with air, which is used in places with hard winters, since the water is admitted to the pipes only when the air is released from them. The alternate-wet-and-dry sprinkler system can be water-filled in warm weather and air-filled in frosty weather. It is used in temperate countries with occasional frost.

3. House Drainage System

As we have already learned, the function of a drainage system in a building is to remove safely and quickly sanitary sewage, industrial wastes and rainwater. As a rule, the various kinds of wastes are disposed of through drainage systems of their own. It would, therefore, be quite to the point to speak of sanitary, industrial and storm-water drainage systems.

(1) Sanitary Drainage System

A sanitary drainage system (Fig. 7-16) consists essentially of plumbing fixtures 3 (which receive and discharge water, liquid, or water-borne wastes), fixture traps 2 (which maintain a water seal against gases, air and odours escape) and drainage piping 1.

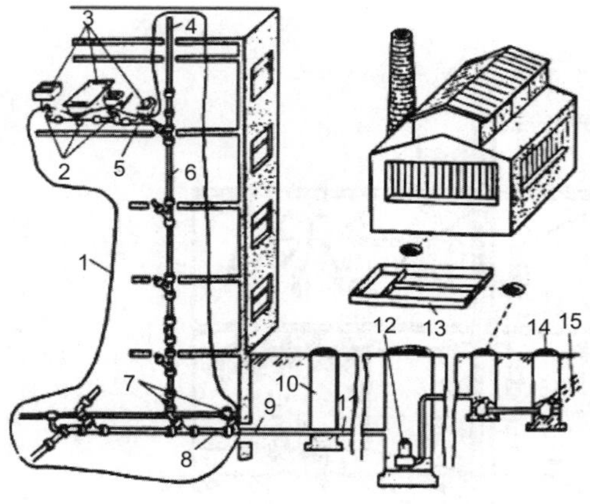

Fig. 7-16 Schematic diagram of sanitary drainage system[11]

1-Drainage piping; 2-Fixture traps; 3-Plumbing fixtures; 4-Vent stack; 5-Fixture branches; 6-Soil stack; 7-Clean-outs; 8-Building drain; 9-Building sewer; 10-Manhole; 11-Public sewer; 12-Lift pumping plant; 13-Water-conditioning plant; 14-Manhole; 15-Sub-main or trunk sewer

The drainage piping collects sewage from the plumbing fixtures and carries it off through a building drain 8 into a building sewer 9. The building sewer discharges through a manhole 10 into a public sewer 11. The flow from the public sewer into a manhole 14 of a sub-main or a trunk sewer 15 is usually by gravity. If the bottom of the manhole 14 is at a higher elevation than the outlet of the building sewer 9, a lift pumping plant 12 is included in the sewer system. Industrial wastes that may not be discharged directly into the public sewer need to be pretreated at a water-conditioning plant 13. Plumbing fixtures are open-top receptacles which receive and discharge liquid or water-borne human wastes.

(2) Industrial Drainage System

The industrial drainage system is mainly used to remove sewage and waste water produced by industrial enterprises in the process of production. In order to facilitate the sewage treatment and comprehensive utilization of waste water, industrial waste water seriously contaminated by chemical impurities (organic matter, heavy metal ion, acid, alkali, etc.) and mechanical impurities (suspended matter and colloids) in the production process shall be treated and purified until it reaches the discharge standard before being discharged, in addition, the less contaminated waste water in the production process (only when the water temperature rises) can be treated simply and then put back into production for reuse.

(3) Storm-water Drainage System

The storm-water drainage system of a building is to organize the discharge of roof rainwater to the outdoor in time. Otherwise, roof water will overflow everywhere and even lead to roof damage and leakage and affect people's lives.

The storm-water drainage system can be divided into internal drainage system and external drainage system according to whether there are rainwater pipes inside the building. The system with rainwater pipes inside the building and rainwater hopper on the roof (a device that directs rainwater from the roof of the building into the rainwater drainage system) is an internal drainage system, otherwise, it's an external drainage system.

Engineering experience shows that the roof rainwater collection method gives priority to the gutter form, while the rainwater hopper is in the gutter. Gravity semi-pressure flow systems are generally suitable for the internal drainage system of the building roof and the external drainage system of the long gutter. The internal drainage systems of plant, warehouse and public building with large roofs are suitable to adopt siphon pressure flow system, and it is advisable to use gravity-free flow system for eaves drainage system. Balcony rainwater should install independent drain system and cannot be connected to the roof rainwater system.

7.4 Urban Gas Pipeline and Heat Pipeline Installations

7.4.1 Urban Gas Pipeline Network

1. Classification and Selection of Urban Gas Pipeline Network System

(1) Classification of Gas Pipeline

The city gas pipeline network system is composed of crisscrossing gas pipelines with different pressure functions. The gas pipeline can be classified according to the laying mode and pipe network shape of gas transmission pressure.

China's natural gas industry has grown rapidly in all aspects of upstream, midstream and downstream, aiming to build a natural gas production, supply, storage and marketing system characterized by domestic supply, diversified import sources, perfect pipeline layout, complete gas storage and peak regulating facilities, rational gas structure, and safe and reliable operation.

With the application of gas transmission in China and the reference to the gas experience of foreign cities, gas pipelines in Chinese cities and towns are divided into seven grades according to the gas design pressure P (MPa) (Table 7-4).

The gas pipeline can be divided into buried pipeline and overhead pipeline according to the laying mode. According to the shape of the pipe section, it can be divided into ring pipeline network and branch pipeline network.

Municipal gas pipeline network systems can be divided into first-level system, secondary system, high-medium-low pressure three-level pipe network system and multiple-level pipeline

Design specification for urban gas **Table 7-4**

Name		Pressure(MPa)
High-pressure gas pipeline	A	$2.5<P\leq 4.0$
	B	$1.6<P\leq 2.5$
Sub-high pressure gas pipeline	A	$0.8<P\leq 1.6$
	B	$0.4<P\leq 0.8$
Medium pressure gas pipeline	A	$0.2<P\leq 0.4$
	B	$0.01\leq P\leq 0.2$
Low pressure gas pipeline		$P<0.01$

network system according to different pressure levels.

The first-level system refers to the pipeline network with only one pressure level of low or medium pressure to distribute and supply gas, which is not commonly used at present. The secondary system is a pipe network system composed of two pressure classes. The medium-low pressure secondary pipe network system is more commonly used, which is suitable for small and medium-sized towns with large gas supply area, large gas supply volume and uneconomical low-pressure mode. The three-level pipe network system is suitable for the occasions where the gas supply volume is large or the gas supply range is large with a long distance. Four or more pressure networks form a multiple-level pipeline network system.

(2) Composition of Gas Transmission and Distribution System

The urban gas transmission and distribution system generally consists of gas stations, gas pipeline networks, gas storage equipment, pressure regulating facilities, management facilities and monitoring systems. It is the basic responsibility of gas pipeline companies to ensure that the natural gas pipeline system can meet the customers' demand to the greatest extent.

2. Urban Gas Pipeline Wiring

Underground gas pipelines should be laid along urban roads, usually in sidewalks or green belts. Factors contributing to the accident of urban buried gas pipeline network are numerous with complex interactions. Hence, it is essential to identify dependent relationships among various accident-causing factors in qualitative and quantitative ways. The following situations should be taken into account when determining the wiring of gas pipelines at different pressures:

(1) The pressure of the gas in the pipe;

(2) The density and layout of other pipelines underground;

(3) The volume of street traffic and the structure of the pavement;

(4) The number of users connected with the pipeline and the gas usage;

(5) Soil properties, corrosion properties and freezing line depth;

(6) Reliability index of gas pipeline network.

For pipeline network systems, reliability indicators are used to describe three features of system reliability:

(1) Reliability, which in its narrowest definition, describes the ability of the system to perform prescribed functions within a specified time under specified conditions.

(2) Robustness, which describes the ability of the system to withstand interference.

(3) Maintenance, which describes the ability of the system to resume normal operations after a system failure.

3. Prospect of New Gas Technology

With the rapid development of natural gas industry, China is marching into the era of natural gas. In order to meet the strong demand of natural gas consumption, China has built a more complex gas pipeline network, with higher transmission pressure and larger diameter. A long-distance pipeline often has a diameter of more than 1200mm and the maximum design pressure up to 12MPa. The flow of a single flowmeter is more than 10,000m^3/h. In addition, the national pipeline network company is under preparation for establishment. All these will raise new and higher requirements for the management and technology of natural gas measurement. The natural gas measurement technology, as the basis of the natural gas industry, will be further developed and upgraded during the 14th Five-Year Plan and even longer period, mainly in the following aspects:

(1) The measuring tools used for natural gas trade handover will change from verification to calibration, and field measurement will change to focus on the in-service inspection and remote diagnosis of flowmeters, series application of mainstream flowmeters and

verification flowmeters, as well as performance evaluation of field metering systems.

(2) The number of gas flow detection stations will gradually increase to support the natural gas development. So far, 3 primary standard equipment traceability chains have been formed. The specific natural gas metering stations include Chengdu, Nanjing, Wuhan, Guangzhou, Urumqi, Beijing Caiyu, Chongqing Yulin and Tarim; Shenyang and Yunnan substations will be built. Now, the annual verification capacity is equivalent to the capacity of more than 3000 flowmeters.

(3) China will improve its standard systems, and actively lead the formulation and revision of international standards. China will actively undertake and participate in the formulation and revision of international standards, in order to enhance China's international influence in the field of natural gas standards. The international standards include the guide for flowmeter selection in natural gas trade, the research on natural gas measurement performance evaluation, and the channel flowmeters for measuring natural gas flow.

(4) Flow measurement instruments will be localized. In China, only the orifice flowmeter reaches the international advanced level, while the high-accuracy ultrasonic flowmeter and gas chromatography are imported. It is necessary to localize the latter instruments and improve the long-term stability of the instruments in line with the international advanced level.

7.4.2 District Heat Supply Network

1. Development of District Heating Systems

The so called 1st generation of district heating networks was introduced in the late 19th century in the U.S., using steam as a heat carrier. These systems were replaced by 2nd generation networks, which use pressurized hot water with supply temperatures over 100℃ as a heat carrier. Since space heating in traditional residential buildings has generally been ensured by heat transfer from radiators at about 80℃ supply temperature, district heating networks with lower supply temperatures were therefore established in the market starting from the 1970s, i.e., the so called 3rd generation of district heating systems, which are currently and widely applied: the total number of systems has been estimated at 80,000 systems worldwide, of which about 6000 systems (7.5%) in Europe. These 3rd generation of district heating systems are characterized by using pressurized hot water as a heat carrier but with lower temperatures (below 100℃) than 2nd generation systems.

The higher efficiency in energy production due to centralization in comparison with decentralized systems has a drawback in distribution losses, which are usually in the range of 10%~30% and in the costs related to distribution pipeline insulation. Furthermore, high temperature-sustaining materials must be used for the network, including heat exchangers, valves and instrumentations. Moreover, integration in the network is suitable only for high temperature sources, thus limiting the possibility of renewables exploitation. Finally, in the last few years the improvement in building energy performance and in the retrofitting of building stock contributes to a generalized decrease in heating demand. This fact produces an increase of the relative distribution losses and a decrease of the overall cost effectiveness of existing district heating networks.

A further reduction of heat supply temperatures has been identified as one of the possible solutions. This so called 4th generation of district heating networks is a novel and quite recent district heating concept developed in Denmark and characterized by lower temperatures. The temperature decrease in the network can (1) lead to an increase of efficiency of district heating systems due to reduction of heat losses and (2) can allow easier integration of cost-efficient renewable energy technologies (like solar energy or geothermal heat pumps) which are not linked to combustion processes. A further division of 4th generation systems can be made between so called low temperature district heating (LTDH) and ultra-low temperature district heating (ULTDH). LTDH networks are usually characterized by supply temperatures in the range of 50~70℃, while ULTDH networks have supply temperatures below 50℃, without a common definition of lower limit. In ULTDH networks, decentralized micro-heat power stations should be used to boost the domestic hot water (DHW) temperature as needed. Another advantage of 4th generation systems is the possibility of using cheaper materials like plastic for pipes and other network components, also including a relevant decrease of insulator thickness and related costs.

Although the idea has only recently been developed, a considerable amount of literature can be found on 4th generation or low temperature district heating. Qualitative benefits and drawback analysis have been assessed, including the evaluation of how 4th generation of district heating systems can be integrated into the smart city concept. Further specific analysis has been carried out to identify how technical, legislative and economic barriers can be overcome. Case studies from different countries have also been reported. Nevertheless, the development of 4th generation of district heating systems is characterized by two strong limitations: The first is that the demand side of a district heating network consists of buildings built at different times. This means that they have different thermal loads and, in particular, they use different indoor space heating distribution systems, which require diverse inlet temperature set points. However, heating terminals of existing buildings are usually oversized compared to real needs. Therefore, lower distribution temperatures may be acceptable in some cases and can cover a fraction of the seasonal thermal load. The application of 4th generation of district heating systems is thus mainly suitable in new and/or renovated buildings. Secondly, 4th generation as well as traditional district heating systems have low compatibility with district cooling networks. Some authors suggested creating a bidirectional low temperature network. The warmer pipe has temperatures between 12℃ and 20℃, while the colder pipe has 8~16℃. In the case of a heating demand, the circulation pump of the building withdraws water from the warm line, uses it in a heat pump to reach temperatures suitable for space heating and then discharges the cooled water to the cold line. In the case of a cooling demand, the system works in the other direction. The system requires a complex regulation of both the energy supply network and the end-users' substations and does not solve the problem of requiring two pipelines (one for heating and one for cooling). Finally, a great challenge in district heating networks is the hygienic preparation of DHW if supply temperatures fall within a certain range (25~55℃): to avoid the origin and the proliferation of legionella in a centralized DHW system, minimum temperatures must be guaranteed.

2. The Principle of the Layout of the District Heating Pipeline Network

The layout principle of heating pipeline network is to consider the distribution of heat load, the location of heat source, the relationship with all kinds of above-ground and underground pipelines, buildings, gardens and green space, hydrology, geological conditions and other factors under the guidance of urban construction planning, and determine by technical and economic comparison. The following basic principles shall be observed in determining the plane position of the heating pipeline.

(1) Technically Reliable

The lines of heating pipelines should go as far as possible in areas with flat terrain, good soil quality and low underground water level. Meanwhile, factors such as quick elimination of possible faults and accidents, safety of maintenance personnel, feasibility of construction and installation should also be considered.

(2) Economically Reasonable

The layout of the main line is short and straight, and the layout of the main line passes through the heat load concentration area. Attention should be paid to the proper arrangement of piping accessories such as valves on the line, as this will involve the location and number of examination rooms, and the number should be reduced as much as possible.

(3) Less impact on the environment and the coordination

Heat supply lines should cut through the main lines of communication. Generally they are parallel to the road center line and should be laid outside the roadway as far as possible. Normally the pipeline should run along only one side of the street.

3. Pipeline Network Laying Method and Structure

The laying of heating pipelines is divided into two categories: overground laying and underground laying.

Overground laying is to lay heating pipes on some independent or truss type supports on the ground, so it is also called overhead laying. Overhead laying is suitable for areas with high groundwater level, large annual rainfall, collapsible loess or corrosive soil, high density of existing underground facilities along the pipeline, and too much earthwork during underground laying. It is mostly used in urban fringe, non-residential areas and industrial areas. According to the different height of the support structure, it can be divided into low support laying, middle support laying and high support laying.

Underground laying does not affect the city appearance and traffic, so it is a widely used laying method for urban central heating pipelines. It is divided into two kinds: trench laying and directly buried laying. Trench laying is to lay pipes in underground trench. Trench is the enveloping structure of underground pipeline laying. Direct buried pipe is directly buried in the soil. In hot water heating pipeline network, direct buried pipe has been widely used at home and abroad.

4. Accessories and Structures

(1) Pipe Flex Compensator

In order to prevent deformation or damage of pipelines caused by thermal elongation or thermal stress when heating pipelines, it is necessary to set a telescopic compensator on pipelines to compensate thermal elongation of pipelines, so as to reduce the stress on the pipe wall and the acting force on the valve parts or bracket structure.

There are many types of compensators used in heating pipelines. The main types are pipe natural compensation, square compensator, corrugated compensator, sleeve compensator and spherical compensator. The first three make use of compensating the deformation of the material to absorb heat elongation, and the last two make use of the displacement of the pipe to absorb heat elongation.

(2) Pipe support

The support of the heating pipe is the main member between the supporting structure and the pipe. It supports the pipe or restricts the deformation and displacement of the pipe, bears the force generated by the pipe, and transmits the force to the supporting structure.

There are two types of supports used in heating pipelines: movable supports and fixed supports. The movable support is the pipe support which allows the pipe and supporting structure to have relative displacement. According to its structure and function, it can be divided into sliding support, rolling support, hanging support and guiding support. Fixed support is a kind of support that does not allow relative displacement between the pipe and the supporting structure. It is mainly used to divide the pipe into several compensation pipe segments for heat compensation, so as to ensure the normal operation of the compensator. The most commonly used is the metal structure of the fixed support, there are ring-curved fixed support and baffle-type fixed support.

7.5 Urban Flood Control Projects

7.5.1 Levee Projects

The levee is an effective engineering measure to resist flood disasters and reduce disaster losses. An embankment (sometimes considered to be the same as a dike) is designed to control a stream in a defined course or to prevent a body of water from intruding into areas to be protected. River walls and sea walls also serve such purposes. It is an important flood control project that was first widely used in the world.

The levee projects include planning, design, construction, flood prevention, rescue and maintenance management. Levee breach occurs due to the overtopping flow or seepage, which can trigger catastrophic flood damage and result in huge social and economic losses. This section mainly introduces the design and planning of levee projects.

1. Principle and Function of Levee Projects

Frequently happened flood disasters often cause heavy losses to the national economy and people's livelihood. Levees are constructed to protect the relevant area from floods.

(1) Principles for the Planning of Levees

No matter the renovation of the old levee system or the establishment of the new levee system, a comprehensive plan must be carried out according to the short-term and long-term flood control requirements, combined with the specific local conditions, and following the general principles as follows.

1) The levee projects should be incorporated into the plan for the comprehensive utilization of water resources in the river basin, and strive to achieve the most effective and economical purpose of flood control with the cooperation of other flood control measures.

2) Not only must the technical and economic aspects of the levees be taken into account, but also landscape and urban environment issues must be considered.

3) Set different flood control standards and levee sections according to different rivers and different sections.

4) The impact of the city levees on the upstream and downstream reaches of the river must be carefully studied, to detect undesirable consequences.

(2) Types of Levees

1) Levees can be divided into earth levees, earth-rock mixed levees, stone levees, reinforced concrete levees, etc., according to the materials of the levees.

2) According to their positions and functions, levees can be divided into flood dike, seawall dike and canal dike. Among them, flood dike can be divided into river dike, lake dike, reservoir dike, and flood storage and detention dike. Seawall dikes are built at the seashore to protect against high tides and wind waves; The canal dike is built on both sides of the channel to carry water diversion or drainage.

2. Design and Construction Technology of Levee Projects

(1) Selection of Embankment Lines

When selecting the embankment line, multi-scheme demonstrations and comparisons should be made according to the trend of the river, the evolution characteristics of the river and the socio-economic conditions of both sides of the bank.

1) The embankment line should be as parallel as possible to the flood flow, avoiding sharp bends and local protrusions;

2) The embankment line should not be too close to the river channel, so as to avoid the evolution of the river bed or the collapse of the river bank, which may endanger the safety of the embankment;

3) The embankment line should be arranged in the area where the land is occupied with few demolished houses, and the cultural relics should be avoided. At the same time, it should be conducive to flood control and

project management;

4) The layout of lake and seawall lines should avoid direct attack by strong winds or storm tides;

5) The layout of urban flood control levees should be coordinated with municipal facilities;

6) The levee projects should use the existing embankment and favorable topography to build on the relatively stable beach with good soil quality. Beach land of appropriate width should be reserved, and weak foundations, deep water zones, ancient river channels and strong permeable foundations should be avoided.

(2) Levee Spacing and Levee Roof Elevation Determination

The distance between levees and the elevation of the levees shall be determined according to the flood control standards in different areas, and then the distance between the levees along the river shall be arranged according to the principles of the selection of the levees.

The elevation of levee top is the sum of the designed flood level and the superelevation of levee top, which should be calculated by the following formula

$$Y=R+e+A \quad (7\text{-}5)$$

where Y— the elevation of levee top;

R— the wave climbing design;

e — the height of wind obstructing water surface design; for seawall, when the designed high water level includes the surface height of wind shove, it is not calculated separately;

A— the security enhancement.

(3) Smart Levee Monitoring System

The mechanisms that cause failures in levees are divided into two main categories: structural failure including damage to the embankment from physical disturbance and failure due to the hydraulic forces such as underseepage, overtopping or wave erosion, piping and liquefaction. Under such circumstances, the construction of smart levee has gradually received attention from all countries.

Smart levees are increasingly considered as an important part of a future flood protection infrastructure enabling real-time monitoring of flood embankments and early warning of approaching disasters. A flood early warning and decision support system needs to rely on advanced middleware services that provide urgent computing capabilities, so that results of resource-intensive data analyses and simulations, required for decision-making in emergencies, are delivered in a timely fashion.

Fig. 7-17 presents a high-level overview of a smart levee monitoring and flood decision support system. The system's three main functions are reflected in the user interface layer: real-time monitoring of the environment (water level, temperature and pore pressure inside the levee, etc.), flood decision support, and management of controlled flooding experiments using the experimental smart levee (specific to integrated smart leeve monitoring and flood decision support system). Here we focus on the decision support aspect implemented in middle layers responsible for workflow (business logic) and execution management (execution middleware), also called the central system. The subsystems connected to the central system are as follows.

7.5.2 Flood Diversion Storage Projects

1. Planning and Design of Flood Diversion Area

The flood diversion and storage area is an important main part of the projects. It is generally selected on the banks near the river reach where the flood discharge capacity is less than the safe discharge volume. Based on the flood control planning of the river basin and the river reach, and combined with comprehensive utilization and comprehensive treatment, the task and layout of flood diversion and storage project can be determined according to local conditions.

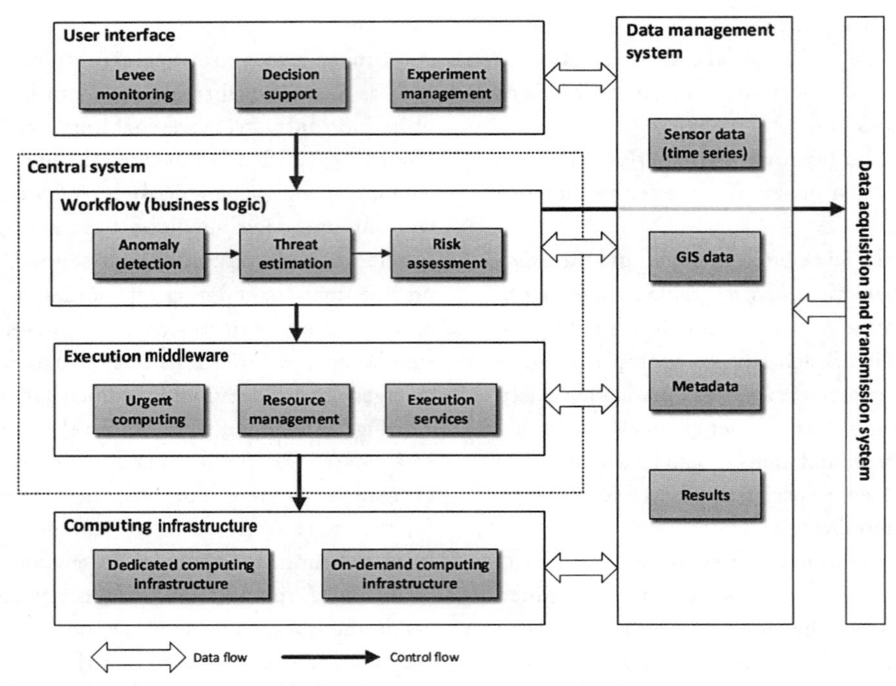

Fig. 7-17 Overview of subsystems and components of a smart levee monitoring and flood decision support system[50]

Flood diversion and storage areas are divided into flood storage areas and flood detention areas according to their different flood regulation functions. Some flood diversion areas first act as flood storage and then detention. Flood diversion area is generally composed of flood avoidance safety facilities, alarm communication system, flood entry facilities and discharge facilities in the dike area of flood diversion area. The following factors should be considered in the selection of flood diversion area:

(1) It should be built in low-lying areas as far as possible to ensure that the flood storage volume is large, the submergence loss is small, and the amount of dike engineering is small;

(2) As close as possible to the protected embankment section, the flood control function can be maximized;

(3) Separate discharge methods should be selected according to local conditions, and flood inlet and discharge should be controlled separately.

2. Layout of Hydraulic Structures

Flood entry facilities in flood diversion and storage areas can be divided into controlled, semi-controlled and uncontrolled facilities. Controlled facilities are controlled by entrance gate. When the flood rises to or near the guaranteed water level, the guaranteed water level is generally regarded as the flood diversion water level. The semi-control is the flood diversion area with a reserved gap at the inlet, and the flood diversion level should be slightly lower than the guaranteed level. Uncontrolled engineering is imported, and the water level of flood diversion is generally lower than the guaranteed water level of flood control.

Controlled flood entry facilities are generally proposed to set up a flood entry gate at the inlet, and the flood discharge can be controlled by the flood entry gate. The study indicates that if gate is adjusted appropriately, it will greatly improve and ease the high water level risk. When selecting gate address, the following points should be considered:

(1) The location of the gate should be in the upper reaches of the protected area, but it is generally not allowed to be in the lower reaches of larger cities;

(2) The gate location is generally selected in a relatively stable curve or straight section;

(3) The diversion angle between the axis of the gate hole and the direction of the river flow should not be too large.

The function of the flood discharge facilities in the flood discharge area is not only to discharge storage water quickly, but more importantly to undertake multiple flood discharge tasks throughout the flood season, so that the flood discharge facilities can be used for multiple times in the flood discharge area. There are two kinds of flood discharge facilities in flood diversion and storage areas: sluice gate and temporary scraper. The selection of facilities can be based on the frequency of use and investment benefits.

Flood control to mitigate hazardous flood waters in rivers and watershed is of vital importance for inundation prevention, flood risk management and water resource management. By operating in-stream hydraulic structures (e.g. reservoirs, dams, floodgates, spillways, etc.), peak flood discharges and high water stages in channels during storms can be reduced so that overflow and overtopping on levees, as well as the resulting inland flooding and inundation are eventually prevented. In case of emergency when flood waters are predicted to exceed capacities of river reaches, controls of flood water diversion/withdrawal through flood gates, diversion channels, or deliberate levee breaching can quickly mitigate flood water stages over the target reaches and even the entire watershed. Among them is the optimal flood control that can give the best cost-effective flood control schedule to minimize the risk of hazardous flood waters.

3. Practical Case of Water Conservancy Projects

The Dujiangyan Irrigation System , which remains in use after 2000 years, was designed and constructed by famous flood designers Li Bing and his son. "In the one side to take water for irrigation, in another side to transport sediment for flood management" reflects part of Li Bing's flood management theory applied in the Dujiangyan Irrigation System. The flood management policy aimed to divide the stream flow to reduce flood risk and to irrigate fields.

(1) The main part of this system is Yuzui or Fish Mouth, shown as "A" in Fig. 7-18. It is named for its conical head that is said to resemble the mouth of a fish. It is an important structure that divides the water into inner and outer streams. The division for the inner stream is normally 40%~60% of flow during floods. The inner stream carries the river's flow into the irrigation system, and the outer stream drains the rest, flushing out much of the silt and sediment.

(2) Feishayan or Flying Sand Weir shown as "G" in Fig. 7-18 is about 200m wide and is designed to direct the water from the inner stream to the outer stream. This component cleans water by drawing out the large sediment to reduce the water level and also ensures against flooding by allowing the flow of the water to drain from the inner to the outer stream. Li Bing's original weighted bamboo baskets have been replaced with modern reinforced concrete weirs.

(3) Baopingkou or Bottle-neck Mouth, shown as "I" in Fig. 7-18, is the final and main part of this system. It conveys clean water to the irrigation channel and works as a check gate, creating whirlpool flow that carries away excess water over Flying Sand Weir or the narrow entrance near Bottle-neck Mouth between Lidui Park (J) and Renzi Levee (H) (Fig. 7-18) to ensure against flooding.

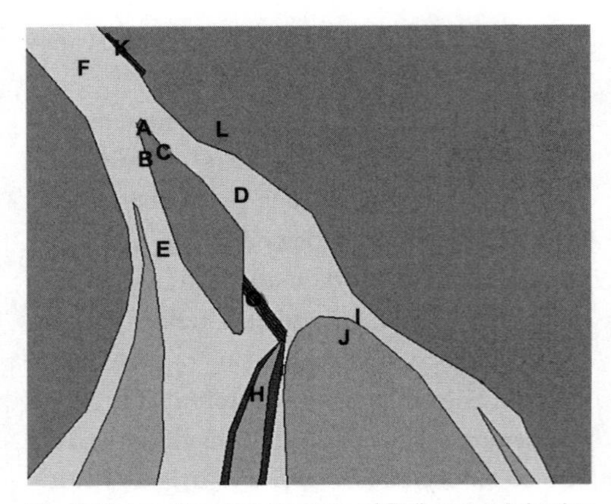

Fig. 7-18 Location and structure of Dujiangyan Irrigation System[57]
A-Fish mouth; B-Outer river dike (Jingang Dike); C-Inner river dike (Jingang Dike); D-Inner river; E-Outer river; F-Mingjiang River; G-Feishayan (Drainage Dam); H-Renzi Levee; I-Bottle-neck Mouth; J-Lidui Park; K- Baizhang Levee (Dike); L-Erwang Temple

7.5.3 Stormwater Management System

1. Foreign Urban Flood Management Philosophy

In the past few decades, there has been growing awareness of the value of stormwater as a resource to be factored into urban development. This has been driven by various trends including rising populations and increased water demand, increased environmental awareness, risk of storm damage exacerbated by climate change, and growth in urban areas and related impervious surfaces. There has been a parallel emergence in many countries of more sustainable paradigms for urban stormwater management including water-sensitive urban design (WSUD) in Australia, sustainable urban drainage systems (SUDS) in Britain, and low-impact development (LID) in North America. As opposed to conventional drainage approaches, which treat stormwater as a nuisance to be removed from the urban area as quickly as possible, the sustainable management of stormwater sees it as a multi-functional resource with many potential benefits for society and the environment if managed wisely.

(1) Stormwater best management practices (BMPs) are designed to control surface runoff in urban areas by means of detention/retention, infiltration and filtration. Other than the filtration/flow-through type BMPs which do not change very much, the local hydrological conditions, the remaining BMPs can generally be classified into the following two types according to their primary hydrological functions: the storage/retention-based BMPs and the infiltration-based BMPs. The best management practices consist mainly of delay and retention devices, infiltration equipment and wetlands. These infrastructures are composed of structural facilities and non-structural measures. Structural measures are devices for temporary storage and treatment of rainwater; Non-structural measures refer to technical treatment of stormwater runoff that reduces pollution levels through natural regulation.

Low-impact development was first introduced in Maryland as a means to mitigate the effects of increased impervious surfaces, though some individual techniques were already in place before the term "low-impact development" was coined. LID aims to reduce stormwater management costs by considering a site's natural features in the design. LIDs are generally very cost-effective, are climate-resilient and suitable for urban developments which plan to be sustainable.

(2) Sustainable stormwater management (SSWM) is closely related to urban planning and landscape design. SUDS include different technologies and techniques used to drain storm water/surface water in a manner that is more sustainable than conventional solutions. To further promote SSWM, evidence is needed based on context-specific work of the interconnections between SSWM measures and benefits to human beings and nature, including improved local quality of life (blue-green landscapes, better urban microclimate), reduced municipal expenses (on irrigation), and contribution to ecological goals.

(3) Water sensitive urban design (WSUD)

is a recent planning and design philosophy in Australia primarily used to minimise the hydrological impacts of urban development on the surrounding environment. Oswaldo once pointed out that the concept of water sensitive urban design means completely changing the logic of water supply system adjustment according to the needs of urban adaptive methods, reducing the huge impact brought by urbanization, and at the same time giving full play to the potential of landscape and urban living groundwater.

(4) Urbanization has put pressure on existing stormwater systems, yielding high flood rates, degraded urban aquatic habitat and low water quality in lakes and rivers. Cities increasingly rely on green infrastructure (GI) as stormwater solutions. GI can be defined as the concept of (semi-) natural structures strategically structured in networks and characterized by their multi-functionality. Green infrastructure used in urban interventions can improve the percolation, transpiration and evaporation of rainwater, reduce the heat island effect and create a better climate framework. In addition to playing a role in stormwater management, green infrastructure can help reduce flooding and improve air quality.

2. Domestic Urban Flood Management Philosophy

(1) President Xi Jinping announced in December 2013 a plan to decrease the impacts of flooding events in China, as a response to repeated serious flooding occurring annually. The main aim was to transform current cities into sponge cities by upgrading the existing urban drainage infrastructure and utilizing more naturally inspired drainage systems. It was believed that this, would reduce the magnitude and frequency of flooding events. This specific program was inspired by LID in the US, SUDS in the UK, and WSUD in Australia.Water environment and water ecology are systematic problems across scales and regions. It is the core of sponge cities to build water ecological infrastructure across scales and combining multiple specific basic technologies from the perspective of ecosystem services.

The sponge city implementation process consists of four phases. Phase 1 is analyzing regional context including water issues and existing water management to identify the demand for sponge city implementation. The next phase is developing scenarios based on climate change scenarios, population growth scenarios and water demand scenarios. Phase 3 indicates the selection and development of modelling software to simulate the performance of sponge city measurement. The final phase is the planning and implementation of sponge city.

Sponge city is a new urban construction concept, as well as an urban planning and design method. It is also a technology that integrates landscape design, architectural design and urban design. The concept must extend from land planning and design to the back garden of each family, the roof of the building and even the interior of the building.

(2) Smart city refers to the integration of urban systems and services by using various information technologies or innovative concepts, so as to improve the efficiency of resource utilization and optimize urban management and services. Smart levees are increasingly considered as an important part of a future flood protection infrastructure enabling real-time monitoring of flood embankments and early warning of approaching disasters. Many new smart technologies have been proposed to improve the operations, performance and capabilities of existing or new infrastructures by adding sensing, controls, communications and computing. New tools can help capture information, analyze data and control infrastructure systems. Many different tools have been tested in water management to monitor, control and manage water volume and quality, and to support decision-making practices in water distribution networks.

In summary, an integrated system of flood monitoring, forecasting and warning system, includes weather prediction, flood monitoring and forecasting, flood dispatching and flood control discussion, has play a vital role in basin flood control.

(3) Rain sewage diversion means the separate discharge of rainwater and sewage, rainwater is directly discharged through the pipeline network, and sewage can only be discharged after being treated and meeting the drainage requirements. Rain sewage diversion is advantageous for the concentration of rainwater utilization and wastewater treatment, ensuring the processing efficiency of wastewater treatment plant.

3. Opportunities and Challenges of Stormwater Management Systems

Floods are among the world's most devastating natural disasters, causing immense damage and accounting for a large number of deaths worldwide. Good flood management policies play an extremely important role in preventing floods.

Integrated water resources management is an overall design strategy aiming at the hydrological cycle, which takes various water sources into consideration comprehensively, makes them meet different requirements or terminal demands, and establishes appropriate hydrological levels to meet future demands. Development of effective flood management policies in the future will require consideration of both historical flood management policies and the future climate conditions.

There is still a long way to go in our efforts to mitigate the problems associated with urban stormwater management (e.g. waterlogging) in Chinese cities, which continues to be a very complex and challenging issue. With the anticipated impacts of climate change, the increasing trend (i.e. more frequent and greater magnitude) of regional hydroclimatic extremes (especially floods) is very likely to continue, and even accelerate in the future. As a result, more meticulous and rational actions and policies should be developed and undertaken, for achieving further development to improve the sustainable urban stormwater management in Chinese cities and cities around the world more broadly.

7.6 Municipal Waste Treatment Projects

7.6.1 Municipal Solid Waste

1. Classification of Municipal Solid Waste

Municipal solid waste (MSW) refers to the solid waste generated from the daily life of urban residents or activities providing services for urban daily life. MSW includes durable goods, non-durable goods, containers and packaging, food wastes and yard trimmings, and miscellaneous inorganic wastes. Sometimes it is also referred as garbage.

MSW can be divided into three categories: biodegradable, non-biodegradable and inert components. Inert component refers to the inorganic component in garbage which is not harmful to the environment, such as concrete, macadam, brick, tile, etc. In China, the physical components of MSW typically include food waste, paper, textile, rubber, plastic, glass, metals, wood and miscellaneous inorganic wastes (e.g., stones, ceramics and ashes). The percentage of content of MSW in different continents shows noteworthy difference, as shown in Table 7-5.

2. Properties of Municipal Solid Waste

(1) Physical Properties

The physical properties of MSW mainly include density, moisture content, water holding degree, permeability, etc. And the physical properties of municipal solid waste

Composition of MSW from different continents Table 7-5

Continent	Organic (%)	Paper (%)	Plastic (%)	Glass (%)	Metal (%)	Others (%)
Europe	35	21	10	6	4	24
Asia	51	14	11	4	4	16
South America	51	15	11	3	3	17
North America	42	22	13	3	6	14
Australia	52	18	8	5	4	13
Africa	56	9	10	3	3	19

are directly related to its composition. Table 7-6 and Table 7-7 show weight and moisture content for typical wastes.

(2) Chemical Properties

The chemical properties of MSW affect the choice of treatment methods, particularly in terms of energy recovery and incineration. It includes proximate analysis, ash melting point, elemental analysis, energy value, etc. Table 7-8 shows examples for typical proximate analysis values of wastes.

Table 7-9 and Table 7-10 show composition content and elemental analysis.

(3) Biological Properties

The main biological characteristics of MSW are that almost all organic components can be converted into gases and relatively stable organic and inorganic solid materials by biology (Table 7-11 and Table 7-12).

3. Environmental Impacts of Municipal Solid Waste

With the rapid development of China's economy and the acceleration of urbanization, the pollution of municipal solid waste is becoming more and more serious. It is necessary for us to pay attention to its environmental pollution, which mainly includes the pollution to soil, air, water and human beings.

(1) Soil

Firstly, soil is home to many bacteria,

Typical specific weight values of wastes Table 7-6

Component	Specific weight (density)(kg/m^3)	
	Range	Typical
Food waste	130~480	290
Paper	40~130	89
Plastic	40~130	64
Yard waste	65~225	100
Glass	160~480	194
Tin can	50~160	89
Aluminum	65~240	160

Typical moisture contents of wastes **Table 7-7**

Component	Moisture content (%)	
	Range	Typical
Food waste (mixed)	50~80	70
Paper	4~10	6
Plastic	1~4	2
Yard waste	30~80	60
Glass	1~4	2
Mixed demolition combustible	4~15	8
Mixed construction combustible	4~15	8
Sawdust	10~40	20
Wood (mixed)	30~60	35
Mixed agricultural waste	40~80	50

Typical proximate analysis values of wastes (%) **Table 7-8**

Type of waste	Moisture	Volatiles	Carbon	Ash
Mixed food	70.0	21.4	3.6	5.0
Mixed paper	10.2	75.9	8.4	5.5
Mixed plastic	0.2	95.8	2.0	2.0
Yard waste	60.0	32.3	7.3	0.4
Glass	1.0~4.0	—	—	96.0~99.0
Residential municipal solid waste	21.0	52.0	7.0	20.0

Chemical compositions of typical municipal solid waste **Table 7-9**

Composition	Content
Water	28.0%
Carbon	25.0%
Oxygen	21.0%
Hydrogen	3.3%
Glass-ceramic	9.3%
Metal	7.2%
Ash	5.5%
Sulfur	0.1%
Nitrogen	0.5%
Others	0.1%

Typical data in elemental analysis (% by weight) **Table 7-10**

Type	C	H	O	N	S	Ash
Mixed food	73.0	11.5	14.8	0.4	0.1	0.2
Mixed paper	43.3	5.8	44.3	0.3	0.2	6.1
Mixed plastic	60.0	7.2	22.8	—	—	10.0
Yard waste	46.0	6.0	38.0	3.4	0.3	6.3
Refuse-derived fuel	44.7	6.2	38.4	0.7	<0.1	9.9

Data on the biodegradable fraction of selected organic waste components based on lignin content

Table 7-11

Component	Volatile solids, percent of total solids(%)	Lignin content, percent of volatile solid(%)	Biodegradable fraction
Food waste	7.0~15.0	0.4	0.82
Newsprint	94.0	21.9	0.22
Office paper	96.4	0.4	0.82
Cardboard	94.0	12.9	0.47
Yard waste	50.0~90.0	4.1	0.72

Calculation of biodegradable fraction of municipal solid waste **Table 7-12**

Component	Percent of municipal solid waste (%)	Biodegradable fraction
Paper and paperboard	37.6	0.50
Glass	5.5	0
Ferrous metal	5.7	0
Aluminum	1.3	0
Other nonferrous metal	0.6	0
Plastic	9.9	0
Rubber and leather	3.0	0
Textile	3.8	0.50
Wood	5.3	0.70
Other materials	1.8	0.50
Food waste	10.1	0.82
Yard trimming	12.8	0.72
Miscellaneous inorganic	1.5	0.80

fungi and other microorganisms that form an ecosystem with their surroundings. Under the action of rainwater carrier, solid waste is easy to penetrate into the soil layer and kill microorganisms in the soil layer, thus destroying the balance of soil ecosystem and changing the soil quality, which has a bad impact on the soil ecosystem. Secondly, a large amount of urban household garbage is piled up in a lot of land areas. Although garbage is mostly piled up in suburbs far away from the city, the occupied farmland area is not small, which exacerbates the problem of farmland shortage to some extent.

(2) Air

Some solid wastes that are stacked contain fine particles and dust which can fly in the wind, into the atmosphere and spread to far away places. They can produce pollution to the atmosphere, and these dust and small particles fly into the atmosphere, which will make the visibility of the atmosphere greatly reduced, causing haze. Some organic solid wastes can be decomposed by microorganisms under appropriate temperature and humidity, releasing harmful gases, poisonous gases or odors, causing regional air pollution and even explosion.

(3) Water

Dumping MSW into water bodies will directly pollute water bodies, seriously endanger the living conditions of aquatic organisms and affect the full utilization of water resources. Dumping solid wastes into water bodies would also reduce the effective surface area of rivers and lakes, reducing their capacity for drainage and irrigation. In addition, solid wastes can enter rivers and lakes with surface runoff or wind migration into water bodies, which will bring toxic and harmful substances into water bodies, kill aquatic organisms and pollute human drinking water sources. Solid wastes deposited on land or simply buried in landfills will produce leachate after impregnation by rainwater and decomposition of the wastes themselves, which will increase the content of heavy metals and organic pollutants in groundwater and reduce the water quality.

(4) Human Beings

All these effects will eventually be harmful to human life. When a large amount of garbage is exposed to the outside, mosquitoes, flies, rodents and other animals will breed, laying a hidden danger for the spread of diseases. In the process of municipal solid waste being piled up or buried, a large number of acid and alkaline pollutants will be produced. In case of rainy days, the harmful components contained in the garbage can easily enter the water body and soil. These harmful ingredients will be concentrated in the plant, once ingested, it will pose a great threat to people's health and life safety.

7.6.2 Municipal Refuse Disposal Area

With the advancement of urbanization, the improvement of people's living standard and the development of social economy, the output of municipal solid waste is increasing year by year. During the disposal, different treatment methods should be selected in combination with the characteristics of solid waste to achieve harmless treatment of garbage. Common treatments include landfill, incineration and composting.

1. Landfill Treatment for Municipal Refuse

(1) What is Landfill

A landfill site is a site for the disposal of waste materials by burial, either by virtue of the natural topography or by means of artificial construction, in an underground space, compassed, covered and sealed.

There are two landfill methods for solid waste: direct landfill method and sanitary landfill method. Direct landfill means that the collected solid waste is directly placed in a reasonable space, and after sealing and compaction, the organic components of the waste are used for degradation reaction to minimize the pollution capacity of the waste. Sanitary landfill method refers to the treatment of solid waste in a specific space, with the help of seepage prevention, soil covering, etc., to reduce the pollution of waste to the water environment and atmospheric environment.

The landfill method has the advantages of simple treatment, low cost, wide application and large processing capacity. However, this method also has some disadvantages, such as a wide range of environmental pollution, easy pollution of groundwater resources, large land possession, methane gas generation, etc.

(2) Social and Environmental Impact of Landfill

Leachate—Some pollutants will leak from landfills and pollute groundwater nearby. The results indicate that 96 kinds of pollutants have been detected in groundwater near landfills distributed throughout China, and 22 kinds of pollutants were considered to be the most important.

Dangerous gases—Rotting food and other decaying organic waste allows methane and carbon dioxide to seep out of the ground and up into the air. Methane is a potent greenhouse gas, and can itself be a danger because it is flammable and potentially explosive.

Infections—Poorly run landfills may become nuisances because of vectors such as rats and flies which can cause infectious diseases. The occurrence of such vectors can be mitigated through the use of daily cover.

2. Incineration Technology for Municipal Refuse

(1) What is Incineration

Incineration is a waste treatment process that involves the combustion of organic substances contained in waste materials. Incineration of waste materials converts the waste into ash, flue gas and heat. An incinerator is a furnace for burning waste. Modern incinerators include pollution mitigation equipment such as flue gas cleaning.

The advantages of this kind of garbage disposal method are as follows: less floor space, low site requirements, short operation cycle, innocuity and completeness. At the same time, the heat produced by burning can be used for heating and power generation. However, a range of air pollutants, particularly dioxins, are produced in MSW incineration and can be released into the atmosphere in significant quantities if the incinerator and the flue gas cleaning system are not properly designed and operated.

(2) Gaseous Emissions

Dioxin and furan—Dioxins are chlorine-containing organic compounds containing two or one oxygen bond to two benzene rings. Dioxins can enter the human body through the skin, respiratory tract, digestive tract, etc., and are highly toxic organic chemicals containing chlorine. The amount through food, especially lipids and through the digestive tract into the human body accounts for more than 90%. They accumulate in the fat and liver, to a certain extent that will cause a lot of adverse effects. The effects of dioxins on the body can be summarized into three aspects: reduced immune function, changes in reproductive and genetic functions and susceptibility to malignant tumors.

CO_2—Carbon dioxide is an inorganic substance, which in room temperature is a colorless odorless gas, and non-toxic, slightly more dense than air that can dissolve in water. Carbon dioxide has a heat preservation effect. When the content of carbon dioxide in the atmosphere increases, there will be a greenhouse effect, leading to climate warming, sea level rise, land desertification and so on, affecting human survival.

3. Composting for Municipal Refuse

(1) What is Composting

Using the natural process of decay to change organic wastes into a valuable humus-like material is called compost. A compost pile needs decomposers (mainly bacteria and fungi), food for the decomposers (mainly organic materials), the right amount of air, water and warmth.

Composting method can be divided into two treatment methods: aerobic method and anaerobic method. It has the advantages of effectively reducing the amount of garbage, realizing the reuse of resources and high utilization rate of resources. However, the disadvantages are that the treatment mode is relatively simple, the application scope is narrow, the treatment is not thorough, the occupation is large, the cycle is long, the sanitation condition is poor and so on.

(2) Materials for Decomposers

The compost workers will thrive if a balanced diet is given. Composting will be most rapid if the decomposers are fed a mix of carbon-rich and nitrogen-rich materials. Carbon-rich organic wastes are known as "browns", such as leaves, straw, paper, sawdust, animal bedding mixed with manure. Nitrogen-rich organic wastes are known as "greens", such as vegetable scraps, coffee grounds, grass clippings, manure. In particular, the materials containing inorganics that are hard to degrade, toxic organics and heavy metals should be avoided.

4. New Treatments for Municipal Refuse

Based on the three industrialized technologies of municipal refuse disposal—landfill, incineration and composting, a batch of forward-looking, pioneering and exploratory technologies have emerged gradually through years of development and technological update.

(1) Anoxic Pyrolysis Gasification

Pyrolysis is to take advantage of the thermal instability of organic matter and make the organic matter with large molecular weight pyrolyze under the condition of hypoxia, which is converted into fuel gas and liquid (oil and grease) with small molecular weight. It involves the change of chemical composition and is irreversible.

Gasification is a process that converts organic or fossil fuel-based carbonaceous materials into carbon monoxide, hydrogen and carbon dioxide. This is achieved by reacting the material at high temperatures ($>700℃$), without combustion, with a controlled amount of oxygen and/or steam. The resulting gas mixture is called syngas (from synthesis gas) or producer gas and is a fuel itself.

Pyrolysis and gasification recover energy (oil or fuel gas) that can be stored and transported from waste. In pyrolysis and gasification, the volatile matter influences composition and yield of the gas products. The higher the volatile matter content is, the more gas pyrolysis and gasification will generate. However, due to the large types of waste, large changes, complex components and the need for stable and continuous operation, the technical and operational requirements are very strict and the treatment cost is high. So far, the industrialized pyrolysis or gasification technologies in the world are very limited, especially in the treatment of urban garbage.

(2) Anaerobic Digestion

Anaerobic digestion refers to the process in which organic matter stabilizes under anaerobic conditions through the catabolism of anaerobes and facultative bacteria, releasing methane and carbon dioxide at the same time. Anaerobic digestion is one of the effective ways to treat organic solid waste, which can produce clean energy biogas and high quality organic fertilizer.

In China, anaerobic digestion is widely used in high concentration of organic wastewater, sludge, feces and agricultural straw treatment. However, municipal waste is different, especially at the present stage, mixed collection of municipal waste, the composition and properties of complex, its substrate fermentation conditions, microbial growth conditions and catabolism and so on still need to be further studied. With the active development of garbage classification, the separate collection of organic and inorganic garbage has a positive effect on the application and promotion of anaerobic digestion.

(3) Plasma Gasification

Plasma is a state in which matter exists,

along with solid, liquid and gas, known as the fourth state of matter. The morphology and properties of plasma are strongly affected by external magnetic field, and there is abundant collective movement. Therefore, the energy of plasma is especially concentrated and it has extremely high electric heating efficiency (over 85%). The high temperature generated can reduce all-the-hard-to-reduce and insoluble materials.

Plasma gasification is a new waste incineration method, which uses high pressure electric arc to produce higher temperature than the surface of the sun. At these high temperatures, anything can go into a gas or a liquid and it's actually called a plasma. Its treatment process is the decomposition and recombination process of waste materials. Its working principle is to ionize the air to produce plasma through a strong electric arc in a confined space, and then in another oxygen-deprived confined space, the plasma to be produced will conduct ultra-high temperature heating for municipal solid waste. In the absence of oxidation, the inorganics in the garbage mixture are vitrified rapidly, and the resulting harmless slag can be used as building materials.

Plasma gasification uses an external heat source to gasify the waste, resulting in very little combustion. Almost all of the carbon is converted to fuel gas. Plasma gasification is the closest technology available to pure gasification. Because of the temperatures involved, all the tar, char and dioxins are broken down. The exit gas from the reactor is cleaner, and there is no ash at the bottom of the reactor. As the temperature is more than 1200℃, organic matter including infectious virus, bacteria and other toxic and harmful substances are all decomposed, the gas and ash produced are harmless. However, it uses electricity as the energy source, thus the economic cost is high.

Many domestic scientific research institutions have also started this research, and developed some plasma waste disposal systems to treat high-risk waste, such as medical waste, discarded tires and e-waste. They have achieved some results, but there is still a gap between marketization and industrialization.

(4) Bioreactor Landfill

Bioreactor landfill is a solid waste disposal landfill system that uses the biological functions of enzymes or organisms to carry out biochemical reactions in vitro. Bioreactor landfill can be divided into three types according to different operation modes: anaerobic type, aerobic type and quasi-aerobic type.

The anaerobic bioreactor landfill adopts the leachate recharging measures, which has the advantages of accelerating the stabilization of landfill waste, reducing the leachate concentration, and recycling and utilizing biogas, but its leachate ammonia nitrogen concentration is very high and COD degradation is slow in the later stage. The aerobic bioreactor landfill is filled with air while recharging leachate to maintain the state of aerobic reaction inside the landfill, which greatly speeds up the stabilization process of the landfill, but forced air blowing will increase the operating cost. Quasi-aerobic bioreactor landfill maintains local aerobic status of the landfill site by means of natural ventilation. Compared with the anaerobic type, it has a faster stable rate, lower ammonia nitrogen concentration in leachate, and no need for ventilation equipment and energy consumption. However, the methane content in the directly discharged gas is still relatively high, which is easy to cause secondary pollution.

Different from traditional landfill leachate recharging, bioreactor landfill leachate recharging is controllable, which provides an optimal living space for the mass reproduction of microorganisms. Therefore, it can achieve a faster degradation rate and realize rapid stabilization. Leachate recharging can promote the degradation of organic compounds in garbage, shorten the time of marsh-producing and increase the effective storage capacity of landfill site.

There are still some uncertainties in the bioreactor landfill technology in terms of durability, compactness, structural characteristics, oxidation-reduction environment and cost-benefit analysis. In addition, due to the limitations of the characteristics of anaerobic landfills, the leachate cannot be completely eliminated by recharging, and the leachate after recharging has a high ammonia nitrogen content, which still needs further treatment before discharge. Therefore, the bioreactor landfill technology still needs further research and improvement.

7.6.3 Municipal Garbage Collection and Transportation

1. Municipal Garbage Collection Options

Waste collection, often divided into "primary" and "secondary" services, is the process of picking up waste from residences or collection points, loading into vehicles and transporting them to locations for processing, transfer or disposal.

Primary collection is the means by which waste is collected from its source (dwellings and commercial premises) and transported to community storage, transfer points or even disposal sites. Secondary collection is the collection of solid waste for the second time, such as from community collection points, prior to its transport (often as part of a collection round by larger vehicles) to a transfer station, treatment facility or disposal site.

There are two types of waste collection: general and detailed. Each has different options and all have advantages and disadvantages, as shown in Table 7-13 and Table 7-14.

2. Municipal Garbage Transportation Equipment

Garbage transportation equipment refers to the equipment used for short-distance transportation after the collection of urban household garbage. The most important feature of these devices is that they can move easily in the residential areas and adapt to the road conditions of local urban residents.

(1) Dump Garbage Truck

Dump garbage truck is the most commonly used collection truck in China. It is generally divided into hood-type dump truck and sealed dump truck. The dump garbage truck is usually equipped with a forklift, which is convenient for mechanical loading on the upper part of the car and suitable for fixed container collection operation.

Comparison of general waste collection options **Table 7-13**

Collection option	Concept	Advantage	Disadvantage
Mixed collection	All kinds of municipal solid wastes are collected together without any treatment	Simple and easy to operate, low operating cost	It reduces the purity and reuse value of useful materials and increases the difficulty of municipal solid waste disposal
Separate collection	This collection method is according to the composition of municipal solid waste classification	It can improve the purity and quantity of recycled materials, reduce the amount of garbage that needs to be treated, and is conducive to the recycling and reduction of urban household garbage	The work process is complex and difficult, which requires a certain amount of economic strength. Publicity, active cooperation of residents and necessary conditions for classification and collection are required

Comparison of detailed waste collection options **Table 7-14**

Collection option	Operation	Comments
Door to door collection	Waste collector knocks on each door or rings doorbell and waits for waste to be brought out by residents	Convenient for residents, little waste on streets. Residents must be available to hand waste over. Not suitable for apartment buildings because of the amount of walking required
Block collection	Collector sounds horn and waits at specified locations for residents to bring waste to the collection vehicle	Economical, less waste on streets. No permanent container or storage to cause complaints. If all family members are out when the collector comes, waste must be left outside for collection, it may be scattered by wind, animal and waste pickers
Kerbside collection	Waste is left outside in a container and picked up by passing vehicles, or swept up and collected by sweepers	Convenient, no permanent public storage. Waste that is left out may be scattered by wind, animals, children or waste pickers
Backyard collection	Collection labor enters houses to remove wastes	Very convenient for residents, no waste on streets, the most expensive system. Cultural beliefs, security considerations or architectural styles may prevent laborers from entering properties
Dumping at designated location (Drop off centers)	Residents and other generators put their waste inside a stationary container which is emptied or removed	Low capital costs. Loading the waste into trucks is slow and unhygienic. Waste is scattered around the collection point. Adjacent residents and shopkeepers protest about the smell and appearance
Shared container	Residents and other generators put their waste inside a same container which is emptied or removed	Low operating costs. If containers are not maintained, they quickly corrode or are damaged. Adjacent residents complain about the smell and appearance

(2) Movable Bucket Garbage Truck

The movable bucket garbage truck is used as the movable open type storage container and is usually placed in the garbage collection point. It is also called multi-purpose vehicle, which is generally used for towing container collection.

(3) Side Mounted Sealed Garbage Truck

The inside of the side mounted sealed garbage truck is equipped with a hydraulic drive lifting mechanism, which is used to lift the supporting circular garbage can. The garbage can on the ground can be lifted to the top of the car and tipped over from the inverted entrance. The inverted entrance of the trash can have a cover, which opens and closes with the dumping action of the bucket.

(4) Rear-mounted Compressed Garbage Truck

The rear compartment of the rear-mounted compression garbage truck is equipped with a feeding inlet and a compression push plate device. Usually the height of the entrance is low, which can adapt to the residents of the elderly and children. At the same time, because of the compression push plate, it can adapt to large volume, small density of garbage collection.

3. Municipal Garbage Transfer Station

Municipal garbage transfer station is an important urban sanitation infrastructure, as well as a hub connecting residents' daily household garbage and garbage treatment plants and it plays a decisive role in maintaining urban environmental sanitation and ensuring public health.

Selecting appropriate locations for MSW management facilities, such as transfer stations, is an important issue in rapidly developing regions. Multiple alternatives and evaluation attributes need to be analyzed for finalizing the locations of these facilities. Multi-attribute decision-making (MADM) approaches are found to be very effective for ranking several potential locations and hence selecting the best among them based on the identified attributes. Some researchers are trying to improve and perfect these approaches , so that they can take more factors into consideration in the decision-making and be more holistic.

The traditional municipal garbage transfer station not only pollutes the environment, but also has serious odor and peculiar smell, breeding a large number of mosquitoes and flies, making the transfer station site selection a difficult problem.

The new environment-friendly garbage transfer station adopts environment-friendly compression equipment and has physical deodorization and closed operating system, which eliminates the odor, peculiar smell, mosquitoes and flies and secondary pollution generated in the process of garbage storage and treatment. Thus, the influence of the garbage station on the surrounding environment is completely eliminated, and the comprehensive operating cost is effectively reduced, realizing the scientific, environmental protection and humanized all-weather operation.

7.7 Conclusions

Since municipal engineering projects are concerned with public infrastructure and services provided by local government, they play a key role in improving the community's health and quality of life. Municipal engineering projects include the design, planning, construction and maintenance of streets, pavements, bicycle paths, public parks and related urban public facilities (street lighting, as well as street furniture and fixtures such as benches, bus shelters, litter bins, traffic control devices, playground equipment and road signs). The term "municipal engineering projects" also covers sanitary and storm sewer systems, municipal solid waste management and disposal facilities. Civil infrastructure (conduits and access chambers) related to utility services (water supply, electrical distribution and telecommunication networks) are also included within this term.

Engineers have been involved in the planning, design and construction of municipal water treatment systems for about 200 years. The last 30 or 40 years, however, have been a time of dramatic changes in the interrelationship between water quality and public health because of increases in scientific understanding and growing human impacts on water sources. As a result, the modern water treatment engineer faces an increasingly complex array of challenges, competing issues and compromises that must be balanced to successfully design a water treatment system. The overall impact of these complexities is a need for engineers to have a solid grasp on the scientific and fundamental principles underlying water treatment processes, rather than designing solely from the perspective of applying previously successful practices. Some of these complexities faced by water treatment engineers include:

(1) Modern water quality management practices must protect against and provide treatment for a wider array of potential sources of microbial contamination. Management does not only cover the utilization of water, it may also apply to the operation of the system to get the most efficient use of equipment to produce a better product at a better price or to get the supply system to last longer.

(2) Modern water treatment must balance the need to provide disinfection to prevent waterborne illness with the need to restrict disinfection to minimize chronic health effects.

(3) Water treatment practices must consider the impact of water distribution on water quality and balance the objectives at the plant effluent with the objectives at the point of use.

(4) Future water management practices must balance the level of water quality achieved with the actual use of the water, potentially supplying drinking water separately from water for other uses.

(5) Future water treatment practices must evaluate water treatment strategies from a holistic perspective that considers all benefits and impacts to the community, environment and society.

The issues make it clear that water treatment engineering continues to evolve. At the same time, the public's expectations for water quality have never been higher. An integration of past strategies and progressive tactics are essential as new challenges continue to surface and the fundamental mission expands.

Exercises

7-1 Please refer to the relevant information to understand the current new classification of road system.

7-2 Please contact the actual list of various bridges common in life.

7-3 At present, the rail transit system develops rapidly, please talk about its influence on the public transportation in our country.

7-4 Please talk about the views on the future development of urban public transport in combination with what you have learned.

7-5 What are the two basic layout forms of water supply system? Try to compare their strengths and weaknesses.

7-6 Please try to refer to the relevant information and talk about the depth treatment technology of drinking water commonly used in China at present.

7-7 In order to make the drainage system design meet the specification and reduce cost as much as possible, what principles should follow as far as possible?

7-8 What methods are often used in sewage treatment? If you were asked to design a sewage treatment plant, how would you arrange the sewage treatment process?

7-9 Please combine the knowledge of water treatment technology and house water supply and drainage engineering, and consider how to make buildings effectively use rainwater for reuse.

7-10 Try to analyze the problems that may be encountered in the design, construction and specification of the urban gas pipeline network.

7-11 Put forward some constructive opinions on the standardized construction of urban gas pipeline network.

7-12 Please talk about the understanding of heating network layout principle.

7-13 What else do you think can be done to further optimize the central heating system?

7-14 Combining actual conditions, try to introduce a new idea of flood control and corresponding flood control construction briefly.

7-15 Discuss how to establish the flood control security system of the whole society according to the national conditions.

7-16 Try to compare the similarities and differences between domestic and foreign stormwater management models.

7-17 Based on the current situation of flood management in China, discuss the problems that the sponge city concept may encounter in the future and put forward some feasible solutions.

7-18 What are the main pollutants in municipal waste? What harm do they each do?

7-19 How is methane produced in landfills? How to prevent methane gas explosion?

7-20 Why do yard and kitchen wastes compost? If you build it, what materials should be needed or avoided?

7-21 Please briefly describe the three reaction stages of anaerobic digestion and list the advantages and disadvantages of this waste disposal method.

7-22 What is the impact of the ongoing garbage classification on garbage collection and disposal method?

References

[1] BUCHAN N. Editorial: International federation of municipal engineering[J]. Proceedings of the Institution of Civil Engineers-Municipal Engineer, 2010, 163(4): 201-202.

[2] JENKINSON I. Briefing: Municipal engineer—the silver anniversary[J]. Proceedings of ICE-Municipal Engineer, 2009, 162(2): 65-68.

[3] LIU B, YAN L, WANG Z. Reclassification of urban road system: integrating three dimensions of mobility, activity and mode priority[J]. Transportation Research Procedia, 2017, 25: 627-638.

[4] Ministry of Housing and Urban-Rural Development of the People's Republic of China. Code for Design of Urban Road Engineering CJJ 37-2012[S]. Beijing: China Architecture & Building Press, 2012.

[5] WANG S G, YU D X, KWAN M D, et al. The impacts of road network density on motor vehicle travel: an empirical study of Chinese cities based on network theory[J]. Transportation Research Part A: Policy and Practice, 2020, 132: 144-156.

[6] ZHOU X, ZHANG X. Thoughts on the development of bridge technology in China[J]. Engineering, 2019, 5(6): 1120-1130.

[7] AN Y H, CHATZI E, SIM S H, et al. Recent progress and future trends on damage identification methods for bridge structures[J]. Structural Control and Health Monitoring, 2019, 26(10).

[8] BAO X. Urban rail transit present situation and future development trends in China: overall analysis based on National Policies and Strategic Plans in 2016–2020[J]. Urban Rail Transit, 2018, 4(1): 1-12.

[9] ZHAI W, HAN Z, CHEN Z, et al. Train–track–bridge dynamic interaction: a state-of-the-art review[J]. Vehicle System Dynamics, 2019, 57(7): 984-1027.

[10] TM A. Water Supply, Water Treatment[M]. Washington.D.C.: Departments of the Army and the Air Force, 1985.

[11] WANG C L, GUAN D, MI H R. Professional English on Water Science and Engineering[M]. Harbin: Harbin Engineering University Press, 2016.(in Chinese)

[12] YAN X S, LIU S Q. Water Supply and Sewerage Pipeline Network System(3rd Edition)[M]. Beijing: China Architecture&Building Press, 2014.(in Chinese)

[13] JIANG Y J, GOODWILL J E, TOBIASON J E, et al. Comparison of ferrate and ozone pre-oxidation on disinfection

byproduct formation from chlorination and chloramination[J]. Water Research, 2019, 156: 110-124.

[14] LU Z D, SUN W J, LI C, et al. Effect of granular activated carbon pore-size distribution on biological activated carbon filter performance[J]. Water Research, 2020, 177: 115768.

[15] BABBITT. Water Supply Engineering[M]. New York: McGraw-Hill Book Co., 1949.

[16] FORSTER C F. Wastewater Treatment and Technology[M]. London: Thomas Telford, 2003.

[17] STEEL E W. Water Supply and Sewerage[M]. New York: McGraw-Hill Book Co., 1960.

[18] ATIMTAY A T, SIKDAR S K. Security of Industrial Water Supply and Management[M]. Dordrecht: Springer Science+Business Media B.V, 2011.

[19] CHANG H, DUAN J. Natural gas measurement technology system and its prospect in China[J]. Natural Gas Industry B, 2020, 7(4): 370-379.

[20] Ministry of Construction, PRC. Code for Design of City Gas Engineering GB 50028-2006[S]. Beijing: China Architecture & Building Press, 2006.

[21] CHEN Q, ZUO L, WU C, et al. Supply adequacy assessment of a gas pipeline system based on the Latin hypercube sampling method under random demand[J]. Journal of Natural Gas Science and Engineering, 2019, 71: 102965.

[22] LI F, WANG W, DUBLJEVIC S, et al. Analysis on accident-causing factors of urban buried gas pipeline network by combining DEMATEL, ISM and BN methods[J]. Journal of Loss Prevention in the Process Industries, 2019, 61: 49-57.

[23] AI M. Discussion on issues regarding the reliability of large-scale oil and gas pipeline network systems[J]. Oil & Gas Storage Transportation, 2013, 32(12): 1265-1270.

[24] LI M F, ZHENG H L, XUE X D, et al. Reliability evaluation and management of Petro China's large-scale system of natural gas pipeline networks[J]. Journal of Natural Gas Geoscience, 2019, 4(5): 287-295.

[25] PELLEGRINI M, BIANCHINI A. The innovative concept of cold district heating networks: a literature review[J]. Energies, 2018, 11(1): 236.

[26] VESTERLUND M, TOFFOLO A. Design optimization of a district heating network expansion, a case study for the Town of Kiruna[J]. Applied Sciences, 2017, 7(5): 488.

[27] DANIELEWICZ J, ŚNIECHOWSKA B, SAYEGH M A, et al. Three-dimensional numerical model of heat losses from district heating network pre-insulated pipes buried in the ground[J]. Energy, 2016, 108: 172-184.

[28] ÇOMAKLı K, YÜKSEL B, ÇOMAKLı Ö. Evaluation of energy and exergy losses in district heating network[J]. Applied Thermal Engineering, 2004, 24(7): 1009-1017.

[29] KEÇEBAŞ A, ALI ALKAN M A, BAYHAN M. Thermo-economic analysis of pipe insulation for district heating piping systems[J]. Applied Thermal Engineering, 2011, 31(17-18): 3929-3937.

[30] NUSSBAUMER T, THALMANN S. Influence of system design on heat distribution costs in district heating[J]. Energy, 2016, 101: 496-505.

[31] TERESHCHENKO T, NORD N. Importance of increased knowledge on reliability of district heating pipes[J]. Procedia Engineering, 2016, 146: 415-423.

[32] WINTERSCHEID C, HOLLER S, DALENBÄCK J. Integration of solar thermal systems in existing district heating systems[J]. Energy, 2017, 116: 158-169.

[33] CHRISTENSON M, MANZ H, GYALISTRAS D. Climate warming impact on degree-days and building energy demand in Switzerland[J]. Energy Conversion and Management, 2006, 47(6): 671-686.

[34] WANG H, CHEN Q. Impact of

climate change heating and cooling energy use in buildings in the United States[J]. Energy and Buildings, 2014, 82: 428-436.

[35] ØSTERGAARD D, SVENDSEN S. Space heating with ultra-low-temperature district heating—a case study of four single-family houses from the 1980s[J]. Energy Procedia, 2017, 116: 226-235.

[36] YANG X, LI H, SVENDSEN S. Evaluations of different domestic hot water preparing methods with ultra-low-temperature district heating[J]. Energy, 2016, 109: 248-259.

[37] LUND H, WERNER S, WILTSHIRE R, et al. 4th generation district heating (4GDH): Integrating smart thermal grids into future sustainable energy systems[J]. Energy, 2014, 68: 1-11.

[38] TUNZI M, ØSTERGAARD D, SVENDSEN S, et al. Method to investigate and plan the application of low temperature district heating to existing hydraulic radiator systems in existing buildings[J]. Energy, 2016, 113: 413-421.

[39] ZIEMELE J, GRAVELSINS A, BLUMBERGA A, et al. Combining energy efficiency at source and at consumer to reach 4th generation district heating: economic and system dynamics analysis[J]. Energy, 2017, 137: 595-606.

[40] BALDVINSSON I, NAKATA T. A feasibility and performance assessment of a low temperature district heating system—a North Japanese case study[J]. Energy, 2016, 95: 155-174.

[41] BRAND M, SVENDSEN S. Renewable-based low-temperature district heating for existing buildings in various stages of refurbishment[J]. Energy, 2013, 62: 311-319.

[42] BÜNNING F, WETTER M, FUCHS M, et al. Bidirectional low temperature district energy systems with agent-based control: Performance comparison and operation optimization[J]. Applied Energy, 2018, 209: 502-515.

[43] YANG X, LI H, SVENDSEN S. Decentralized substations for low-temperature district heating with no Legionella risk, and low return temperatures[J]. Energy, 2016, 110: 65-74.

[44] ZENG J, XU Q, NING Y, et al. Pipe network optimization in district cooling/heating system: a review[C]. 2019 International Conference on Robots & Intelligent System (ICRIS), 2019.

[45] Ministry of Housing and Urban-Rural Development of the People's Republic of China. Design Code for Heating Ventilation and Air Conditioning of Civil Buildings GB 50736—2012[S]. Beijing: China Architecture & Building Press, 2012.

[46] CHEN X, HUANG D, CHEN X, et al. Risk assessment for dangerous sections of the levees: a case study in Guangdong Province, China[J]. Ocean & Coastal Management, 2020, 185: 105061.

[47] WU S B, YU M H, WEI H Y, et al. Non-symmetrical levee breaching processes in a channel bend due to overtopping[J]. International Journal of Sediment Research, 2018, 33(2): 208-215.

[48] TUCCI C E M, VILLANUEVA A O N. Flood control measures in União da Vitoria and Porto União: structural vs. non-structural measures[J]. Urban Water, 1999, 1(2): 177-182.

[49] Ministry of Housing and Urban-Rural Development of the People's Republic of China. Code for Design of Levee Project GB 50286—2013[S]. Beijing: China Planning Press, 2013.

[50] BAYOUMI A, MEGUID M A. Wildlife and safety of earthen structures: a review[J]. Journal of Failure Analysis & Prevention, 2011, 11(4): 295-319.

[51] RAHIMI S, WOOD C M, COKER F, et al. The combined use of MASW and resistivity surveys for levee assessment: a case study of the Melvin Price Reach of the Wood River Levee[J]. Engineering Geology, 2018, 241: 11-24.

[52] BALIS B, BARTYNSKI T, BUBAK M, et al. Smart levee monitoring and flood decision support system: reference architecture and urgent computing management[J]. Procedia Computer Science, 2017, 108: 2220-2229.

[53] BALIS B, KASZTELNIK M, BUBAK M, et al. The urban flood common information space for early warning systems[J]. Procedia Computer Science, 2011, 4: 96-105.

[54] LEONG S H, KRANZLMÜLLER D. Towards a general definition of urgent computing[J]. Procedia Computer Science, 2015, 51: 2337-2346.

[55] BALIS B, KASZTELNIK M, MALAWSKI M, et al. Execution management and efficient resource provisioning for flood decision support[J]. Procedia Computer Science, 2015, 51(1): 2377-2386.

[56] XUE L Q, HAO Z C, LIU X Q, et al. Numerical simulation and optimal system scheduling on flood diversion and storage in Dongting Basin, China[J]. Procedia Environmental Sciences, 2012, 12: 1089-1096.

[57] DING Y, WANG S S Y. Optimal control of flood diversion in watershed using nonlinear optimization[J]. Advances in Water Resources, 2012, 44: 30-48.

[58] LUO P P, HE B, TAKARA K, et al. Historical assessment of Chinese and Japanese flood management policies and implications for managing future floods[J]. Environmental Science & Policy, 2015, 48: 265-277.

[59] CAO S Y, LIU X N, ER H. Dujiangyan Irrigation System — a world cultural heritage corresponding to concepts of modern hydraulic science[J]. Journal of Hydro-environment Research, 2010, 4(1): 3-13.

[60] FLETCHER T D, SHUSTER W, HUNT W F, et al. SUDS, LID, BMPs, WSUD and more — The evolution and application of terminology surrounding urban drainage[J]. Urban Water Journal, 2014, 12(7): 525-542.

[61] BARBOSA A E, FERNANDES J N, DAVID L M. Key issues for sustainable urban stormwater management[J]. Water Research, 2012, 46: 6787-6798.

[62] HERING J G, INGOLD K M. Water resources management: what should be integrated?[J]. Science, 2012, 336(6086): 1234-1235.

[63] WANG J, GUO Y. Dynamic water balance of infiltration-based stormwater best management practices[J]. Journal of Hydrology, 2020, 589: 125174.

[64] ECKART K, MCPHEE Z, BOLISETTI T. Performance and implementation of low impact development — a review[J]. Science of the total Environment, 2017, 607-608: 413-432.

[65] FLECHER T D, ANDRIEU H, HAMEL P. Understanding, management and modelling of urban hydrology and its consequences for receiving waters: a state of the art[J]. Advances in Water Resources, 2013, 51: 261-279.

[66] POUR S H, WAHAB A K A, SHAHID S, et al. Low impact development techniques to mitigate the impacts of climate-change-induced urban floods: current trends, issues and challenges[J]. Sustainable Cities and Society, 2020, 62: 102373.

[67] GOULDEN S, PORTMAN M E, CARMON N, et al. From conventional drainage to sustainable stormwater management: beyond the technical challenges[J]. Journal of Environmental Management, 2018, 219: 37-45.

[68] LA ROSA D, PAPPALARDO V. Planning for spatial equity — a performance based approach for sustainable urban drainage systems[J]. Sustainable Cities and Society, 2020, 53: 101885.

[69] MORISON P J, BROWN R R. Understanding the nature of publics and local policy commitment to water sensitive urban design[J]. Landscape and Urban Planning, 2011, 99(2): 83-92.

[70] ANDO A W, CADAVID C L, NETUSIL N R, et al. Willingness-to-volunteer and stability of preferences between cities: estimating the benefits of stormwater

management[J]. Journal of Environmental Economics and Management, 2020, 99: 102274.

[71] VAN OIJSTAEIJEN W, VAN PASSEL S, COOLS J. Urban green infrastructure: a review on valuation toolkits from an urban planning perspective[J]. Journal of Environmental Management, 2020, 267: 110603.

[72] RUBINATO M, NICHOLS A, PENG Y, et al. Urban and river flooding: comparison of flood risk management approaches in the UK and China and an assessment of future knowledge needs[J]. Water Science and Engineering, 2019, 12(4): 274-283.

[73] NGUYEN T T, NGO H H, GUO W S, et al. A new model framework for sponge city implementation: emerging challenges and future developments[J]. Journal of Environmental Management, 2020, 253: 109689.

[74] MENG T, HSU D. Stated preferences for smart green infrastructure in stormwater management[J]. Landscape and Urban Planning, 2019, 187: 1-10.

[75] WANG K, CHU D Y, YANG Z H. Flood control and management for the transitional Huaihe River in China[J]. Procedia Engineering, 2016, 154: 703-709.

[76] YANG M, SANG Y, SIVAKUMAR B, et al. Challenges in urban stormwater management in Chinese cities: a hydrologic perspective[J]. Journal of Hydrology, 2020, 591: 125314.

[77] CHENG H F, HU Y. Municipal solid waste (MSW) as a renewable source of energy: current and future practices in China[J]. Bioresource Technology, 2010, 101(11): 3816-3824.

[78] HAN Z Y, MA H N, SHI G Z, et al. A review of groundwater contamination near municipal solid waste landfill sites in China[J]. Science of the total Environment, 2016, 569-570: 1255-1264.

[79] ZHOU H, MENG A H, LONG Y Q, et al. An overview of characteristics of municipal solid waste fuel in China: physical, chemical composition and heating value[J]. Renewable and Sustainable Energy Reviews, 2014, 36: 107-122.

[80] MOUNTOURIS A, VOUTSAS E, TASSIOS D. Solid waste plasma gasification: equilibrium model development and exergy analysis[J]. Energy Conversion and Management, 2006, 47(13-14): 1723-1737.

[81] YADAV V, KALBAR P P, KARMAKAR S, et al. A two-stage multi-attribute decision-making model for selecting appropriate locations of waste transfer stations in urban centers[J]. Waste Management, 2020, 114: 80-88.

[82] GANGOLELLS M, CASALS M, FORCADA N, et al. Predicting on-site environmental impacts of municipal engineering works[J]. Environmental Impact Assessment Review, 2014, 44: 43-57.

[83] HOWE K J, HAND D W, et al. Principles of Water Treatment [M]. Hoboken: John Wiley & Sons, 2012.

[84] STEPHENSON D. Water Supply Management[M]. Baton Rouge: Kluwer Academic Publishers, 1998.

Chapter 8
Hydraulic Engineering

8.1 Overview

Water plays a major role in human perception of the environment because it is an indispensable element. More importantly, human life is totally dependent upon water. Hydraulic engineering is the science of water in motion and its interactions with the surrounding environment. From a technological viewpoint, hydraulic engineering is defined as the application of fluid mechanics principles to problems dealing with the collection, storage, control, transport, regulation, measurement and use of water. The fundamental task of hydraulic engineering is to adapt the actual natural water regime of rivers, lakes, underground waters, seas, etc. for the purpose of a worthwhile and economical water utilization. For fulfilling this purpose, hydraulic structures are usually designed and constructed. Large projects such as a major dam providing for flood control, power generation, navigation, irrigation and recreation involve many aspects of hydraulic engineering. Smaller projects, such as an individual bridge across a river, the measurement of the flow rate in a small irrigation ditch, the protection of a stream bank from erosion, or the sizing of a water supply pipe or storm drain, require different hydraulic principles. This chapter starts by defining hydraulic engineering and describes briefly the historic development of hydraulic engineering. We will show how the evolution of hydraulic engineering is related to civil engineering, which has been shaped by changes in human beings. General hydraulic structures are categorized according to their assignment, and the features and functions are also presented. The creation of hydraulic structures are generally discussed, including investigation, design, construction and operation. Hydraulic engineering encompasses a broad portion of the field of civil engineering. The hydraulic engineer must collaborate with the structural engineer since many hydraulic engineering projects involve a variety of different types of structures.

8.2 Historical Development of Hydraulic Engineering

Earliest uses of hydraulic engineering were to irrigate crops and date back to the dawn of human civilization. Controlling the movement and supply of water for growing food has been used for many thousands of years. The remote history of hydraulic structures concerning when and where irrigation systems and dams were first constructed is not very clear. However, study on ancient Egypt, Iraq (Babylonia), Iran (Persia), India, Sri Lanka (Ceylon), China, Greece and Rome does confirm that such hydraulic works in these lands were begun thousands of years ago.

The first water-related projects involving the rudiments of hydraulic engineering date back to the first civilizations. The Sumerian Civilization along the Euphrates and Tigris Rivers created a network of irrigation canals dating from 4000 B.C. According to Durant, this system "was one of the greatest achievements of Sumerian Civilization, and certainly its foundation." Although the first canals were surely used for irrigation, their obvious use for navigation was apparent as well. Thus, canal as well as river navigation was important to the early Middle-Eastem Civilizations. The degree of this importance may be gauged by the frequent association, and even identification of the canals and rivers.

Although much of the water diversion into the canals depended on manual labor, it was only natural that dams or other low control structures would also be used for this purpose. With the added advantage, some storage could be gained thereby. The first dam is certainly lost in antiquity. The oldest known dam, and the oldest such structure remaining in existence, is the Saddel-Katara or Dam of the Pagans built between 2950 B.C. and 2750 B.C. The dam was constructed across a valley some 20 miles south of Cairo. The crest length was 348ft and it had a maximum height of 37ft. The lack of sediment accumulation behind the ruins of the stone-faced structure indicates that it probably was overtopped and washed out within a few years of its closure.

The legend that King Menes dammed the Nile at an even earlier date is unsubstantiated and, considering the magnitude of such an undertaking, it is most unlikely. What was more likely accomplished, itself no mean feat, was the protection of the city of Memphis from the waters of the Nile by a levee along the river.

Civilizations whose very existence depended upon a single water supply, like the Nile, learned to both respect and study the river behavior. Nilo–meters to measure the stage of water level have been in use from as early as 3000 B.C. to 3500 B.C. Not only did these gauges provide a means of recording the annual rise and fall of the Nile, but also they were used for flood control. As the flood season arrived, rowers moving in the direction of the current and more rapidly than the flood wave could bring river forecasts to the downriver regions. At the Roda gauge just north of Cairo, water level records have been kept, since 620 A.D.

The use of groundwater from springs or dug wells dates back to antiquity, with recurring references in the *Old Testament*. The most notable groundwater collection system was the qanats of Persia, which dates back to 3000 years before. The qanats were a system of nearly horizontal tunnels that served the combined functions of infiltration galleries and water transmission tunnels. They provided an ideal solution in arid regions in that they collected the water over large areas and in addition kept the water cool thereby reducing evaporation. The construction procedure began when an initial well was dug in a water-bearing region, followed by a second well some distances away. The two would be connected by a tunnel and the process would continue from vertical shaft to vertical shaft. Qanats extended to lengths of 25 to 28 miles, with typical slopes of 1ft to 3ft per 100ft, and depths as great as 400ft.

The Greek Civilization brought innovations in many areas that have come down to use mainly through the writings of Archimedes and Hero of Alexandria. The principles of hydrostatics and flotation postulated by Archimedes are taught in introductory fluid mechanics and Archimedean screw pumps are still used today. Hero's surviving manuscripts express an understanding of flow measurement and the role of velocity, a concept that perhaps the Romans never understood. In Pneumatica, Hero describes the construction of pumps for fire fighting and the use of hydrostatics and hydraulic forces to open temple doors automatically and operate other devices.

If the Greeks were the innovators and thinkers, the Romans were the builders and engineers. This was partly due to Greek abhorrence of any association with manual labor, which left all aspects of engineering entirely to slaves. The Romans, on the other hand, were organizers and developers. Engineering became a highly respected profession, including in its ranks the Emperor Hadrian.

All told, 11 aqueducts served the city of Rome, discharging an estimated 84 million gallons per day, or about 38 gallons per day per capita. Evidence of over 200 Roman aqueducts serving more than 40 cities has been found. The outstanding example is probably the famous Pont du Gard, which supplied water to Nimes in what is now southwest France. The sewer constructed to drain the Roman Forum, the Cloaca Maxima, is still in service.

Although great strides were not made during the Dark and Middle Ages, nevertheless, progress related to hydraulic engineering continued. More bridges were built, and wind and water power came into ever-increased usage. The Industrial Revolution in the 18th century and the industrialization that followed greatly increased the need for engineering and the scope of engineering. Hydraulic engineering as we know it today advanced rapidly during this period. When the principles of fluid mechanics were utilized in the 20th century, progress became even more accelerated. Our understanding of hydraulic concepts is far greater than that of our professional ancestors, which only serves to increase our admiration of their accomplishments. Since the 1950s, the increase in the number and height of dams was accelerated around the whole globe, and dams of several types (gravity, arch, multiple arch, zoned earthfill or rockfill) continued to mount toward new height until exceeding 250m. New dam height and volume records were set and broken in quick succession.

New dam type, material, construction method, etc., had been developed fast. Some of them had limited influences, and the others, particularly on the aspects of material and construction, have profound influences until today. Three events are the most worthwhile to be noted for this significant historical period, namely the invention of roller compacted concrete (RCC) dams, fast development of rockfill dams, and comprehensive design theory of super-arch dams. These three dam types are prevalent in the construction of modern high dams, particularly on the southwest areas in China.

8.3 Types of Hydraulic Engineering Structures

Civil engineering structures carried out for solving specific water utilization tasks are called hydraulic structures, while the applied science dealing with their general theory, design, construction and operation is hydrotechnics or hydraulic engineering.

The fundamental task of hydraulic engineering and hydraulic engineering structures is to adapt the actual natural water regime of rivers, lakes, underground waters, seas, etc. for the purpose of a worthwhile and economical water utilization for the needs of the water utilization branch and for the protection of the environment from the harmful effects of water. Another task is the creation of artificial watercourses and reservoirs when there is a shortage of natural ones. For fulfilling the above-mentioned tasks, main hydraulic engineering structures are distinguished as water retaining structures, water conveying structures and special-purpose structures.

(1) Water Retaining Structures

Dams are typical water retaining structures that affect closure of the stream and create heading-up afflux. Made of various materials, dams fall into soil and/or rockfill embankment, concrete, reinforced concrete, masonry and wooden, of which, the first two are the most prevalent nowadays.

1) Concrete Dams

By the structural features, concrete dams are termed as gravity (massive), buttress and arch. A gravity dam is a concrete structure resisting the imposed actions by its weight and section without relying on arch. In its common usage, the term is restricted to solid masonry or concrete dam which is straight or slightly curved in plan. A buttress dam depends principally upon the water weight in addition to the concrete weight for stability. It is composed of two major structural elements: a water-supporting deck of uniform slope and a series of buttresses supporting the deck. An arch dam (shown in Fig. 8-1) is always curvilinear in plan with its convex side facing headwater. On its vertical cross section, the dam is a relatively thin cantilever that is slightly curved. An arch dam transmits a major part of the imposed actions to the canyon walls mainly in the form of horizontal thrusts from it abutments.

2) Embankment Dams

Embankment dams (shown in Fig. 8-2) are massive fills of natural ground materials composed of fragmented particles, graded and compacted, to resist seepage and sliding. The friction and interlocking of particles bind the material particles together into a stable mass rather than by the use of a cementitious

Fig. 8-1 Victoria Dam in Sri Lanka
Source: https://structurae.net/en/structures/victoria-dam-1985.

Fig. 8-2 Mica Dam in Canada
Source: http://www.ctgpc.com.cn.

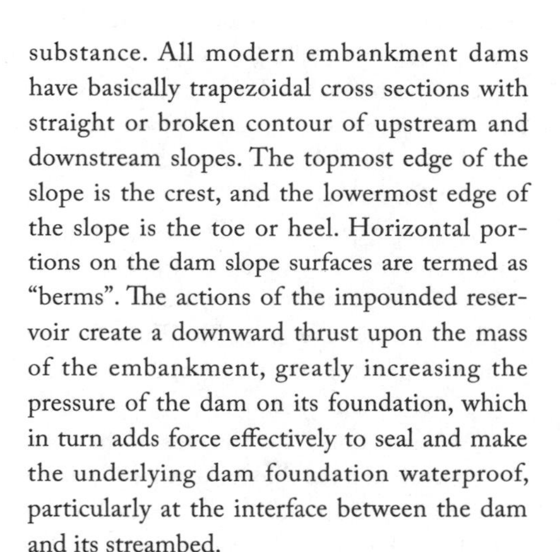

substance. All modern embankment dams have basically trapezoidal cross sections with straight or broken contour of upstream and downstream slopes. The topmost edge of the slope is the crest, and the lowermost edge of the slope is the toe or heel. Horizontal portions on the dam slope surfaces are termed as "berms". The actions of the impounded reservoir create a downward thrust upon the mass of the embankment, greatly increasing the pressure of the dam on its foundation, which in turn adds force effectively to seal and make the underlying dam foundation waterproof, particularly at the interface between the dam and its streambed.

(2) Water Conveying Structures

Water conveying structures are artificial channels cut in the ground and made of either ground materials such as soil and rock (e.g., canals and tunnels) or artificial materials such as concrete and metal (e.g., aqueducts, flumes, siphons, pipelines). By means of these structures, water is conveyed in appropriate quantities for utilization of various purposes—for instance, for obtaining hydroelectric power in turbines of hydroelectric power plants (shown in Fig. 8-3), for water supply for inhabited places and the industry, for irrigation and for draining water from surfaces, which are to be dried out, etc. Water conveying structures can also be utilized for water navigation, and to encompass overflow and spillway structures, which serve for drawing off excess water from impounding reservoirs and its transfer downstream of the water retaining structure. Fig. 8-3 shows the Queenston-Chippawa Power Canal in Canada.

(3) Special-purpose Structures

Special-purpose structures are accommodated in a hydraulic project to meet the requirements of:

1) Hydroelectric power generation, inclusive power plants, forebays and head ponds, surge towers and shafts, etc.;

2) Inland waterway transportation, inclusive navigation locks and lifts, berths, landings, ship repair and building facilities, timber handling structures and log passes, etc.;

3) Land reclamation, inclusive sluices (head works), silt tanks, irrigation canals, land draining systems, etc.;

4) Water supply and waste disposal (sewerage), inclusive water intakes, catchment works, pumping stations, cooling ponds, water treatment plants, sewage headers, etc.;

5) Fish handling, inclusive fish ways, fish locks and lifts, fish nursery pools, etc.

Fig. 8-3. Queenston-Chippawa Power Canal in Canada
Source: https://www.flickr.com/photos/3336/135951184/.

Fig. 8-4 Three Gorges Dam in China
Source: https://news.sina.com.cn/c/p/2006-05-22/20199934972.shtml.

The Three Gorges Dam is the largest hydroelectric dam in the world shown in Fig.8-4. It is located in the middle of the Three Gorges on the Yangtze River, in Hubei Province of China. The 135.266 billion RMB mega project was formally approved in 1992 by the National People's Congress. Construction began in 1994 and was complete in 2009.

The dam consists of three parts: the dam itself, the ship locks and the shiplift. The dam is 182m high and 1.3 miles across. It consists of a center spillway to let water over the dam, and each side houses the hydroelectric turbine generators, twenty-six in all. The right side of the dam is fully operational, while the left side of the dam is scheduled to be fully operational by 2009. The generators will be able to produce as much power as eighteen nuclear power plants, about 84.6 billion kilowatt hours. The ship locks are located on one side of the dam. There are two in total, one for upstream traffic and one for downstream traffic. There are five locks, or stages, used to transport large ships, twelve at a time, over the dam. Both ship locks are fully operational. The third part of the dam is the shiplift, an elevator used to raise smaller ships over the dam. The shiplift has been scheduled for operation in 2009.

The Chinese government has four goals for the Three Gorges Dam project:

1) Flood control: The history of the Yangtze River includes many devastating floods over the centuries, killing thousands of people and causing millions of dollars in damages. The dam will reduce the impacts of flooding since it will have a flood control capacity of 22.15 billion cubic meters.

2) Power generation: The use of hydroelectric turbine generators will reduce China's dependency on coal, a hydro carbon that produces greenhouse gases. The Three Gorges Dam will produce about 84.6 billion kilowatt hours of clean energy annually.

3) Navigation: The presence of the dam, the reservoir and the ship locks will allow large ships to travel up and downstream for the first time. Ships from Chongqing will be able to transport goods all the way to the sea in Shanghai.

4) Tourism: Since the Three Gorges Dam project is the largest hydroelectric dam in the world, it is expected to be popular among tourists visiting China.

8.4 Design and Construction of Hydraulic Structures

8.4.1 Basic Stages in the Creation of Hydraulic Structures

The process of creation of hydraulic structures, especially dams, is long, troublesome, painstaking and costly. What precedes this process is a water utilization design for the hydraulic system. This phase is highly significant, since the water economies are being set increasingly more complex tasks, caused by increasingly large demands made on water resources as a result of increased consumption of water, reduction of available water, as well as increased demands in relation to the efficiency of hydraulic systems.

A water utilization design helps in selecting a solution to the water utilization problem. The water utilization design is a very important phase in the realization of a hydraulic system and its separate parts, since by having such a design, one can define the strategic orientation—where possible mistakes cannot in practice be eliminated later on—irrespective of the quality of the hydraulic part of the design.

After elaboration of the water utilization design, we come to the realization of a specific hydraulic structure. The process can be generally divided into four stages:

(1) Investigation. In the first stage, it is necessary to obtain all the necessary data on natural conditions of the region and the place of the more immediate location of the structure. Here, there belongs data on the topography of the terrain, its geological structure, seismicity of the ground, hydrological conditions of the catchment area and the watercourse, climatic particularities of the region, availability of local materials, economic and water utilization conditions and needs, etc.

(2) Design. In this stage, on the basis of the design task and the requirements that have been set up with the water utilization assignment for the future water regime of the structure, and taking into consideration the data obtained from investigations, we establish the dimensions of the structure and the structural elements, then work out the necessary plans and schemes, determine the methods of construction, as well as the necessary plant and equipment. Thus, it is possible to obtain the economic indicators for a structure that has been anticipated for construction.

(3) Construction. The third stage encompasses all the works concerning the organization of the construction of the structures, as well as the mechanization of the process of execution of the works. Then comes the construction of the structures, dismantling of the construction plant and equipment, dismantling of the temporary structures that have served in the course of the construction and, in the end, handing over the structure for operation.

(4) Operation. The last stage encompasses all the works regarding the operation, i.e. service of the structure—management of its operation, supervision of the fulfilment of the requirements specified by the design, as well as the condition of the structure and equipment. What follows is an analysis of results from surveillance measurements, periodic new analyses of static and dynamic stability, employment of the most sophisticated methods and newly-acquired parameters, checking the rate of flow, i.e. capacity of the overflow structures on the basis of new knowledge and criteria, regular maintenance, and capital overhaul of the structure.

8.4.2 Investigation for Design and Construction of Hydraulic Structures

Before designing hydraulic structures, it is necessary to conduct the following investigative works:

(1) Topographic investigation works. By means of the topographic investigation works, one can obtain: 1) general regional maps; 2) ordinance maps; 3) levelling (elevation network) and a tachometrically-surveyed longitudinal profile of the watercourse; 4) geodetic location plans; 5) detailed tachometric survey photographs of longitudinal and cross-profiles in the zone of the most important structures.

(2) Engineering-geological and hydro-geological investigation works, the aim of which is to throw light on the geological structure of the region of a hydraulic scheme, the physical and mechanical properties of its minerals, hydro-geological conditions, and to determine occurrences of natural materials for construction of the structures.

(3) Geotechnical investigation works, by means of which we can determine the elastic and deformational characteristics of the rock materials at the dam site and in the foundations of some more significant structures, as well as of the rock materials which will be used for construction of the dam's body; they are effectuated through field and laboratory investigations.

(4) Geomechanical investigation works,

which determine the characteristics of non-rock (soil) materials in the foundation, as well as in the borrow pits for local earth materials. For their realization, field and laboratory equipment has been developed, as well as precisely described methods and procedures.

(5) Hydrological investigation works, which are useful for studying the hydrological regime of a watercourse planned for utilization. At the same time, an investigation is performed of the climatic regime of the area in order to obtain data on the atmospheric precipitation, air temperature and its variations, direction, duration of winds, etc.

(6) Constructional output investigation works, which make it possible to obtain data that are necessary in the realization of the construction works, for instance, for conditions leading to connection of the construction site with the existing traffic network, for sources of supply of construction materials, water and electric power, conditions for settling down, supply and the living accommodation of workers, etc.

(7) Other investigation works, which help clarify the existing water utilization of the considered water resource. Then comes abundance of the river with fish and conditions for fishing, kinds of soil and vegetation in the region in which the hydraulic structure might be set up, as well as basic economic branches (industries) in the zone in which the planned hydraulic scheme will have an influence.

Actually, the scope of necessary investigative works varies depending on the size and importance of a hydraulic scheme, i.e., on the structure, the design phase for which the investigative works are carried out, the size of the catchment area and future storage space, the character and structure of the ground of the future impounding reservoir, the geological structure of the dam site, the possible type of planned dam, and its appurtenant structures.

On the scope of investigative works that will be undertaken, an influence can also be the degree of previous investigation of the location for the needs of the subject structure, or else for some other designs, accomplished or not. In utilizing old data, it is necessary to verify its credibility, the way it has been obtained, as well as the methods applied. Site investigative works are, as a rule, expensive; however, when it is a question of hydraulic schemes with a dam, most often they do not extend to a significant part of the total investment expenditure. On the other hand, mistakes made in investigations and their insufficient scope, can lead to considerable failures in the realization of a hydraulic scheme with a dam, and that is a sufficient reason for not to save on them without an especially convincing reason.

8.4.3 Contents for Design and Construction of Hydraulic Structures

Based on data obtained from investigations that have to be accomplished with a water utilization structure, we move to work out a design, which implies a variety of well-thought-over and coordinated activities such as would lead to the realization of the specific concept. The following major constituent elements come into the design for a hydraulic scheme with a dam for example:

(1) Elaboration and definition of the future hydrological and water utilization regime of the hydraulic structure, which includes determination of characteristic elevations of the tail-water and headwater, determination of area of submersion, volume of the impounding reservoir, etc.;

(2) Selection of types, schematic structure of the works and materials for their construction and equipping, way of founding and structural interventions in the foundation and, finally, constructing the works;

(3) Hydraulic dimensioning of the structures and seepage calculations for the watertight elements, structure and foundation;

(4) Static and dynamic dimensioning and control, which verify the strength and stability of the designed structure and a foundation for the effects of all possible forces;

(5) Execution of solution and selection of surveillance instrumentation (monitoring) of the structure in the course of construction and operation;

(6) Elaboration of methods for organization of works with a time schedule, i.e. program and progress schedule, period of completion of structures, etc.;

(7) Preparation of technical specifications for construction and control of quality of materials and their placement;

(8) Preparation of *Bill of Quantities and Estimate* for the works along with determination of the total cost and the technical-economic indicators for the structure.

In addition, in cases of very large and significant structures, especially if they are not executed in the most favorable ground conditions—and that often happens in hydraulic engineering—within the framework of the design it is also possible to include a part of the scientific and investigative works for proving the justification of the selected structure and dimensions of the structure, if by means of the usual methods and existing theories, it cannot be proved.

The process of designing a hydraulic structure is usually carried out in three phases. The first phase consists of preparing a basic design for the watercourse, which is a study by means of which we set up a technical concept for a complex utilization of the water of the considered watercourse. In general, the objective of the first phase is to present the designed hydraulic structure, approximately indicate its dimensions, the scope of the works and the quantity of necessary materials, necessary equipment, cost of the structure and, most importantly, to prove the technical possibility, as well as the economic and commercial usefulness and justifiability of the structure. On the basis of that study, the structure can be included in the plans for further elaboration and financing.

The second phase involves working out a preliminary design for the structure, in which, on the basis of more detailed data from the investigative works and detailed requirements, we determine the water utilization regime, as well as the type and basic details of the structure. It contains more accurate hydraulic, seepage and static calculations and checks, elaborate drawings, and it gives methods for the organization and execution of the works. In addition, a more detailed *Bill of Quantities and Estimate* of works is carried out. The preliminary design should be a basis for precise determination of the investment value of the structure, in order to be able to serve as preparation of an investment program, and for closing the financial structure for construction of the structure, that is to say for making the decision for construction.

The third phase implies preparation of a main and constructional design, in which, on the basis of additional investigative works and more accurate data, final hydraulic, seepage, static and dynamic calculations are made, and detailed execution drawings for all of the structure and structural elements are prepared. Also, this phase finally elaborates the methods for organization of the works, anticipates the necessary plant and equipment, issues technical specifications for construction and control of quality and placement of materials, anticipates instruments and methods for surveillance (monitoring) of the structures in the course of construction and operation, performs detailing, and makes a more accurate *Bill of Quantities*.

On the basis of this worked out and accepted main design, a building permit is obtained so that the construction of works can commence. In this, owing to the complexity of hydraulic structures, especially dams, the relatively large territory on which they extend, which renders impossible investigative works in detail to define the characteristics of the foundation, the complexity of the conditions for construction (presence of water, need for diverting the watercourse, etc.), almost always implies the need for the design to go on even after the completion of the main design. Namely, following the opening of the construction pit and the commencement of construction, there often occur conditions that, to a large or small extent, differ from anticipated ones. Because of that, a need is imposed for performing alterations and amendments to the design. That is why designers should be in touch with everything that happens during the realization of the main design, in order to be able to react and respond in the right manner and at the right moment.

The above-described three phases of design occur with all large and more significant hydraulic structures. In the cases of small structures, the design can be performed in two phases, in which the first one and the second one are usually combined. In some cases, it is possible to go directly into preparation of the main design, if field conditions enable an experienced engineer, on the basis of analysis of data from investigative works, field prospecting and small analysis, the ability to give an evaluation of the justification of the construction, type and structure of the works.

8.5 Case Study—Gravity Dams in General

8.5.1 General

Gravity dams may be the earliest water retaining structure in the human history. Their name comes from the Latin word gravitas, which means weight, since they resist the action of the basic force—the horizontal static pressure of water—by means of their self-weight, which creates a force of resistance against sliding across the foundation.

Basically, gravity dam is a kind of solid concrete structure that resists horizontal water pressure of reservoir to maintain its stability against sliding by both the friction related to the concrete weight and the cohesion between the dam concrete and the foundation rock. The flared or inclined upstream dam face enables it to make use of a part of water weight to improve its stability against sliding.

Gravity dams built in earlier times had almost trapezoidal or even rectangular profiles in cross section. Later, with the development of design theory, more economical profiles of curvilinear or polygonal appeared. Nowadays, almost all modern concrete gravity dams are triangularly configured in cross section.

Generally, gravity dams are constructed on a straight axis, but may be slightly curved or angled to accommodate the specific site conditions. A gravity dam is conventionally divided into independent dam monoliths cut by transverse joints (Fig. 8-5), each of them can be looked at as a cantilever fixed on the foundation. At any horizontal section of dam monolith, there is a moment M attributable to the water pressure P, which leads to tensile stress at the upstream and compressive stress at the downstream (Fig. 8-6b). Since the tensile strengths of concrete and rock are very low, the tensile stresses within the dam body and at the dam/foundation interface are not permitted. Therefore, the compressive stress induced from the dam weight W and the water weight Q as well as the uplift U (Fig. 8-6c) should be larger enough to offset the tensile stress due to the water pressure P (Fig. 8-6d).

The modern solid gravity dam shown in Fig. 8-5 has been widely practiced attributable to the following advantages:

(1) With solid concrete structure, water may overflow the gravity dam or be bleed-off through

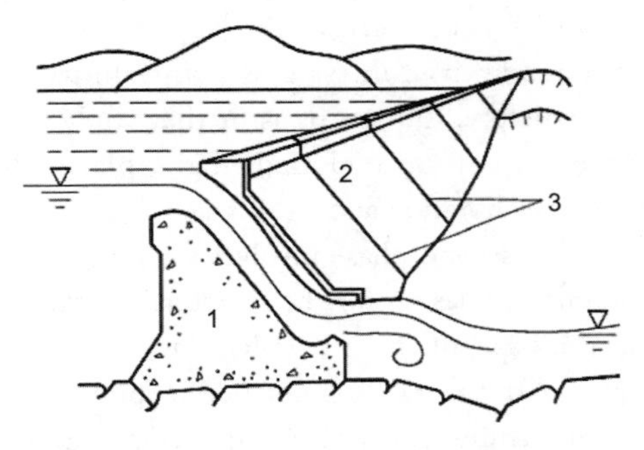

Fig. 8-5 A gravity dam on riverbed [3]
1-Overflow dam monolith; 2-Non-overflow dam monolith; 3-Transverse joints

the bottom outlets (permanent or temporary) during the service period and construction period.

(2) Compared to the arch dam and buttress dam, the sectional profile of the gravity dam is simpler, which facilitates the mechanized concrete placement and simplifies the formwork technique.

(3) The gravity dam possesses high degree of safety in any climatic conditions inclusive of harsh winter.

(4) Regarding the foundation rock quality requirement, the gravity dam has moderate adaptability, which is higher than the arch and buttress dams, but lower than the embankment dam.

(5) The engineers have plenty of experience and expertise in the design and construction of gravity dams, which guarantees higher reliability, longer service life and lower maintenance cost.

The main disadvantages with gravity dams are cited as follows:

(1) Due to the huge bulk of dam concrete and relatively lower stress level, the strength of the concrete material is not fully exploited.

(2) Also due to the huge bulk of dam concrete, the dissipation of the heat from hydration process and thermal flow is difficult, which may give rise to cracking and in turn, deteriorate the monolithic and strength of the dam.

(3) Due to the larger surface of dam base compared to the arch and buttress dams, the gravity dam is exerted by larger uplift which offsets a portion of the dam weight and is not favorable to the dam stability.

With respect to the unit construction cost, the gravity dam is more costly than the embankment dam, but less costly than the arch and buttress dams.

To make the most of the advantages and to remedy the weaknesses of gravity dams, the design and construction techniques have been developing in recent decades, by which the most important advances achieved are given as follows:

(1) Use of inclined or flared upstream dam face to improve the stability;

(2) Use of different concrete grades in the zoned dam body to save the cement;

(3) Use of roller compacted concrete (RCC) technique to save the cement and to raise the construction speed.

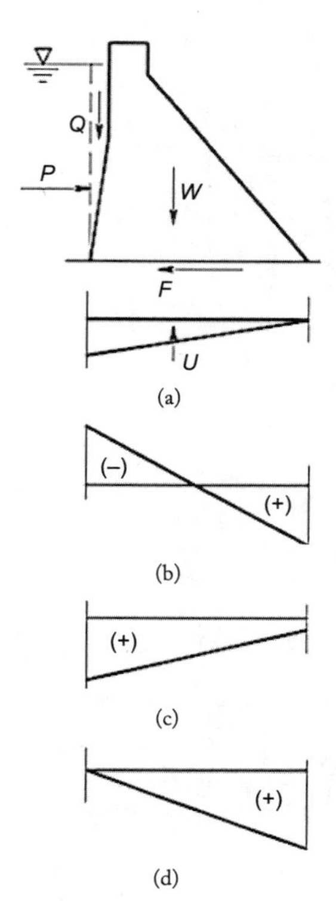

Fig. 8-6 Working principle of gravity dam [3]

8.5.2 Design Theory and Profile of Gravity Dams

In the mid of the 19th century, the cantilever theory was firstly proposed for the profile design of gravity dams by Sazilly—a French engineer. Nowadays, this theory becomes classical termed as "gravity method" and is still widely exercised in the gravity dam design, which includes following principles:

(1)The resultant of all forces exerting above any horizontal plane through a dam intersects that plane inside the middle third kernel to maintain compressive stresses.

(2)It is safe against sliding on any horizontal or near-horizontal plane within the dam at the base, or on any rock seam in the foundation.

(3)The allowable stresses in the dam concrete

or in the foundation rock shall not be exceeded.

Generally, the principles of non-tensile stress and stability against sliding are the major two conditions to ensure the safety of the gravity dam. The analysis according to these principles leads to a theoretic (basic) profile of triangular cross section. Actually, the prevalent practical profile of modern gravity dam is revised by this triangular basic profile.

8.5.3 Layout of Gravity Dam Projects

Basically, a gravity dam project consists of overflow dam monoliths, non-overflow dam monoliths, abutment piers, guide walls and crest works. Fig. 8-7 shows the layout of a most typical and simple gravity dam project, which consists of the non-overflow dam monoliths on the right and left banks, the overflow spillway dam monoliths on the main river stream. The power plant of dam toe type is located to the left riverbed.

The layout of gravity dam depends on the comprehensive consideration concerning the topographic and geographic conditions as well as the accommodation requirements for the other structures (e.g., hydropower plant and ship lock or lift). The layout also should pay attention to the shape harmoniousness of the dam monoliths. Where there are significant differences in the topographic and geographic conditions with different dam monoliths, it is advisable to use different downstream dam slopes while to keep a consistent upstream dam slope.

In order to make the general layout of a project more economically rational, safe and reliable, modern trends are to use solid gravity dams, particularly on deep gorge sites in mountainous areas, which are mainly guided by the following considerations:

(1) Dam safety is the paramount concern and the quick reservoir drawing down or emptying under emergent situations that should not be neglected. In addition to crest

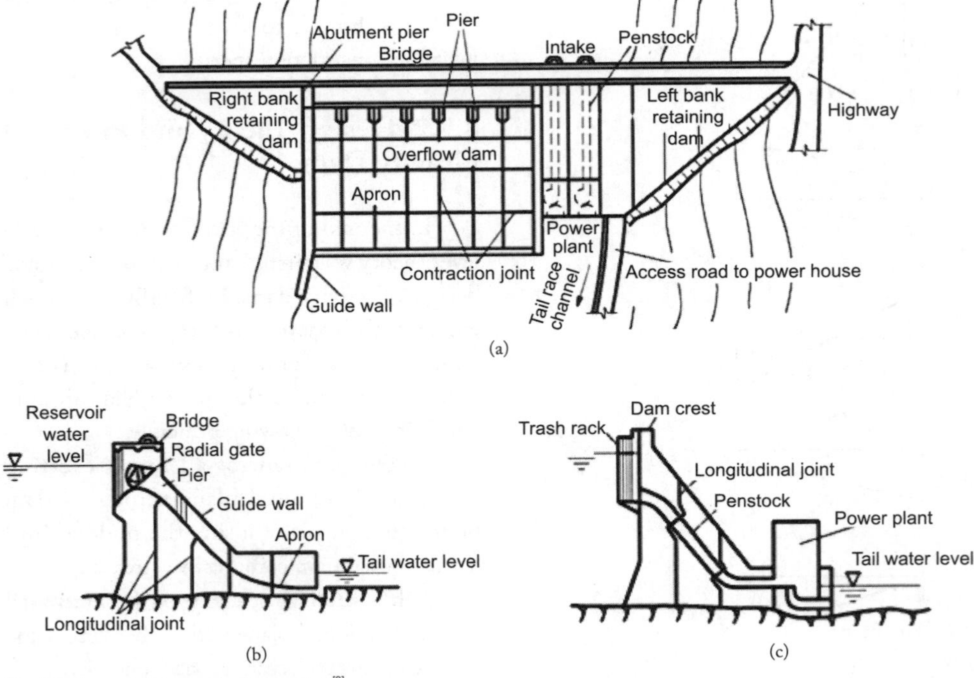

Fig. 8-7 Layout of a gravity dam project[3]
(a) Plan; (b) Profile of spillway dam monolith; (c) Profile of non-overflow dam monolith

spillway, intermediate and bottom outlets have to be installed for controlling the reservoir's water level as flexibly as possible.

(2) It is necessary to provide sluices of certain-scale as bottom outlets for discharging flood water and flushing silt, to keep effective reservoir storage capacity and to get rid of sediments in front of power plant intakes.

(3) With respect to construction diversion and schedule, the solid gravity dam performs well by the flexible combination using open channels and bottom outlets of large diameter.

(4) With a solid gravity dam, penstocks and powerhouse can be flexibly arranged. Power plant inside the dam or at the toe are all good alternatives, and power intakes even may be installed in the piers of surface outlets.

8.5.4 Main Design Tasks for Gravity Dams

(1) Overall Layout

Overall layout is accomplished by the selection of dam site and axis, the alignment of the dam monoliths, and the other works along the dam axis in the whole project.

(2) Flood-releasing Design

Flood-releasing design comprises the study on the flood-releasing methods and their combination, the positioning and sizing of spillway, the flood routing, the computation of the design and the catastrophe (check) flood levels, the energy dissipation of high velocity flow as well as the downstream river protection against scouring, the discharge atomization and bank-slope protection, the gate operation and control scheme, etc.

(3) Profile Design

Profile design is tightly related to the study on the loads and load combinations. The optimization of dam profile is obtained mainly by means of stability and stress analysis (static and dynamic). The results of stability and stress analysis also provide guidelines for the dam foundation treatment and appurtenant design.

(4) Appurtenant and Miscellaneous Design

The appurtenant and miscellaneous design comprises of concrete zoning with respect to corresponding concrete grades and performance indices (physical and mechanical property, thermal property, permeability property, freezing resistance property, abrasion resistance property, corrosion resistance property, cement–water ratio, cement type and consumption, aggregate and cement admixture), transportation system, gallery system and lift, anti-seepage and drainage systems in dam body and foundation, lighting and ventilating systems within the dam body, piers and bridges on the dam, guide walls and handrails, etc.

(5) Foundation Treatment Design

In the foundation treatment design, the major attentions are focused on the issues such as the requirements for the foundation excavation, anti-seepage, drainage and grouting; the study on the seeping, deforming, piping and argillization of foundation rock under the action of water and dam; the treatment for the fractures, faults and seams in the foundation. The deformation and stability of the abutment slopes are also studied.

(6) Instrumentation Design

Instrumentation system is installed to monitor the structural integrity and to check the design assumptions, as well as to forecast the performance of the foundation and dam body during the construction and service periods.

(7) Construction Design

The design works conducted in this stage comprise the following: the selection of lift thickness, transverse and longitudinal joints, time interval between lifts, maximum allowable placing temperature of concrete, surface insulation and post cooling scheme; to put forward technical requirements for the construction and to specify the concrete placement method including construction equipment; to complete the design of main temporary structure for construction and general construction schedule; and the selection of construction diversion scheme and the design of diversion works (e.g., cofferdam and diversion bottom outlets). Care should also be called at the issues with regard to the traffic and downstream water supply during the construction period.

8.6 Conclusions

This chapter presents some general aspects of hydraulic engineering. The historical developments of hydraulic engineering demonstrate that mankinds are making use of water resources by hydraulic structures (such as dams, tunnels, canals, etc). The fundamental task of hydraulic engineering and hydraulic structures is to adapt the actual natural water regime of rivers, lakes, underground waters, seas, etc. for the purpose of a worthwhile and economical water utilization for the needs of the water utilization branch and for the protection of the environment from the harmful effects of water. For fulfilling these tasks, hydraulic structures are well-thought-over designed to their functions. The considerations of design, construction and other aspects of hydraulic engineering structures are also generally discussed. Case study of design and construction of a gravity dam shows the general creation process. It's clear that hydraulic engineering encompasses a broad portion of the field of civil engineering. The hydraulic engineer must collaborate with the structural engineer since many hydraulic engineering projects involve a variety of different types of structures.

Exercises

8-1 What is hydraulic engineering and what is the main task?

8-2 How many types of hydraulic structures and what are they?

8-3 What are characteristics of gravity dams?

8-4 Describe the basic stages in creation and use of hydraulic structures and its main task.

8-5 Describe some common hydraulic structures around us and list some basic features.

References

[1] PRASUHN A L. Fundamentals of Hydraulic Engineering [M]. New York: Oxford University Press, 1996.

[2] HOUGHTALEN R J, AKAN A O, HWANG N H C. Fundamentals of Hydraulic Engineering Systems (5th Edition.) [M]. New York: Prentice Hall, 2016.

[3] CHEN S H, CHEN M L. Hydraulic Structures [M]. New York: Springer Berlin Heidelberg, 2015.

[4] TANCHEV L. Dams and Appurtenant Hydraulic Structures (2nd Edition) [M]. Florida: CRC Press, 2014.

[5] NOVAK P, MOFFAT A I B, NALLURY C, et al. Hydraulic Structures (4th Edition)[M]. Florida: CRC Press, 2007.

[6] JANSEN R B. Dams and public safety, a water resources technical publication [J]. Water and Power Resources Service, 1980.

[7] JAMES C S. Hydraulic Structures [M]. New York: Springer, 2020.

[8] HAGER W H, BOES R M. Hydraulic structures: a positive outlook into the future [J]. Journal of Hydraulic Research, 2014, 52(3), 299-310.

[9] MARRIOTT M. Civil Engineering Hydraulics [M]. New Jersey: John Wiley & Sons, 2009.

[10] JIA J S. A technical review of hydro-project development in China [J]. Engineering, 2016, 2(3): 302-312.

Chapter 9
Ocean Engineering

9.1 Overview

Ocean engineering is a relatively new engineering discipline that is involved with mankind's need to use natural ocean energy such as wind, waves, temperature, currents, and mineral resources beneath the sea surface, provide a food source, accommodate recreational activities, transport goods and people, provide alternative space for living quarters and facilities, further understand oceanic processes, and develop engineering concepts for protecting the land from various ocean meteorological processes. Ocean engineering may be defined as the application of engineering principles to the analysis, design, development and management of systems that must function in water environments such as oceans, lakes, estuaries and rivers. Ocean engineering is actually a combination of multi disciplines that are applied to the ocean environment such as mechanical, electrical, civil, chemical engineering techniques. Man's utilization of the oceans as a resource stimulates the need for structures that must operate in the complex ocean environments. Currently, the major use of offshore structures is for the exploration and production of oil and gas. Therefore, a brief explanation of some terminology and procedures for drilling and producing oil and gas is given, followed by the types of offshore system. Special concerns lay on the corrosion protection methods. The technical challenges are formidable and sustained research efforts are essential.

9.2 History of Ocean Engineering

Although engineers have been engaged with engineering applications in the ocean since before the beginning of 20th century, the academic discipline of ocean engineering only surfaced at some universities in the late 1960s and early 1970s. As a consequence, engineers educated in ocean engineering are relatively new. The development of ocean engineering was fueled by exploration of the underwater environment, development of offshore gas and oil, and the continued need for coastal protection and port expansion. The development of offshore oil and gas fields by the various oil and gas companies (e.g., Amoco, Arco, British Petroleum, Conoco, Esso, Exxon, Mobil and Texaco) in the Gulf of Mexico North Sea and Persian Gulf has been tremendous, and opportunities for ocean engineering applications have prospered at the same time. Other offshore developments have occurred offshore Alaska, Canada, Brazil, Mexico, China, Africa, India, Australia and Indonesia. Contaminated sediments in ports have created new engineering problems related to maintenance dredging in ports and the related placement of dredged materials that are necessary to allow ships to continue accessing the facilities. The development of the nation's coastlines and ports as centers of trade and recreation continues to expand.

(1) Coast

Protection of coastlines and beaches from erosion and flooding has always been a concern of engineers. In 1950, the beach erosion board was first established in the United States to protect the nation's coastlines. A major activity that occurs worldwide is beach nourishment, which is placing beach material back on beaches after severe erosion over many years or because of severe storms such as hurricanes. These beach nourishment projects, such as what occurred at Port Royal Sound, South Carolina, the United States (Fig. 9-1), are necessary to protect the land from flooding and wave action, provide beaches for recreation and protect wetlands where a diverse

(a) (b)

Fig. 9-1 Before (a) and after (b) beach restoration at Port Royal Sound, South Carolina
Source: https://hiltonheadislandsc.gov/projects/beachrenourish/2016beachrenourish/home.cfm.

marine habitat exists. Coastlines are protected by many different manmade coastal protection structures that include seawalls, breakwaters, revetments, groins and submerged berms.

In the 1960s, port, harbor and marina development rose sharply. Large ports contribute to a strong economy and increased commerce and trade vital to all nations. Safe and navigable entrance channels are critical to ports and harbors, and the construction and maintenance of channel jetties and breakwaters have provided safe passage to these important trading ports. Recreational boating and fishing that occurs along coastlines and in coastal bays and estuaries, inland lakes, and rivers also require the development of small boat marinas for support. Commercial fishing and the seafood industry require a port and harbor infrastructure to support these very important activities.

Development of ports and harbors requires dredging of the bottom sediments; this requirement led to the initiation of the Dredging and Dredge Material Disposal Research Program spearheaded by the U.S. Army Corps of Engineers in the 1970s. This program was followed by the Dredging Research Program in the 1980s and the current Dredging Operations and Environmental Research Program. The need for dredging is worldwide so that ports and harbors can remain open for commercial and military ships, submarines and other water-borne crafts. New ships have greater drafts, and consequently, it is necessary to further deepen the entrance channels to the ports (e.g., 17m). Large dredges, such as hydraulic cutterhead and sea-going hopper dredges, and mechanical clamshell, dippers and backhoe dredges, are used to deepen the channels and subsequently maintain the channel depths. The placement of the dredged material is also an important engineering activity. In some cases, the dredged material can be beneficially used or placed in an environmentally safe manner in specific placement areas upland or offshore.

(2) Offshore

Trends in the ocean engineering field have paralleled the Offshore Industry, whose center in the United States is in Houston, Texas. However, the first offshore exploration for oil was in 1887 off the coast of California in a few feet of water. In 1910, an oil well was drilled in Ferry Lake, Louisiana. Internationally, the first wells were drilled in Lake Maricaibo, Venezuela in 1929, and the Gulf of Mexico followed with the development of the Creole Field in 4.3m of water off the coast of Louisiana. Shallow water wells continued the slow development, and in 1959, Shell installed a platform in 30.5m of water off Grand Isle, Louisiana. The Persian Gulf and the North Sea experienced oil finds and subsequent offshore platform development starting in 1960. Development in the 1970s was explosive, and offshore platforms and drilling advanced into deeper water at a rapid rate. In 1973, the North Sea was the site of the first concrete gravity platform, which is a concrete structure that is built on land, floated to the site, and sunk to the bottom. The Hondo Platform was installed by Exxon in 259.1m of water off the California Coast near Santa Barbara, and in 1978, Shell placed its Cognac Platform in 312.5m of water in the Gulf of Mexico. Exxon installed the first guyed tower, Lena, in the Gulf of Mexico during 1983. A guyed tower is a slender, bottom supported tower that is laterally braced by cables. The following year, 1984, Conoco placed the first tension leg platform in the North Sea in 147.9m of water.

In 1988, Shell installed the Bullwinkle Fixed Platform in 412m, and 5 years later (in 1993), Shell installed the Auger Tension Leg Platform in a water depth of 852m in the Gulf of Mexico. A brief history of offshore platform installations is illustrated in Fig. 9-2. As of 1984, there were 16 gravity structures in the North Sea in depths ranging from 70m to 152m water depth, and other gravity platforms were installed offshore Brazil and the Baltic Sea. McClelland and Reifel report that over 3500 offshore structures have been placed in offshore waters of over 35 nations and nearly 98% of them are steel structures supported by piles driven into the sea floor. The 1990s experienced the push to deeper waters (>610m) with the installation of tension leg platforms, and floating production systems were being used to produce oil in marginal fields. New platform concepts continue to be proposed with the goal to reduce the cost of production and to be able to work in greater and greater water depths since large oil reserves have been found in very deep water depths (>1829m).

(3) Underwater Systems

Underwater habitats, diving equipment, submarines and subsea completion equipment are examples of underwater systems that ocean engineers are researching, developing, designing and operating to advance man's use of the ocean environment. The first manned underwater habitat for saturation diving, Man-in-the-Sea I, was developed and tested in 1962 by E. A. Link. Since then, over 65 underwater habitats have been built worldwide.

Divers use many different types of breathing equipment to assist in their exploration and work in the underwater environment. Prior to the use of compressed air, breath-hold divers developed goggles, snorkels and fins to improve their diving efficiency. Self-contained underwater breathing apparatus (SCUBA) has been around since the 1500s, but the double hose Cousteau-Caglan aqualung that was developed in 1943 and sold worldwide over the subsequent 10 years started the common use of SCUBA for research and recreational diving. The need to conserve the breathing gas used for mixed gas breathing (e.g., helium/oxygen mixtures) resulted in the development of the semiclosed breathing apparatus which recirculates the exhaled breathing gas through a carbon dioxide absorber. This system was followed by the development of the closed circuit breathing apparatus that totally contains the breathing gas (i.e., no breathing gas

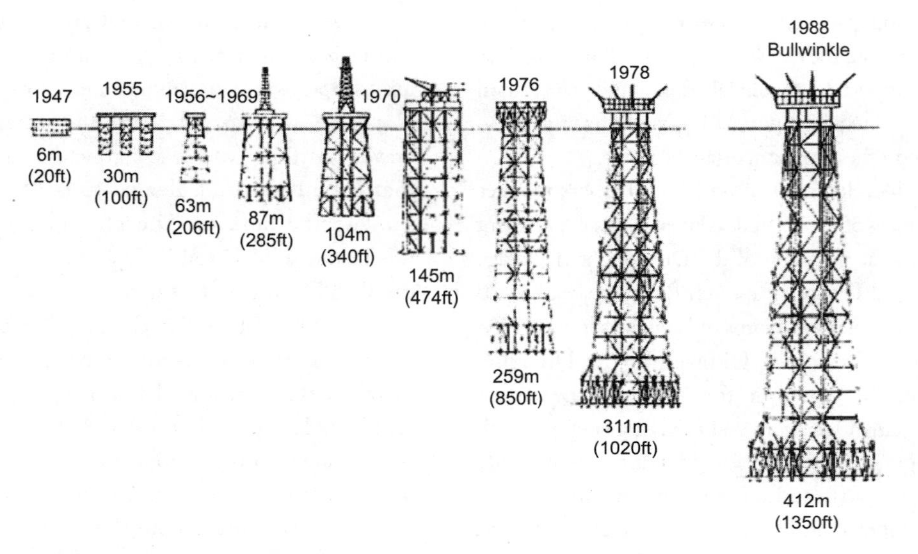

Fig. 9-2 Progression of fixed platforms and their installed water depth [7]

leaves the system).

Submarines are important military undersea vehicles, and they were first used in World Wars I and II. The U-boats of Germany were formidable weapons in the sea that patrolled the shipping lanes and disrupted shipping and supply routes. The German and allied submarines were typically 91.5m long and could work to depths of near 122m, and they were powered by diesel engines on the surface and electrical batteries underwater. In the late 1960s and early 1970s, larger and faster submarines were constructed that used nuclear power, inertial navigation systems and oxygen generating equipment so that the submarines could stay beneath the water for nearly unlimited time and could travel under ice caps without having to surface. The *USS* Nautilus was the world's first nuclear powered submarine.

A remotely operated vehicle (ROV) is an unmanned underwater system consisting typically of a propulsion device, closed circuit television and mechanical or electro-hydraulic manipulator. The vehicle is controlled from a surface vessel through an umbilical; video pictures and data are also transmitted through the umbilical and viewed on the surface vessel. The first ROV to gain fame was CURV that was developed by the U.S. Navy and used to recover a hydrogen bomb resting on the sea floor at a depth of 869m off the coast of Spain in 1966. ROVs are being used in the offshore industry, military applications and scientific investigations. The future of ROVs is very bright, and these vehicles are a valuable tool for the ocean engineer in a wide variety of underwater applications.

Generally, the equipment associated with subsea systems is necessary for the production of oil and gas from subsea wells. When the oil and gas fields are marginal, subsea production technology is more economical than conventional platform production techniques. Marginal fields apply to oil reserves of 30~50 million barrels and are typically in shallow water depths of less than 160m. As subsea equipment and technology advance, use in deeper waters is anticipated.

9.3 Ocean Environment

9.3.1 General Information

The surface of earth is approximately 71% covered by ocean. There are five major ocean areas including the (1) Arctic Ocean, (2) Atlantic Ocean, (3) Indian Ocean, (4) Pacific Ocean, and (5) Antarctic Ocean distributed worldwide. For many years, only the first four oceans were officially recognized, and then in the spring of 2000, the International Hydrographic Organization established the Antarctic Ocean, and determined its limits. If the Antarctic Ocean were considered part of the Pacific Ocean, the world ocean area would be subdivided into the Pacific Ocean (46%), Atlantic Ocean (23%), Indian Ocean (20%) and the Arctic Ocean (11%).

The average depth of the ocean is about 3700m. The deepest part of the ocean is called the Challenger Deep and is located beneath the western Pacific Ocean in the southern end of the Mariana Trench, which runs several hundred kilometers southwest of the U.S. territorial island of Guam. Challenger Deep is approximately 11,030m deep. It is named after the HMS Challenger, whose crew first sounded the depths of the trench in 1875. In comparison, the average land elevation is 840m and the highest elevation is 8,848.86m at the top of Mount Everest. The distributions of depths in the oceans and elevations on land are illustrated in Fig. 9-3. Depths shallower than 1000m make up 8.4% of the oceans. Offshore oil and gas platforms are found in water depths shallower than 3000m, which is about 18.2% of the world's oceans.

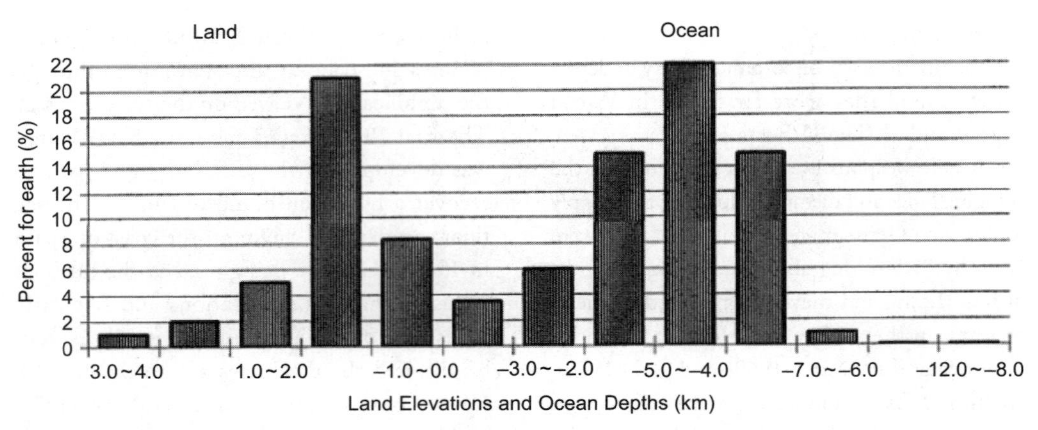

Fig. 9-3 Distribution of depths in the ocean and elevation on land [7]

Approximately 52% of the ocean depths are between 2000m and 6000m.

9.3.2 Ocean Water Properties

The important physical properties of seawater include both "thermodynamic properties" like density, as well as "transport properties" like viscosity. Density in particular is an important property in ocean science because small spatial changes in density result in spatial variations in pressure at a given depth, which in turn drive the ocean circulation. In addition, the mass density of water is a function of the salinity in the water. The temperature plays a major role in the values of these quantities and the values change with the change in the water temperature. The density values not only determine the forces on a structure placed in water, but also the difference in the density in different layers of water with depth may contribute to the internal waves in the deeper region, which has a very important effect on a submerged structure.

Generally, the largest thermocline occurs near the water surface irrespective of the geographic region of an ocean. The temperature of water is the highest at the surface and decays down to nearly constant value just above 0℃ at a depth below 1000m. This decay is much faster in the colder polar region compared to the tropical region and varies between the winter and summer seasons. The variation of salinity is less profound, except near the coastal region. The river run-off introduces enough fresh water in circulation near the coast, producing a variable horizontal as well as vertical salinity. In the open sea, the salinity is less variable having an average value of about 35%. The dynamic viscosity may be obtained by multiplying the viscosity with mass density. For comparison purposes, the values of the quantities in fresh water are also shown. These fresh water values are applicable in model tests in a wave basin where fresh water is used to represent the ocean. In these cases, corrections may be deemed necessary in scaling up the measured responses on a model of an offshore structure.

9.3.3 Ocean Currents

Ocean currents are the continuous, predictable, directional movement of seawater and is mainly driven by the wind effect on the water, variation of atmospheric pressure and tidal effects. Ocean water moves in two directions: horizontally and vertically. Horizontal movements are referred to as currents, while vertical changes are called upwelling or downwelling. Currents in the ocean are an important contributor to many physical, chemical and biological processes that occur in the ocean environment. In ocean engineering, currents create forces on structures, vehicles, shorelines and other systems which must be designed to withstand these forces. The most common categories of current are as follows.

(1) Wind-generated Currents

Wind-generated currents are caused by the direct action of the wind shear stress on the surface of the water. The wind-generated currents are normally located in the upper layer of the water body and are therefore not very important from a morphological point of view. In very shallow coastal waters and lagoons, the wind-generated current can, however, be of some importance. Wind-generated current speeds are typically less than 5 percent of the wind speed.

(2) Tidal Currents

Tidal currents are the strongest in large water depths away from the coastline and in straits where the current is forced into a narrow area. The most important tidal currents in relation to coastal morphology are the currents generated in tidal inlets. Typical maximum current speeds in tidal inlets are approximately at 1m/s, whereas tidal current speeds in straits and estuaries can reach as high as approximately at 3m/s.

(3) Loop and Eddy Currents

Loop current is a part of the Gulf Stream System and loops through the Gulf of Mexico continually. It enters the gulf through the Yucatan Channel and exits through the Straits of Florida oscillating north and south. The phenomena of most concern to deep water operators in the Gulf of Mexico are surface intensified currents associated with the loop current, loop current eddies and other eddies. While reliable information is available regarding the general speed distributions, translation speeds, sizes and shapes of these currents, the details of the velocity distributions and their variability are not well known.

9.3.4 Ocean Wave

Engineering systems that are designed to operate in the ocean must withstand the forces exerted by ocean waves. Consequently, ocean engineers must understand the physical processes and theories describing wave motion. Ocean waves are random in nature. However, larger waves in a random wave series may be given the form of a regular wave that may be described by a deterministic theory. Even though these wave theories are idealistic, they are very useful in the design of an offshore structure and its structural members. The forces related to wind, currents, storm surges and ice are often less important than those related to waves. As waves propagate over the ocean surface, they eventually impact offshore platforms, subsea systems, coastal protection structures and the shoreline that must absorb, reflect or dissipate the wave energy. Ocean engineers must design systems such that the wave forces do not cause it to fail.

Ocean waves are very complex, and many wave periods may be present at a given location and time. Shorter waves are commonly found superimposed on longer waves. In addition, the waves from different directions interact and cause wave conditions that are very difficult to describe mathematically. There are several wave theories that are useful in the design of offshore structures. Regular waves have the characteristics of having a period such that each cycle has exactly the same form. Thus the theory describes the properties of one cycle of the regular waves and these properties are invariant from cycle to cycle. There are three parameters needed in describing any wave theory shown as follows:

(1) Period (T)—the time duration taken for two successive crests to pass a stationary point.

(2) Height (H)—the vertical distance between the crest and the following trough.

(3) Water depth (d)—the vertical distance from the mean water level to the mean ocean floor. For wave theory, the floor is assumed horizontal and flat.

From these parameters, we can get several important quantities in water wave theory. They are:

(1) Wavelength—horizontal distance between successive crests.

(2) Phase speed—the propagation speed of the wave crest.

(3) Frequency—the reciprocal of the period;

(4) Wave elevation—the instantaneous elevation of the wave from the still water level (SWL) or the mean water level (MWL) ;

(5) Horizontal water particle velocity—the instantaneous velocity along x of a water particle;

(6) Vertical water particle velocity (v)—the instantaneous velocity along y of a water particle;

(7) Horizontal water particle acceleration (G)—the instantaneous acceleration along x of a water particle;

(8) Vertical water particle acceleration—the instantaneous acceleration along y of a water particle.

9.3.5 Wind Effects

The wind effect on an offshore structure becomes important when the superstructure is significant. The wind generally has two effects—one from the mean speed and the other from the fluctuation about this mean value. The mean speed is generally treated as a steady load on the offshore structure. For a fixed structure, it is only the mean speed that is taken into account. The effect of the fluctuation of wind about the mean value has a little effect on the fixed structure. However, this is not the case for a floating structure. In this case, the dynamic wind effect may be significant and may not be ignored. It should be noted that even the mean wind flowing over a changing free surface produces a fluctuating load due to the variation of the exposed structure with the wave. This effect is sometimes considered. For a linear wave, this fluctuation may be determined in a simple straightforward manner if the exposed surface is assumed to vary sinusoidally.

The accepted steady wind speeds in a design of an offshore structure are generally taken as the average speed occurring for a period of 1h duration. The steady speeds are considered to be the mean speed measured at a reference height, typically 10m above the mean still water level. A mean wind speed for a 100-year return period should be used in the design, based on the marginal distribution of wind speeds at the specific location. The directionality of the wind may be important in some applications. Wind load on the structure should be treated as a steady component based on the above mean speed. Additionally, a load with a time-varying wind component known as the gust should be calculated, which generates low-frequency motion. The time varying wind is described by a wind gust spectrum.

In ocean engineering, wind loading caused by extreme weather events such as hurricanes or typhoons controls the design of offshore structures, like offshore oil and gas platforms, mobile offshore drilling rigs, coastal bridges. Tropical cyclones are the result of winds rotating at a rate of 119kph or greater. These tropical cyclones are called hurricanes in the western Atlantic Ocean and typhoons in the western Pacific Ocean. The center of a hurricane, where the winds are light and the pressure is the lowest, is called the eye. Typically, the diameter of a hurricane is less than 560km. Fig. 9-4 shows Hurricane Ivan just before it entered the Gulf of Mexico in September 2004, with the eye clearly visible.

Fig. 9-4 Hurricane Ivan in 2004
Source: https://www.weather.gov/mob/ivan.

9.4 Offshore Structures

9.4.1 Introduction

Offshore structures have special characteristics from economic view, where offshore structure platforms are dependent on oil and gas production, which directly affect worldwide investment by oil price. Offshore structures used for the exploration and production of petroleum is being conducted in most continental shelf areas of the world and that can date back to the 1950s. The Gulf of Mexico led the way and was followed by the coastal waters off Mexico and Brazil. Installation of these structures in the extremely harsh environment of the North Sea began in the 1960s. Offshore structures are built worldwide in waters ranging from shallow water to nearly 3000m for producing petroleum reserves. Undersea pipelines are designed and installed to provide a means of transporting energy resources to land.

Ocean resource stimulates the need for structures that must operate in the ocean. Wave and tidal energy systems and the ocean thermal energy conversion systems have been and continue to be pursued for extracting energy from waves, tides and ocean water temperature differences. Wind energy is the newest form of a renewable energy resource, and offshore structures for wind energy harvesting are constructed in shallow waters with blades similar to airplane propellers. The extracted electrical energy is delivered to shore via underwater electrical cables buried in the ocean bottom sediments. The oceans may also be used in the future to support offshore fish processing plants, floating airports and floating communities where land is scarce.

9.4.2 Drilling and Producing Oil Gas

Currently, the major use of offshore structures is for the exploration and production of oil and gas. Therefore, a brief explanation of some terminology and procedures for drilling and producing oil and gas is given. The construction of offshore structures is very expensive, and mobile exploratory drilling rigs are used to drill wells to determine the presence or absence (dry hole) of petroleum at the offshore site. If oil is present in sufficient quantity, then the well is plugged until a permanent production platform is installed.

Offshore wells are drilled by lowering a drill string through a conduit which extends from the drill rig to the sea floor. The drill string (Fig. 9-5) consists of a drill bit, drill collar and drill pipe. The crew sets up the rig and starts the drilling operations. First, from the starter hole, the team drills a surface hole down to a pre-set depth, which is somewhere above where they think the oil trap is located. There are five basic steps for drilling the surface hole:

(1) Place the drill bit, collar and pipe in the hole;

(2) Attach the kelly and turntable, and begin drilling;

(3) As drilling progresses, circulate mud through the pipe and out of the bit to float the rock cuttings out of the hole;

(4) Add new sections (joints) of drill pipes as the hole gets deeper.

(5) Remove the drill pipe, collar and bit when the pre-set depth (anywhere from a few hundred to a couple-thousand feet) is reached.

Once they reach the pre-set depth, they must run and cement the casing—place casing-pipe sections into the hole to prevent it from collapsing in on itself. The casing pipe has spacers around the outside to keep it centered in the hole. In the pump system, an electric motor drives a gear box that moves a lever. The lever pushes and pulls a polishing rod up and down. The polishing rod is attached to a

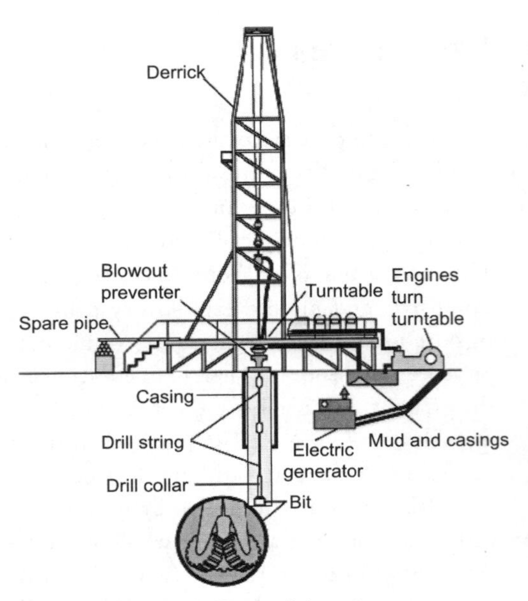

Fig. 9-5 Schematic of offshore oil rig system
Source: https://science.howstuffworks.com/environmental/energy/oil-drilling4.htm.

Fig. 9-6 Jack-up drilling rigs
Source: https://www.treehugger.com/types-of-offshore-oil-rigs-4864111.

Fig. 9-7 Semi-submersible drilling rig
Source: https://www.shippingherald.com/maersk-reactivates-semi-submersible-drilling-rig-in-southeast-asia/.

sucker rod, which is attached to a pump. This system forces the pump up and down, creating a suction that draws oil up through the well.

Some platforms are capable of drilling and producing. The fluids from the well contain a mixture of oil, gas, sand and brine water. This mixture is processed by special equipment before being sent ashore through a pipeline or transported to shore by a tanker. The processing of the well fluid mixture is known as producing, and the equipment used is called the production equipment. Production platforms are designed to support the production equipment.

9.4.3 Types of Offshore Structures

(1) Offshore Drilling System

1) Jack-up Drilling Rig

Jack-up drilling rigs (Fig. 9-6) are a type of self-elevating, mobile platform capable of raising the hull part which is obviously buoyant above the sea level and penetrating its steel legs into the sea-bed for support for deep sea drilling operations. It can drill in water depths of approximately 122m. The name "jack-up" is derived from versatile nature of the legs or jacks which can be suitably jacked above or below the hull accordingly when not in operation and when in operation. These legs may be piled into the sea-bed or may be placed on large footings for better grip and stability. There may be three or four legs for jacking up the buoyant hull above the sea level for operations. After a stipulated time period when the operations are terminated at a place, the legs are dismantled from the sea floor and are jacked above the hull and the entire unit is mobilized to another destination, maybe to some jetty/port or to another drilling site. They are not self-propelled and usually depend on tugs of heavy lift ships for towing from one place to another. Jack-up drilling rigs are used specifically for oil well drilling purposes and has limited or no storage capacity.

2) Semi-submersible Drilling Rig

The semi-submersible (Fig.9-7) is another type of offshore drilling rig. The semi-submersible drilling rig is used in water depths ranging from 91.5m to 3,048.8m, and it is typically towed to the site and moored to the bottom for water depth less than 915m. They are floating installations which rest on four or six pillar-like legs called columns with an equal weight distribution on each. These columns or legs are in turn attached to large basements called pontoons floating on the water surface. These pontoons may be ballasted or de-ballasted accordingly on and off operations. Often these pontoons delve deeper under the water surface and maintain the buoyancy and position of the floating system. Thereafter the operational deck is kept well aloof from the wave disturbance or the rough seas. However due to small water plane area, the structure is sensitive to load variations and must be trimmed accordingly. By the virtue of its equivalent weight distribution and a high draft, it has a greater stability than normal ships. The number of legs, pontoon design, the situation of the risers and drill equipment are decided at pre-design stage. They are generally instrumental in ultra-deep waters where the fixed structures pose a problem. Their position is maintained generally by a catenary mooring system or sometimes in modern structures by dynamic positioning system. These structures are gigantic and may be towed from one location to another by the virtue of a kind of ships called heavy lift ships.

3) Drillship

A drillship (Fig. 9-8) is used to drill oil and gas wells. Typically employed in deep and ultra-deep waters, drillships work in water depths ranging from 610m to more than 3048m. The drilling derrick is usually positioned at amidship, and a moon pool opening is located below the derrick for the drilling operation. Drillships have their own propulsion. A dynamic positioning system is used to keep the drillship over the drilling location. Thrusters are added in the bow of the ship to assist in the positioning of the ship. A local acoustic positioning system or differential global positioning system is used to determine the position of the drillship relative to the well head on the sea floor. Environmental sensors measuring wind, wave, and current and acoustic

Fig. 9-8 Offshore drillship
Source: https://www.rigzone.com/training/insight.asp?insight_id=306&c_id=.

sensors are used to automatically send signals to the ship's thrusters and propulsion system to keep it directly over the well. The conductor pipes, or drilling risers, are flexible to accommodate small inclination angles, and the conductor pipes can be disconnected when the extreme weather approaches.

(2) Offshore Oil and Gas Platforms

Offshore platforms (fixed or floating) are used to recover and process oil and gas that is transported to shore through subsea pipelines and ships such as crude oil tankers and liquefied natural gas vessels. The progression of offshore platforms into deepwater, as illustrated in Fig. 9-9, begins with fixed platforms in up to 460m water depth. Beyond the 460m water depth, floating offshore platforms are used. In the 460m to 915m water depth, compliant towers have been installed, and minitension–leg platforms (TLPs) have been used in water depths between 150m and 1070m. Semi-submersible platforms are considered for depths of 610m to 3048m. Conventional TLPs are considered for water depths from 460m to 1524m. Spar platforms such as the classical and truss spars are considered for the deep waters ranging from 610m to 3048m, similar to the semi-submersible platforms. The use of subsea tiebacks that connect distant wells to fixed or floating production facilities are also in use and expanding in the deeper waters. The floating production storage and offloading (FPSO) is

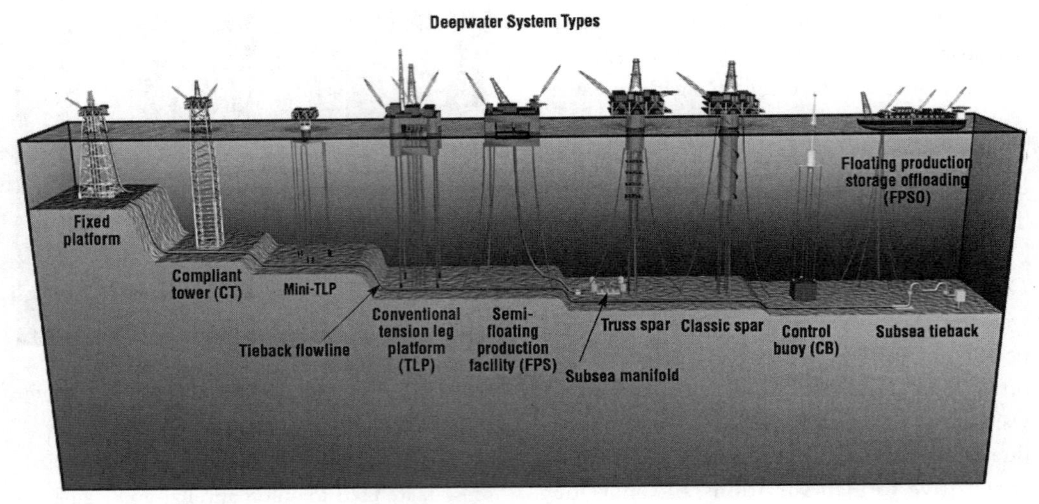

Fig. 9-9 Offshore platforms used for the production of oil and gas
Source: http://www.oil-gasportal.com/oilgas-subsea-production/?print=print.

another production platform used all around the world, especially in offshore locations where there is no pipeline infrastructure. The FPSO transfers the oil to shuttle tankers that bring the oil to shore for offloading at marine terminals.

1) Fixed Jacketed Structure

A fixed jacketed structure (Fig. 9-10) consists of a steel framed tubular structure that is attached to the sea bottom by piles. These piles are driven into the sea floor through pile guides (sleeves) on the outer members of the jacket. The topside structure consists of drilling equipment, production equipment, crew quarters and eating facilities, gas flare stacks, revolving cranes, survival craft, and a helicopter landing pad. Drilling and production pipes are brought up to topside through conductor guides within the jacket framing, and the crude oil and gas travel from the reservoir through the production riser to topside for processing. The produced fluid is then pumped to shore through the export pipeline. The detailed design of the frame varies widely and depends on the requirements of strength, fatigue and launch procedure. Structural members consist of X and K joints, and X and K braced members. The platform phases include design, construction, load-out, launch, installation, piling and hook-up before production begins. The design life of the structure is typically 10 to 25 years. This is followed by the requirement to remove and disposal of the platform once the reservoir is depleted.

Fig. 9-10 A fixed jacketed structure
Source: https://www.engineerlive.com/sites/engineerlive/files/Offshore_Promotional-Image_ist_62362464_OilGasPlatform_L_Original_2%20use.jpg.

2) Concrete/Gravity-based Platform

Concrete structures are purely composed of concrete and is considered the safest mode of offshore industrial operations. They may be directly fixed or molded to the sea floor on a permanent basis. Fixed ones are also known as gravity-based structures (Fig. 9-11). The entire structure is based on a submerged

Fig. 9-11 Gravity-based platform
Source: http://www.chinashipbuilding.cn/info.aspx?id=4605.

Fig. 9-12 Tension-leg platforms
Source: http://www.chinashipbuilding.cn/info.aspx?id=4605.

island-like solid structure which may serve as a foundation structure as well as storage for oil or other by-products. Gravity-based structures are used for water depths between 20m and 300m, and require a preliminary survey of the sea floor before being sunk due to the extreme pressure exerted on the bed. The entire load of the structure acts directly on the subsequent layers of the seabed, eradicating problems even of the extreme wave disturbances or seafloor scour. The design parameters are guided by the number of supporting columns and the diameter of the legs. Constructing is a tedious process where the entire structure or its components are towed and assembled with proper ballasting and de-ballasting measures. They have almost negligible maintenance, high durability and can be also used for greater depths.

3) Tension-leg Platforms

A typical TLP is shown in Fig. 9-12. It has floating facilities which stay afloat and remain in position with the help of specialized steel tubes called tethers or tendons. These are nothing but the supporting legs of the floating platform which by the virtue of their upward tension takes care of the position, loading and the functionality of the extraction system. For all TLPs, there is not much vertical oscillatory motion of the platform no matter how rough the sea conditions or the average wave height may be. This facilitates in tying with the wellheads and the piping system for extraction of oil without much distortion. For all such systems there is always an additional buoyant force which keeps the platform afloat. The topside facilities and the basic parameters like the number of risers have to be fixed at pre-design stage. Generally the platform is manufactured on the shore and is towed to the desired location with the help of tugs. This structure is apt for depths up to 1200m and has limited or no storage facility. As of common usage, these have a high maintenance and surveillance cost of the tethers and under unavoidable circumstances the structure may be prone to irreversible damage.

4) Spar Platform

A spar platform is essentially a buoy-based design. It is a cylindrical shaped floating offshore installation; the cylinder is weighted at the bottom, using a ballast for stabilizing upwards in a vertical position. This cylinder supports the deck at the top, and is moored to seabed with polyester ropes to keep it positioned against waves, tides and current forces. Spar platforms are highly preferred

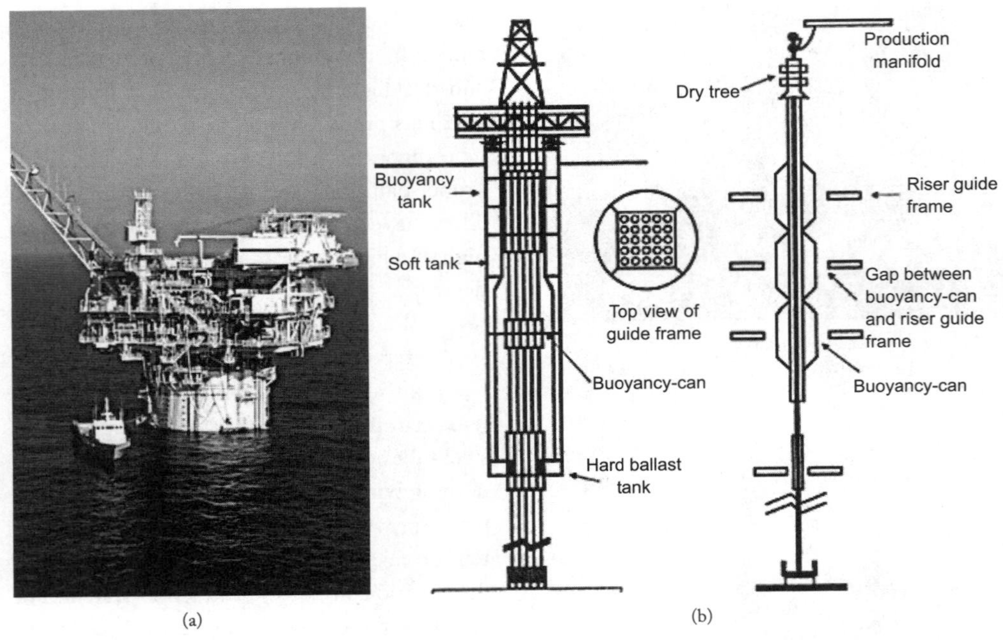

Fig. 9-13 Truss spar platform and schematic of a spar platform
(a) Truss spar platform; (b) Schematic of a spar platform[8]
Source: (a) http://www.chinashipbuilding.cn/info.aspx?id=4605.

structural design for offshore deep-water oil and gas exploration, drilling, processing and storage facility. Commonly there are three spar platform designs: classic spar, truss spar and cell spar. A typical truss spar platform is shown in Fig. 9-13.

Spar platforms have evolved from the conventional "classic" spar that consisted of a full length cylinder with a diameter of 22m to 37m and length of 183m to 213m, as shown in Fig. 9-14. The truss spar followed the conventional spar and that shortened the length of the cylinder and reduced the cost. The plates of the truss provided the necessary damping and maintained the necessary heave characteristics. A cell spar replaces the large diameter cylinder with a bundle of smaller diameter cylinders for easier construction and lower costs. The cell spar uses the truss concept below the bundled cylinders.

(3) Floating Production Storage and Offloading System

A floating production storage and offloading (FPSO) system is a floating facility, usually based on an/a (converted) oil tanker hull (Fig. 9-15). It is equipped with hydrocarbon processing equipment for separation and treatment of crude oil, water and gases, arriving on board from subsea oil wells via flexible pipelines. The FPSO concept allows oil companies to produce oil in more remote areas and in deeper water. Furthermore, it has storage capacity for the treated crude oil and is equipped with an offloading system to transfer the crude oil to shuttle tankers for shipment to refineries, rather than requiring a pipeline to transport oil to shore. One of the typical features of a FPSO is the turret mooring system (Fig. 9-16) , which is usually fitted inside and integrated into the FPSO's hull. A typical example is Bluewater with a core technology. The turret is moored to the seabed with chains, wires and anchors and has bearings allowing free and unrestricted 360° rotation of the FPSO around the turret.

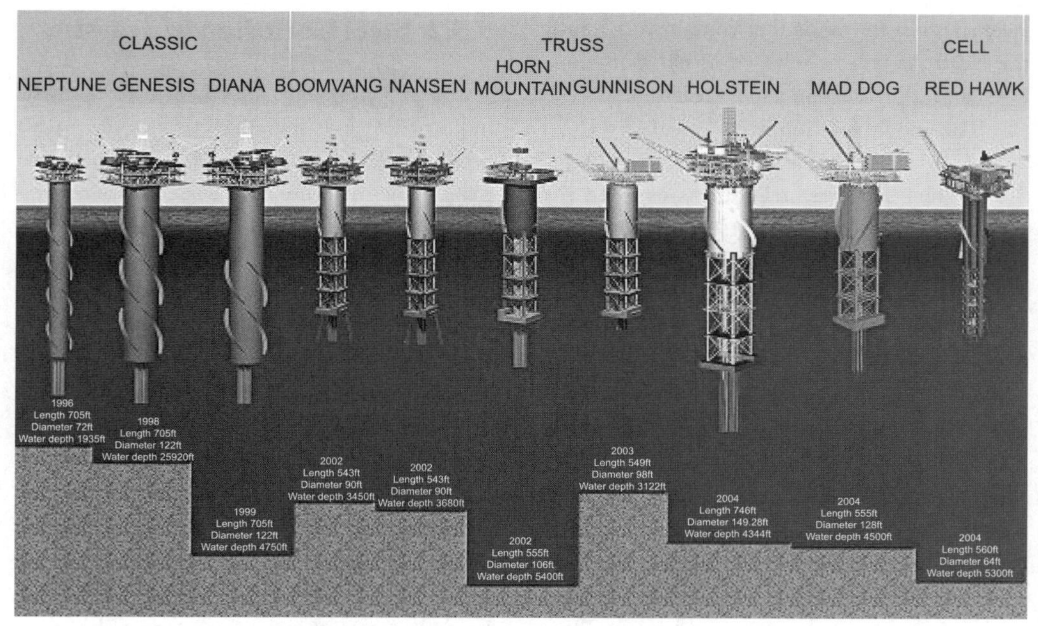

Fig. 9-14 Spar platforms installed from 1996 to 2005 primarily in the Gulf of Mexico [8]

Fig. 9-15 FPSO Glas Dowr offloading to a shuttle tanker
Source: https://www.bluewater.com/fleet-operations/what-is-an-fpso/.

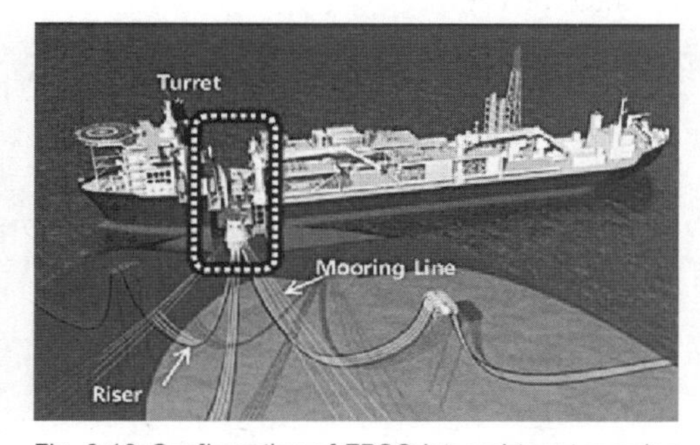

Fig. 9-16 Configuration of FPSO internal turret mooring system and riser [10]

9.5 Corrosion Protection

9.5.1 Introduction

Generally, environmental condition surrounding the structure is one of the key factors leading to corrosion situation. The definition and characteristics of this variable can be quite complex since it can change with time and conditions. Besides, it should be realized that environment usually affects a metal corresponds to the microenvironmental conditions of the local environment at the surface of the metal. It is the reactivity of this local environment that determines the real corrosion damage.

In general, corrosion is signaled by rust appearing on a steel surface. The chemical reactions are the main drivers in the corrosion process due to chloride attack. The corrosion of steel starts in the voids that contain water and the electrons will be released as per Equation (9-1) and it presents an anodic reaction.

The anodic reaction is:

$$Fe \rightarrow Fe^{2+} + 2e^{-1} \quad (9\text{-}1)$$

A steel offshore structure is exposed to saline water throughout its life time. The impact of water on the integrity of materials is thus an important aspect of system management. Since steels and other iron-based alloys are the metallic materials most commonly exposed to water, aqueous corrosion is discussed with a special focus on the reactions of iron (Fe) with water (H_2O).

If the electrons accumulate on another part of the steel but cannot accumulate in huge amounts in the same part of the steel, another chemical reaction takes place as a combination of the electrodes with oxygen and water. This is called the cathodic reaction and is illustrated in Equation (9-2).

The cathodic reaction is:

$$2e^{-1} + H_2O + \frac{1}{2}O_2 \rightarrow 2OH^- \quad (9\text{-}2)$$

The hydroxide ions cause alkalinity and reduce slightly the effect of chlorides. It is important to note from the above equation that water and oxygen are the main causes of the corrosion process.

The anodic and cathodic reactions are the first step in the process of corrosion shown in Equation (9-1). Hydroxide ions (OH^-) react with ferrous iron (Fe^{2+}) resulting from Equation (9-1). This reaction produces ferrous hydroxide which reacts with oxygen and water again and produces ferric hydroxide.

$$Fe^{2+} + 2OH^- \rightarrow Fe(OH)_2 \quad (9\text{-}3)$$

$$4Fe(OH)_2 + O_2 + 2H_2O \rightarrow 4Fe(OH)_3 \quad (9\text{-}4)$$

$$2Fe(OH)_3 \rightarrow Fe_2O_3 \cdot H_2O + 2H_2O \quad (9\text{-}5)$$

From the above chemical reactions, rust films normally consist of three layers of iron oxides in different states of oxidation.

9.5.2 Steel Corrosion in Seawater

Seawater systems are widely used by many industries, such as shipping, offshore oil and gas production, power plants, and coastal industrial plants. The main use of seawater is for cooling purposes, but it is also used for firefighting, oilfield water injection and desalination plants. The corrosion problems in these systems have been well studied over many years, but despite published information on materials behavior in seawater, failures still occur. Most of the elements that can be found on earth are present in seawater, at least in trace amounts.

The concentration of dissolved materials in the sea varies greatly with location and time, because rivers dilute seawater, as do rain and melting ice, however seawater can be concentrated by evaporation. The most important properties of seawater are as follows:

(1) Remarkably constant ratios of the concentrations of the major constituents worldwide;

(2) High salt concentration, mainly sodium chloride;

(3) High electrical conductivity;

(4) Relatively high and constant pH;

(5) Buffering capacity;

(6) The presence of a myriad of organic compounds;

(7) Salinity.

The corrosion of steel in seawater is adequately represented by Equation (9-1), although the process normally proceeds to the precipitation of $Fe(OH)_3$. On clean steel in seawater, the anodic process occurs with greater facility than the cathodic. As a consequence, the corrosion reaction can go no faster than the rate of cathodic, oxygen reduction. The latter usually proves to be controlled by the rate of arrival of the oxygen at the metal surface, which, in turn, is controlled by the linear water flow rate and the dissolved oxygen concentration in the bulk seawater. This may be represented on a polarization diagram as shown in Fig. 9-17. At first, the cathodic kinetics get faster as the potential becomes more

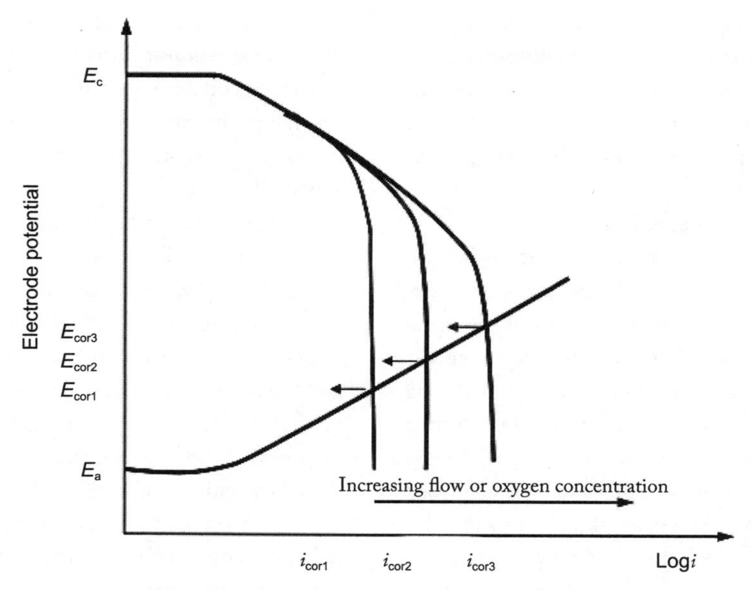

Fig. 9-17 Polarization diagram with increasing oxygen concentration[9]

negative from E_c. This has the effect of depleting the oxygen immediately adjacent to the metal surface, thus rendering the reaction more difficult. Ultimately, a point is reached where the surface concentration of oxygen has fallen to zero and oxygen can then only be reduced when it reaches the surface. Further lowering of the potential cannot increase the cathodic reaction rate, because the kinetics are now governed by potential-independent diffusion processes. Fig. 9-17 shows that the corrosion rate is then equal to this limiting current. The limiting current can be increased by increasing the oxygen flux either by raising the bulk oxygen concentration (the concentration gradient gets steeper) or increasing the flow rate (the oxygen-depleted layer gets thinner). Both serve to increase the corrosion rate as shown in Fig. 9-17. In practice, corrosion products and marine fouling build up on steel as it corrodes in seawater. These generally produce lower corrosion rates. The peak corrosion rate is often attributed to galvanic action between steel in contact with the oxygen-rich surface waters, which is the cathodic area, and the steel at somewhat greater depth exposed to waters of lesser oxygen content, which is the anodic area.

9.5.3 Typical Types of Corrosion Occurring in Ocean Environment

(1) General Corrosion

General corrosion appears as a continuous layer of corrosion over an entire surface area. It occurs more often for objects that are exposed to air, such as piping and plates on exposed structures (e.g., offshore platforms and ships). This type of corrosion is not commonly found when objects are totally submerged. It is easier to design for this type of corrosion.

(2) Galvanic Corrosion

Galvanic corrosion occurs when two dissimilar metals are connected electrically or by a metallic path and are immersed in seawater or some other liquid that acts as an electrolyte. As a result, one of the metals will corrode faster than normal, and the other, more noble metal will corrode more slowly or cease to corrode. Engineers use the process of galvanic corrosion to protect ship hulls by bolting zinc anodes to the steel hulls. It is best to avoid using dissimilar materials, but if dissimilar materials must be used, the two materials should be isolated by means of a nonmetallic material between them. Increasing the distance between two dissimilar metals in a galvanic couple will lower the rate of corrosion. Also, making the anode larger than the cathode will reduce galvanic corrosion. Additionally, situations in which the surface area of the more noble metal (cathode) is larger than the less noble metal (anode) should be avoided.

(3) Intergranular Corrosion

Intergranular corrosion is a microscopic form of corrosion that is caused by the potential difference between the grain boundaries of the metal and grain bodies. When the grain body is anodic to the grain boundaries, corrosion will occur along the boundaries. This is frequently the case in weld zones. Dealloying corrosion can occur in some metal alloys that are susceptible to corrosion when elements in the alloy are attacked. The more active element is removed, the more noble element remains. The result is a porous and weakened structure. This type of corrosion is commonly observed in cast iron immersed in seawater, and it occurs in brass containing more than 15% zinc.

(4) Crevice and Pitting Corrosions

Crevice corrosion can occur in places with a limited amount of oxygen available, such as in slightly open joints (crevices) and under nuts, bolt heads and washers. Crevice corrosion results from the creation of an electrochemical cell between the oxygen-deficient electrolyte inside the crevice and the electrolytic solution outside the crevice. The crevice acts as the anode, and the large surrounding area acts as the cathode. This results in an attack of the anode, which can then cause mechanical failure of the components. The localized nature of the corrosion can cause mechanical failures. Pitting corrosion is similar to crevice corrosion, but it does not require an existing pit or crevice to cause the pits to occur. It produces very small pits or holes in the metal surface. These pits can penetrate a hull and severely weaken the metal structure.

(5) Erosion Corrosion

Erosion corrosion occurs when seawater is flowing at a relatively high velocity and impacts a metal surface. It is often found in the bends and elbows of pipes. Erosion corrosion can be general or localized, and is usually characterized by grooves, waves or rounded holes. Corrosion due to cavitation is also caused by the velocity of seawater, but the mechanism is different. As the fluid flows over the metal surface, vapor bubbles form at the interface (cavitation) as a result of the reduced local pressure. When these vapor cavities collapse in regions of higher pressure, the forces of the water on the metal surface can damage the metal surface or protective coating. This type of corrosion commonly occurs on propellers, hydrofoils and pump impeller blades.

9.5.4 Preventing Corrosion

Corrosion is the gradual deterioration of a material or its properties through a chemical reaction with its environment. The optimal time to consider corrosion prevention is during the design stage, and the worst time is after the existence of corrosion that has been discovered.

Typical requirements and procedures for material selection for corrosion prevention Table 9-1

Requirements	Procedures
Properties (corrosion, mechanical, physical)	Define life of system
Fabrication (constructability, weldability, machinability)	Material life
Compatibility	Reliability (safety, failure consequences)
Maintainability	Availability/delivery time
Data availability	Costs (material, maintenance, inspection)
	Comparison with other corrosion protection possibilities

Several methods of preventing corrosion are now briefly described, including material selection, good design, paint or coating, cathodic protection and inhibitors. By selecting the right material for a particular application, one can avoid or minimize corrosion. However, selecting the most corrosion-resistant material doesn't always work, because there are other requirements that the design must satisfy. Some typical requirements and a few procedures for making proper material choices are tabulated in Table 9-1.

(1) Design

Good design incorporates corrosion protection methods during the design stage, when the engineer can view drawings and determine the possibility of eliminating geometric configurations that are known to cause or accelerate the corrosion process, such as crevices, stagnant areas and stress risers. The design stage is a good time to consider the effects of other corrosion protection methods, but they can also be incorporated at a later time.

(2) Material Selection

The purpose of material selection is to select the best material that will achieve the desired corrosion resistance at the least cost. Corrosion cost, design conditions, safety, reliability and environmental effects must be considered when selecting materials. Additional considerations include material properties, maintainability, compatibility, and ease of fabrication.

(3) Coating

Painting surfaces is a very common type of corrosion control for metallic and nonmetallic surfaces. Four critical elements for using coatings for corrosion control are correct selection, surface preparation, application, and routine inspection and repair. The life of corrosion protection from paint is relatively short (several years), and the repair of worn, scratched, or chipped spots is routine maintenance. Some coatings (long-life paints) have been developed that have a useful lifetime of over 10 years in the marine environment. Coatings are very expensive on a per gallon basis, and they require extensive surface preparation that is also costly. Inorganic zinc is one of the better coatings for above water applications, and its use underwater requires a top-coating.

(4) Inhibitors

The protection of closed systems such as engines, boilers or tanks is usually accomplished with the use of chemical inhibitors. There are five classes of inhibitors: absorption inhibitors (affect anodic and cathodic reactions), hydrogen evolution poisons (affect hydrogen evolution), scavengers (remove oxygen needed for the cathodic reaction), oxidizers (work with iron only), and vapor phase inhibitors. The first four methods use a solution to protect the metal. For example, rust inhibitors are used in automobile coolants to protect the engine and radiator.

(5) Cathodic Protection

Cathodic protection is the most common form of corrosion protection for submerged material. The two types of cathodic protection are called sacrificial anode cathodic protection and impressed current cathodic protection. The impressed current cathodic protection system is a more permanent protection system and requires the use of external electrical power. It is much more complex than the sacrificial anode cathodic protection system and is used for offshore structures and mooring applications.

9.6 Conclusions

This chapter presents a brief introduction of ocean engineering. The historical developments of ocean engineering demonstrate that mankinds are making use of ocean resources by exploration of the underwater environment, development of offshore gas and oil, and the continued need for coastal protection and port expansion. Ocean resources stimulates the need for structures that must operate in the ocean. Therefore, offshore structures are being designed and built to explore the secrets of the ocean. After understanding of harsh ocean environments, it is critically important and necessary to provide corrosion protection for offshore structures. Several common corrosion protection methods are briefly introduced. Human beings are being building offshore structures and developing equipment to make full use of ocean resources. Ocean engineering is the critical link between scientific knowledge and application of that knowledge to work in ocean.

Exercises

9-1 What's ocean engineering? And what is the main task?

9-2 How many types of offshore structures are there and what are they?

9-3 How does an offshore oil rig system work?

9-4 What are anodic reaction and cathodic reaction?

9-5 What is corrosion phenomenon?

9-6 What are typical types of corrosion in ocean?

9-7 What can we do for corrosion protection?

9-8 Describe some common offshore platforms and list their basic features.

References

[1] PICKARD G L, EMERY W J. Descriptive Physical Oceanography: An Introduction [M]. New York: Pergamon Press, 1990.

[2] ALLMENDINGER E E. Submersible Vehicle Systems Design[M]. Jersey City: The Society of Naval Architects and Marine Engineers, 1990.

[3] Goodfellow Associates. Applications of Subsea Systems[M]. Tulsa: PennWell Publishing Co., 1990.

[4] National Oceanic and Atmospheric Administration (NOAA). NOAA Diving Manual (3rd Edition)[M]. Washington, D.C.: Best Publishing Company, 1991.

[5] National Oceanic and Atmospheric Administration (NOAA). NOAA Diving Manual (4th Edition)[M]. Washington, D.C.: Best Publishing Company, 2001.

[6] CHAKRABARTI S. Handbook of Offshore Engineering (2-volume Set) [M]. New York: Elsevier, 2005.

[7] RANDALL R E. Elements of Ocean Engineering(2nd Edition) [M]. The Society of Naval Architects and Marine Engineers, 2010.

[8] KOO B J, KIM M H, RANDALL R E. Mathieu instability of a spar platform with mooring and risers [J]. Ocean Engineering, 2004, 31(17-18): 2175-2208.

[9] EL-REEDY M A. Offshore Structures: Design, Construction and Maintenance [M]. Oxford: Gulf Professional Publishing, 2012.

[10] JO A R, PARK K P, LEE H J. Lug arrangement and dynamic analysis of lifting simulation for underwater installation of structure in asymmetric position[J]. Journal of the Society of Naval Architects of Korea, 2015, 52(4), 283-289.

Chapter 10
Disaster Prevention and Mitigation

10.1 Overview

Our world is extremely vulnerable to natural disasters. Every country is exposed to one or more of a couple of hazards: earthquakes, tornadoes, droughts, floods, landslides, wildfires, tsunamis, volcanoes and hurricanes. Deadliest natural disasters cause the collapse and damage of buildings and structures, disruption of lifeline projects such as transportation and communication, water supply and power supply, and trigger secondary disasters, resulting in a large number of casualties and property losses, leading to serious economic losses. As Wenchuan Earthquake demonstrated in 2008, natural disasters can undo years of development and devastate natural resources in minutes or hours. The main reason of severe losses of casualties and property is due to collapse of civil engineering infrastructures and its facilities. Civil engineering plays a vital role in disaster prevention and mitigation. Almost all disasters are associated with civil engineering. Meanwhile, civil engineering is significantly meaningful and valuable even irreplaceable to resist and mitigate all disasters. Earthquakes, extreme wind and fire needs special considerations in design and construction of buildings since they are more frequent, widespread and more disastrous. In this chapter this aspect of building design and constructions are discussed as well as mitigation measures.

10.2 Definition and Types of Disaster

10.2.1 Definition of Disaster

A disaster is a serious disruption occurring over a short or long period of time that causes widespread human, material, economic or environmental losses which exceeds the ability of the affected community or society to cope with using its own resources. Disasters have been a part of human experience from earliest times and have been significantly impacting human development and civilization. From a modern scientific viewpoint, a disaster is an abrupt event that leads to loss of human lives, properties, resources or environmental wellbeing, exceeding the capacity of the hazard-bearing body, a term that is used herein to refer to any exposure of human society to a hazard. The following four characteristics are inherent to the above definition of disasters: 1) Disasters are consequent to the presence of human beings and communities as hazard-bearing bodies. There would be no disasters if there were no humans; violent changes and movements have occurred since the beginning of the earth, but did not constitute disasters until the appearance of man. 2) Disasters are uniquely expressed in terms of losses by human beings and communities. Such losses are not limited to life and property, but also include natural resources and the environment. 3) There is a threshold of the extent of loss due to an event for it to be considered a disaster. In other words, not all losses causing events are considered disasters. For example, the collapse of an ordinary warehouse, everyday car accidents, and robberies may cause losses, but such are not generally categorized as disasters. While a car accident may constitute a disaster for a family, it is far from a disaster to the city. Moreover, there are always fortunate individuals or families that remain intact during major earthquakes, which may otherwise constitute devastating disasters to local communities. Hence, the aim of disaster mitigation is not necessarily to eliminate loss entirely, but to decrease the loss below the disaster threshold, which is often a more practical strategy. 4) Disaster emphasizes the abruptness of the event from definition. Disastrous events are typically abrupt, such as earthquakes, landslides, aviation accidents and terrorist attacks. However, some disastrous events may occur gradually, such as global warming due to excessive and extended carbon emission, metropolitan smog due to air pollution and the desertification of a forest or prairie. In contrast to such gradual events, which are to some extent expected and observed during their development, abrupt disasters take place suddenly without effective forecast or prediction. Their durations are short, but they cause significant and often lasting consequences. In addition, abrupt and gradual disasters differ in other ways, including in their occurrence mechanisms, consequences and mitigating measures. However, this chapter focuses on abrupt disasters.

10.2.2 Types of Disaster

Generally, disasters are classified into two major categories based on the natures of the related hazards, namely, natural disasters and man-made disasters. Natural disasters are large-scale geological, meteorological or biological events that have the potential to cause severe loss of life or property, which are further classified as:

(1) Geological disasters, which are caused by hazards in the lithosphere, such as earthquakes, landslides, debris flows and volcanic eruptions;

(2) Meteorological disasters, which are caused by hazards in the atmosphere and hydrosphere, such as hurricanes, tornados, droughts, forest fires, heavy rains and floods;

(3) Biological disasters, which are caused by hazards in the biosphere, such as plagues and pests.

As with natural disasters, man-made disasters of traumatic events may also cause loss of life and property. They may also prompt evacuations from certain areas and overwhelm behavioral health resources in the affected communities. Examples include industrial accidents, shootings, acts of terrorism and incidents of mass violence. Man-made disasters can also be classified by hazards into the following three categories:

(1) Disasters caused by technical mistakes, such as nuclear accidents, explosion of dangerous substances;

(2) Disasters caused by human faults, such as fires in buildings, traffic accidents and gas explosions;

(3) Disasters caused by hostile actions, such as wars, riots and terrorist attacks.

10.3 Earthquake Disaster

10.3.1 Basic Concepts of Earthquake

An earthquake is defined as the shaking of the surface of earth resulting from a sudden release of energy in the earth's lithosphere that creates seismic waves. Earthquakes usually happen when two blocks of the earth suddenly slip past one another, as shown in Fig. 10-1. The surface where they slip is called the fault surface or fault plane. The location below the earth's surface where the earthquake starts is called the hypocenter, and the location directly above it on the surface of the earth is called the epicenter.

Earthquakes are quite common natural hazards and occur somewhere around the world everyday. Once every 30 seconds

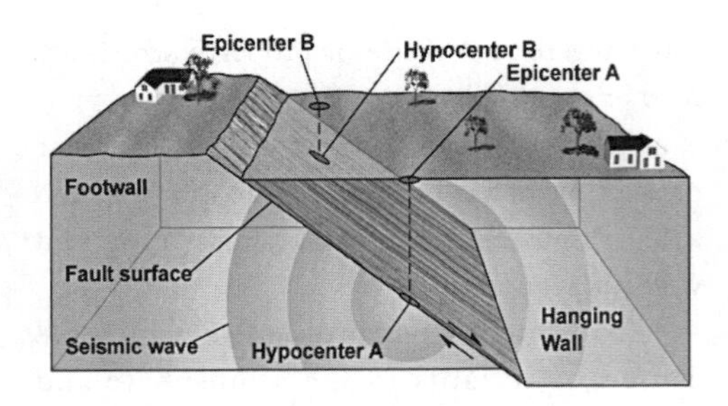

Fig.10-1 Schematic of earthquake due to fault movement
Source: https://geologylearn.blogspot.com/2016/03/what-causes-earthquakes.html.

somewhere in the world the ground shakes. These earthquakes range from quite small events detectable by seismic instrument to great earthquakes that destroy entire cities. Fortunately, the vast majority are considered minor. Actually, the number of lives lost and the amount of economic losses that result from an earthquake depend on "earthquake size", which are measured by earthquake magnitude, shaking intensity and energy release.

(1) Magnitude

Magnitude is a quantitative measure of earthquake size. It is based on the maximum amplitudes of body or surface seismic waves. There is only one magnitude for each earthquake. The time, location and magnitude of an earthquake can be determined from the data recorded by seismometer shown in Fig. 10-2. Generally, there are four types of magnitudes: Richter magnitude M_L, body wave magnitude M_B, surface wave magnitude M_S and moment magnitude M_w. At present, Richter magnitude is widely used all over the world, which is defined by American seismologist Charles F. Richter and the formulation is given by

$$M_L = \log A - \log A_0 \qquad (10\text{-}1)$$

Where A — the maximum seismic wave amplitude recorded on standard Wood–Anderson seismographs located at a distance of 100km from the earthquake epicenter;

A_0 — a calibration factor that depends on distance.

Earthquakes are classified in categories ranging from minor to great, based on their magnitude. These terms are magnitude classes shown in Fig. 10-3. The classification starts with "minor" for magnitudes between 3.0 and 3.9, where earthquakes generally begin to be felt, and ends with "great" for magnitudes greater than 8.0, where significant damage is expected.

(2) Shaking Intensity

The shaking intensity is the measure of shaking at each location, and it varies from place to place, depending mostly on the distance from the fault rupture area. Earthquake intensity measurement is an on-the-ground description. The measurement explains the severity of earthquake shaking and its effects on people and their environment. It is a subjective damage evaluation metric because of its qualitative nature, related to population density, familiarity with earthquake

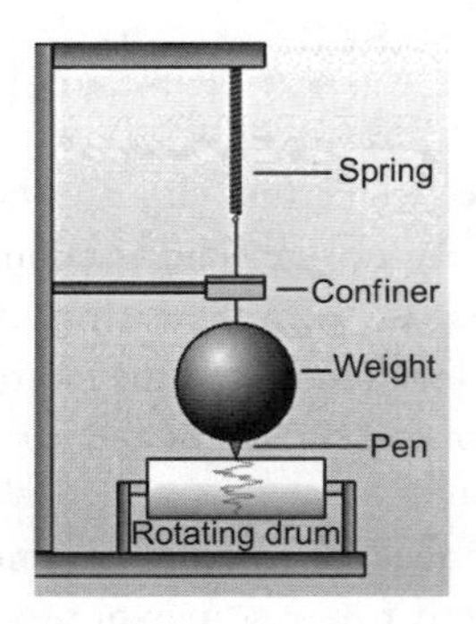

Fig. 10-2 Seismometer
Source: https://www.gcse.com/waves/seismometers.htm.

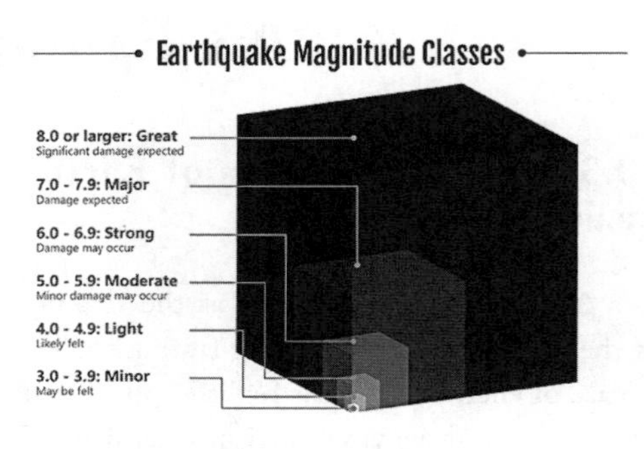

Fig. 10-3 Earthquake magnitude classes
Source: https://www.earthquakeauthority.com/Blog/2020/Earthquake-Measurements-Magnitude-vs-Intensity.

and type of constructions. There are many intensity values for each earthquake that are distributed across the geographic area around the earthquake epicenter, unlike the magnitude which is one number for each earthquake. Discrete scales are used to quantify seismic intensity; the levels are represented by Roman numerals and each degree of intensity provides a qualitative description of earthquake effects. Several intensity scales have been proposed worldwide, such as the Modified Mercalli (MM) shown in Table 10-1.

(3) Energy Release

Another way to measure the size of an earthquake is to compute how much energy it released. When an earthquake occurs, stress

Modified Mercalli (MM) Intensity Scale **Table 10-1**

Intensity	Shaking	Description/Damage
I	Not Felt	Not felt except by a very few under especially favorable conditions
II	Weak	Felt only by a few persons at rest, especially on upper floors of buildings. Delicately suspended objects may swing
III	Weak	Felt quite noticeably by persons indoors, especially on upper floors of buildings. Many people do not recognize it as an earthquake. Standing motor cars may rock slightly. Vibration is similar to the passing of a truck. Duration is estimated
IV	Light	Felt indoors by many, outdoors by few during the day. At night, some will be awakened. Dishes, windows, doors are disturbed; walls make cracking sounds. Sensation is like heavy truck striking building. Standing motor cars are rocked noticeably
V	Moderate	Felt by nearly everyone; many are awakened. Some dishes, windows are broken. Unstable objects are overturned. Pendulum clocks may stop
VI	Strong	Felt by all, many are frightened. Some heavy furniture is moved; a few instances of fallen plaster will occur. Damage is slight
VII	Very Strong	Damage is negligible in buildings of good design and construction; slight to moderate is in well-built ordinary structures; considerable damage is in poorly built or badly designed structures; some chimneys are broken
VIII	Severe	Damage is slight in specially designed structures; considerable damage is in ordinary substantial buildings with partial collapse. Damage is great in poorly built structures. Fall of chimneys, factory stacks, columns, monuments, walls. Heavy furniture is overturned
IX	Violent	Damage is considerable in specially designed structures; well-designed frame structures are thrown out of plumb. Damage is great in substantial buildings, with partial collapse. Buildings are shifted off foundations
X	Extreme	Some well-built wooden structures are destroyed; most masonry and frame structures are destroyed with foundations. Rail is bent
XI	Extreme	Few, if any, (masonry) structures remain standing. Bridges are destroyed. Broad fissures are in ground. Underground pipe lines are completely out of service. Earth slumps and land slips in soft ground. Rails are bent greatly
XII	Extreme	Damage is total. Waves are seen on ground surfaces. Lines of sight and level are distorted. Objects are thrown upward into the air

accumulated in solid rock is suddenly released along fault lines. The energy released when the rocks break along the fault is converted into seismic waves that radiate from the origin. How much energy is involved largely depends on the magnitude of the quake: larger quakes release much more energy than smaller quakes. The Richter Magnitude scale was devised by Charles F. Richter in 1935 to classify local earthquakes in southern California, but has evolved into the most common parameter to describe the size of the quake and hence, its energy and potential of destructive power. Earthquake magnitude can be used to quantify the amount of energy released during fault ruptures. Energy propagating by seismic waves is proportional to the square root of amplitude-period ratios. Magnitude is proportional to the logarithm of seismic energy E. A semi-empirical relationship between surface wave magnitude M_S and E was formulated by Richter and Gutenberg, and is given by

$$\log E = 1.5\, M_S + 11.8 \qquad (10\text{-}2)$$

Fig. 10-4 indicates the correlation between surface wave magnitude and energy released during earthquakes and other events. The number of earthquakes per year is also provided.

10.3.2 Earthquake Effects

Earthquakes are nothing but natural energy release driven by the evolutionary processes of the planet. Earthquakes have caused devastating effects in terms of loss of life and property. Earthquakes, therefore, are and were thought as one of the worst enemies of mankind. The Chi-Chi Earthquake of M_W=7.7 occurred on September 21, 1999. The earthquake resulted in 2386 people dead and missing, and damaged 12,000 buildings and houses, among which 6000 are totally collapsed. The earthquake also damaged infrastructures such as bridges, dams and port facilities, as well as water, electricity and gas lines. In addition, large-scale landslides buried several villages and killed many people. The Gujarat-Kachchh Earthquake with M_W =7.7 occurred on January 26, 2001, about 20km northeast of Bhuj in the Kachchh District of Gujarat State, western India. According to the Gujarat local government, the death toll reached about 13,800 and more than 166,000 people were injured. The earthquake destroyed about 370,000 houses and left approximately 922,000 partially damaged buildings. Total property damage was estimated at about 6

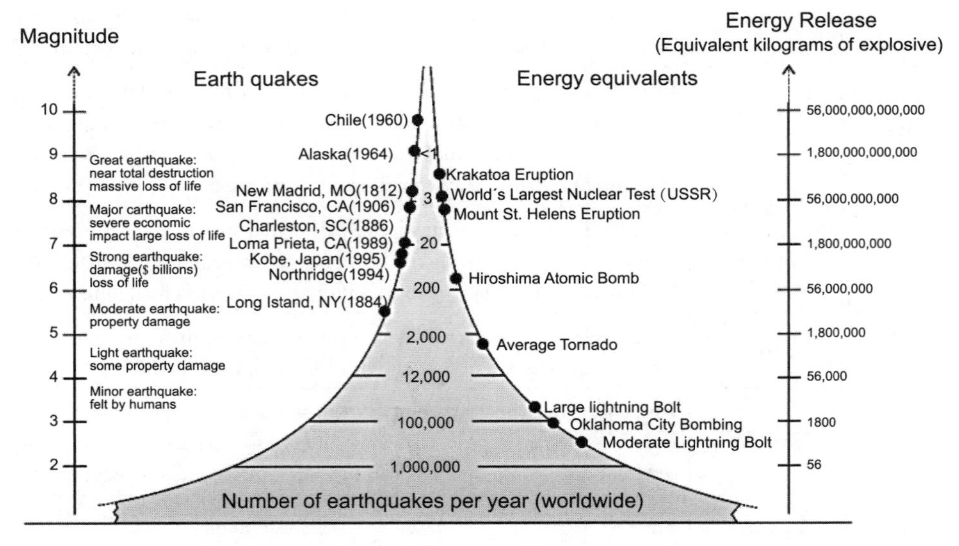

Fig. 10-4 Correlation between magnitude and energy release [9]

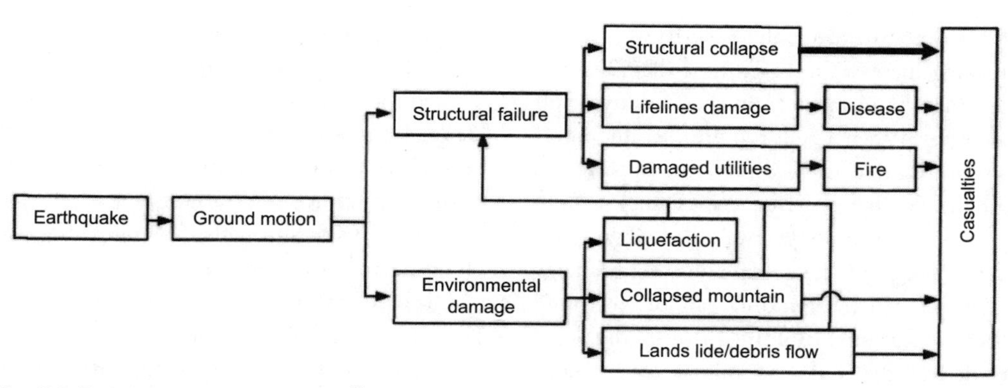

Fig. 10-5 Evolution of earthquake disaster [3]

billion U.S. dollars. In addition to the houses, infrastructures such as roads, port facilities, dams, and lifeline systems such as electricity, water and telecommunications were heavily damaged. On May 12, 2008, the Wenchuan Earthquake of M_W =7.9 struck the western mountainous area of Sichuan Province. This was the largest inland earthquake ever recorded. Ministry of Civil Affairs of the People's Republic of China announced that the earthquake left about 69,277 dead, 17,923 missing, 373,643 injured, and 5.3 million wrecked houses. The estimated direct economic loss exceeded 1.36 billion U.S. dollars.

Actually, the destructive potential of earthquakes depends on many factors. The size of an event (expressed by either intensity or magnitude as described in Section 10.3.1), focal depth and epicentral distance, topographical conditions and local geology are important earthquake characteristics. However, the causes of fatalities and extent of damage depend to a great extent on the type of constructions and the density of population present in the area. Ground shaking is by far the most important hazard resulting from earthquakes. Structural damage, which is a feature of the primary vertical and lateral load-resisting systems, may vary between light damage and collapse. During destructive earthquakes, civil engineering works are initially the hazard-bearing bodies. Their failure or collapse due to insufficient seismic resistance may, however, cause them to become hazards that may cause further casualties and financial loss (Fig. 10-5). The loss of life and property through past earthquakes has actually been primarily attributed to the failure and collapse of buildings. Approximately 80% of the deaths that resulted from the 1995 Kobe Earthquake in Japan were caused by collapsed buildings. Among the 2456 deaths in the city of Kobe, 2221 were dead within 15 minutes of the earthquake. On July 28, 1976, an M_W=7.6 earthquake hit the city of Tangshan, China, resulting in the collapse of 90% of single-storey buildings and 85% of multi-storey buildings in the city, killing at least 250,000 people. The extensive building collapse caused numerous immediate deaths and injuries. We can see that most earthquake-related deaths are caused by the collapse of structures and the construction practices play a tremendous role in the death toll of an earthquake.

It is clear to see that earthquake disasters have attributed to the most significant cause of casualties and financial loss. Actually, the primary effects of earthquakes generally include ground shaking, surface faulting and ground failure.

(1) Ground Shaking

Ground shaking is the most familiar effect of earthquakes. It is a result of the passage of seismic waves through the ground, and ranges from quite gentle in small earthquakes to incredibly violent in large earthquakes.

As a generalization, the severity of ground shaking increases as magnitude increases and decreases as distance from the causative fault increases. When a fault ruptures, seismic waves are propagated in all directions, causing the ground to vibrate at frequencies ranging from about 0.1Hz to 30Hz. Buildings vibrate as a consequence of the ground shaking; damage takes place if the building cannot withstand these vibrations. Buildings can be damaged or destroyed, people and animals have trouble standing up or moving around, due to strong ground shaking in earthquakes. However, you should note that, while many people are killed in earthquakes, none are actually killed directly by the shaking—if you were out in an open field during a magnitude 9 earthquake, you would be extremely scared, but your chance of dying would be zero or damn near it. It is only because that we persist in building buildings, highways, and the like that people are killed; it's our responsibility, not the earthquakes. Fig. 10-6 shows the building that are serious damaged in Wenchuan Earthquake.

(2) Surface Faulting

Surface faulting is the differential movement of the two sides of a fracture at the earth's surface and can be strike-slip, normal and reverse. Combinations of the strike-slip type and the other two types of faulting can be found. Although displacements of these kinds can result from landslides and other shallow processes, surface faulting applies to differential movements caused by deep-seated forces in the earth. Death and injuries from surface faulting are very unlikely, but casualties can occur indirectly through fault damage to structures. Surface faulting, in the case of a strike-slip fault, generally affects a long narrow zone whose total area is small compared with the total area affected by ground shaking. Nevertheless, the damage to structures located in the fault zone can be very high, especially where the land use is intensive. A variety of structures have been damaged by surface faulting, including houses, apartments, commercial buildings, nursing homes, railroads, highways, tunnels, bridges, canals, storm drains, water wells, and water, gas and sewer lines. Damage to these types of structures has ranged from minor to very severe. Fig. 10-7 shows the highway interruption in earthquake of the Pacific Coast of Tokyo.

(3) Ground Failure

Earthquake-triggered liquefaction lateral spreads landsides and any other consequence of shaking that affects the stability of the ground. 1) Liquefaction is not a type of ground failure, it is a physical process that takes place during some earthquakes that may lead to ground failure. As a consequence of liquefaction, clay-free soil deposits, primarily sands and silts, temporarily lose strength and behave as viscous fluids rather than as solids. Liquefaction takes place when

Fig.10-6 Building damage in Wenchuan Earthquake
Source: http://blog.sina.com.cn/s/blog_494677800102vcz3.html.

Fig. 10-7 Highway interruption in earthquake of the Pacific Coast of Tokyo
Source: https://www.nytimes.com/2011/03/12/world/asia/12japan.html.

seismic shear waves pass through a saturated granular soil layer, distort its granular structure, and cause some of the void spaces to collapse. In Niigata Earthquake, three of the buildings (the white ones) have tilted over due to liquefaction shown in Fig.10-8. 2) Lateral spreads involve the lateral movement of large blocks of soil as a result of liquefaction in a subsurface layer. Movement takes place in response to the ground shaking generated by an earthquake. Lateral spreads generally develop on gentle slopes, and usually break up internally, forming numerous fissures and scarps. Damage caused by lateral spreads is seldom catastrophic, but it is usually disruptive. Lateral spreads are destructive particularly to pipelines. In 1906, a number of major pipeline breaks occurred in the city of San Francisco during the earthquake because of lateral spreading. Breaks of water mains hampered efforts to fight the fire that ignited during the earthquake. 3) Landslides are caused by earthquakes both by direct rupture and by sustained shaking of unstable slopes. The most abundant types of earthquake induced landslides are rock falls and slides of rock fragments that form on steep slopes. The size of the area affected by earthquake-induced landslides depends on the magnitude of the earthquake, its focal depth, the topography and geologic conditions near the causative fault, and the amplitude, frequency composition and duration of ground shaking.

Fig. 10-8 Inclined buildings in Niigata Earthquake.
Source: https://depts.washington.edu/liquefy/selectpiclique/nigata64/tiltedbuilding.jpg.

10.3.3 Earthquake Disaster Prevention and Mitigation Strategy

Although earthquakes are destructive and devastating, the greatest losses—both of lives and property—result from the collapse of man-made structures during the violent shaking of the ground. Accordingly, the most effective way to mitigate the damage of earthquakes from an engineering standpoint is to design and construct structures capable of withstanding strong ground motions.

(1) Interpreting Recorded Ground Motions

It is meaningful to explain the characteristics of the recorded ground motions in earthquakes. Such knowledge is needed to predict ground motions in future earthquakes so that earthquake–resistant structures can be designed. Understanding near-source motion can be viewed as a three-part problem. The first part stems from the generation of elastic waves by the slipping fault. The pattern of waves produced is dependent on several parameters, such as fault dimension and rupture velocity. Elastic waves of various types radiate from the vicinity of the moving rupture in all directions. The geometry and frictional properties of the fault critically affect the pattern of radiation from it. The second part of the problem concerns the passage of the waves through the intervening rocks to the site and the effect of geologic conditions. The third part involves the conditions at the recording site itself, such as topography and highly attenuating soils. All these questions must be considered when estimating likely earthquake effects at a site of any proposed structure.

Experience has shown that the ground strong-motion recordings have a variable pattern in detail but predictable regular shapes in general (except in the case of strong multiple earthquakes). An example of actual shaking of the ground (acceleration, velocity and displacement) recorded during an earthquake is given in Fig.10-9. In a strong horizontal shaking of the ground near the fault

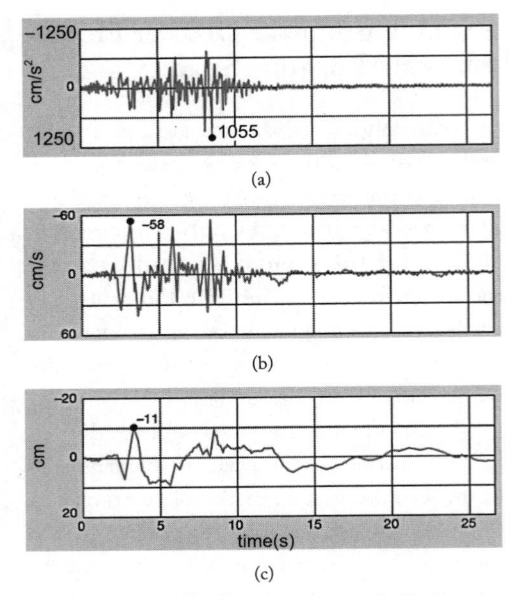

Fig.10-9 Shaking record of San Fernando Earthquake, California 1971
(a) Ground acceleration; (b) Velocity; (c) Displacement
Source: https://www.britannica.com/science/earthquake-geology/Methods-of-reducing-earthquake-hazards.

source, there is an initial segment of motion made up mainly of P waves, which frequently manifest themselves strongly in the vertical motion. This is followed by the onset of S waves, often associated with a longer-period pulse of ground velocity and displacement related to the near-site fault slip or fling. This pulse is often enhanced in the direction of the fault rupture and normal to it. After the S onset, there is shaking that consists of a mixture of S and P waves, but the S motions become dominant as the duration increases. Later, in the horizontal component, surface waves dominate, mixed with some S waves.

(2) Constructing Seismic Hazard Maps

In many regions, seismic expectancy maps are now available for planning purposes. The anticipated intensity of ground shaking is represented by a number called the peak acceleration or the peak velocity. To avoid weaknesses found in earlier earthquake hazard maps, the following general principles are usually adopted:

1) The map should take into account not only the size but also the frequency of earthquakes.

2) The broad regionalization pattern should use historical seismicity as a database, including the following factors: major tectonic trends, acceleration attenuation curves and intensity reports.

3) Regionalization should be defined by means of contour lines with design parameters referred to ordered numbers on neighboring contour lines.

4) The map should be simple and not attempt to microzone the region.

5) The mapped contoured surface should not contain discontinuities, so that the level of hazard progresses gradually and in order across any profile drawn on the map.

(3) Building Earthquake Resistant Structures

To design an earthquake resistant building, engineers need to reinforce the structure and counteract an earthquake's forces. Since earthquakes release energy that pushes on a building from one direction, the strategy is to have the building to push the opposite way. Here are some of the methods used to help buildings withstand earthquakes.

1) Creating a Flexible Foundation

One way to resist ground forces is to "lift" the building's foundation above the earth. Base isolation involves constructing a building on top of flexible pads made of steel, rubber and lead. When the base moves during the earthquake, the isolators vibrate while the structure itself remains steady. This effectively helps absorb seismic waves and prevent them from traveling through a building.

2) Counter Forces with Damping

Damper is a mechanical or hydraulic devices designed to absorb energy to help buildings slow down. This is accomplished in two ways: vibrational control devices and pendulum dampers. The first method involves placing dampers at each level of a building between a column and beam. Each damper consists of piston heads inside a cylinder filled with silicone oil. When an earthquake occurs, the building transfers the vibration energy into the pistons, pushes against the oil. The energy is transformed into heat, dissipating the force of the vibrations. Another damping method is pendulum dampers used primarily in skyscrapers. Engineers suspend a large ball with steel cables with a system of hydraulics at the top of the building. When the building begins the sway, the ball acts as a pendulum and moves in the opposite direction to stabilize the direction. Like damping, these features are tuned to match and counteract the building's frequency in the event of an earthquake.

3) Shield Buildings from Vibrations

Instead of just counteracting forces, researchers are experimenting with ways that buildings can deflect and reroute the energy from earthquakes altogether. Dubbed the "seismic invisibility cloak", this innovation involves creating a cloak of 100 concentric plastic and concrete rings in and burying it at least three feet beneath the foundation of the building. As seismic waves enter the rings, they are forced to move through to the outer rings for easier travel. As a result, they are essentially channeled away from the building and dissipated into the plates in the ground.

4) Reinforcing the Building's Structure

To withstand collapse, buildings need to redistribute the forces that travel through them during a seismic event. Shear walls, cross braces, diaphragms and moment-resisting frames are central to reinforcing a building. Shear walls are a useful building technology that helps to transfer earthquake forces. Shear walls are often supported by diagonal cross braces. These steel beams have the ability to support compression and tension, which helps counteract the pressure and push forces back to the foundation. Diaphragms are a central part of a building's structure. Consisting of the floors of the building, the roof and the decks placed over them, diaphragms help remove tension from the floor and push force to the vertical structures of the building. Moment-resisting frames provide more flexibility in a building's design. This structure is placed among the joints of the building and allows for the columns and beams to bend while the joints remain rigid. Thus, the building is able to resist the larger forces of an earthquake while allowing designers more freedom to arrange building elements.

10.4 Wind Disaster

10.4.1 Basic Concepts and Types of Wind

Wind is the movement of air on a large scale in the atmosphere. The wind is caused by differences in the atmospheric pressure. When a difference in atmospheric pressure exists, air moves from the higher to the lower pressure area, resulting in winds of various speeds. Wind is usually applied to the natural horizontal motion of the atmosphere. Motion in a vertical or nearly vertical direction is called a current. Movement of air near the surface of the earth is three-dimensional, with horizontal motion much greater than the vertical motion. Vertical motion of air is of importance in meteorology but is of less importance near the ground surface. On the other hand, the horizontal motion of air, particularly the gradual retardation of wind speed and the high turbulence that occurs near the ground surface, are of importance in building engineering. In urban areas, this zone of turbulence extends to a height of approximately one-quarter of a mile aboveground, and is called the surface boundary layer. Above this layer, the horizontal airflow is no longer influenced by the ground effect. The wind speed at this height is called the gradient wind speed, and it is precisely in this boundary layer where most human activities are conducted. Therefore, how wind effects are felt within this zone is of great concern.

Wind loading competes with seismic loading as the dominant environmental loading for structures. They have produced roughly equal amounts of damage over a long time period, although large damaging earthquakes occur less often than severe wind storms. On almost every day of the year, a severe wind storm is happening somewhere on earth—although many storms are small and localized. In the tropical oceans, the most severe of all wind events—tropical cyclones (including hurricanes and typhoons)—are generated. Common wind storms that cause disasters in nature include gales, tropical cyclones, tornadoes, etc.

(1) Gales

A gale is a strong wind, typically used as a descriptor in nautical contexts. In the mid-latitudes from about 40 to 60 degrees, the strongest winds are gales generated by large and deep depressions or cyclones. They can be significant contributors to winds in lower latitudes. Gales are usually large in horizontal dimension—they can extend for more than 1000km; so, they can influence large areas of land during their passage—several countries in the case of Europe. They may take several days to pass, although winds may not blow continuously at their maximum intensity during this period. The winds tend to be quite turbulent near the ground, as the flow has adjusted to the frictional effects of the earth's surface over hundreds of kilometres. The direction of the winds remains quite constant over many hours.

(2) Tropical Cyclones

Tropical cyclones are one of the biggest threats to life and property even in the formative stages of their development. A tropical cyclone is a rapid rotating storm originating over tropical oceans from where it draws the energy to develop. It has a low pressure center and clouds spiraling towards the eyewall surrounding the "eye"—the central part of the system where the weather is normally calm and free of clouds. Its diameter is typically around 200km to 500km, but can reach 1000km. A tropical cyclone brings very violent winds, torrential rain, high waves and, in some cases, very destructive storm surges and coastal flooding. The winds blow counterclockwise in the Northern Hemisphere and clockwise in the Southern Hemisphere. Tropical cyclones above a certain strength are given names in the interests of public safety. They are associated with extremely heavy rain which can result in widespread flooding. Cyclones are also associated with damaging or destructive winds and in the most intense systems, surface winds may reach speeds in excess of 300km/h. The combination of wind-driven waves and the low-pressure of a tropical cyclone can produce a coastal storm surge—a huge volume of water driven ashore at high speed and with immense force that can wash away structures in its path and cause significant damage to the coastal environment. The impact of a tropical cyclone and the expected damage depend not just on wind speed, but also on factors such as the moving speed, duration of strong wind and accumulated rainfall during and after landfall, sudden change of moving direction and intensity, the structure (e.g., size and intensity) of the tropical cyclone, as well as human response to tropical cyclone disasters.

(3) Tornado

A tornado is a rapidly rotating column of air that is in contact with both the surface of the earth and a cumulonimbus cloud or, in rare cases, the base of a cumulus cloud. Tornadoes come in many shapes and sizes, and they are often visible in the form of a condensation funnel originating from the base of a cumulonimbus cloud, with a cloud of rotating debris and dust beneath it. Most tornadoes have wind speeds less than 180km/h, are about 80m across, and travel a few miles before dissipating. The most extreme tornadoes can attain wind speeds of more than

480km/h, are more than 3km in diameter, and stay on the ground for dozens of miles. Most tornadoes take on the appearance of a narrow funnel, a few hundred yards (meters) across, with a small cloud of debris near the ground. Tornadoes may be obscured completely by rain or dust. These tornadoes are especially dangerous, as even experienced meteorologists might not see them.

10.4.2 Wind Effects

Wind disasters are responsible for tremendous physical destruction, injury, loss of life and economic damage. According to the statistics, around 80% of natural disaster economic losses in the world are caused by extreme winds and their associated events, i.e., combined effects of wind and flood. With the development of production and construction, the loss caused by wind disaster is increasing over years, especially typhoon disaster, hurricane disaster as well as tornado disaster.

(1) Typhoon Disaster

A typhoon is a mature tropical cyclone that develops between 180° and 100°E in the Northern Hemisphere. This region is referred to as the Northwestern Pacific Basin, and is the most active tropical cyclone basin on the earth, accounting for almost one-third of the world's annual tropical cyclones. Typhoon Hagibis was an extremely violent and large tropical cyclone that caused widespread destruction. It was the strongest typhoon in decades to strike mainland Japan, and one of the largest typhoons ever recorded at a peak diameter of 1529km. It was also the costliest Pacific typhoon on record, surpassing Typhoon Mireille's record by more than $15 billion. It is officially stated that at least 98 people have been confirmed dead, 7 people are missing, with 346 people injured by Typhoon Hagibis. More than 270,000 households lost power across the country. Ten trains of the Hokuriku Shinkansen Line in Nagano City were inundated by flood waters, leading to a loss of $300 million. Insured losses throughout the country are estimated as greater than $15 billion. Typhoon Lekima (shown in Fig.10-10) in 2019 was the costliest typhoon in Chinese history. Striking east China as a super typhoon, according to China Meteorological Administration, Lekima wrought major damage across numerous provinces. In all, the typhoon killed 56 people and left 14 others missing. Damage nationwide exceeds $9.26 billion.

(2) Hurricane Disaster

A hurricane is a type of tropical cyclone or severe tropical storm. A typical cyclone is accompanied by thunderstorms and a counterclockwise circulation of winds near the earth's surface. A hurricane has maximum sustained winds of 120km/h or higher. Hurricanes can cause catastrophic damage to coastlines and several hundred miles inland. Hurricanes can produce winds exceeding 250km/h as well as tornadoes and microbursts. In some hurricanes, wind alone can cause extensive damage such as downed trees and power lines, collapsing weak areas of homes, businesses or other buildings. Additionally, hurricanes can create storm surges along the coast and cause extensive damage from heavy rainfall. The 1944 Great Atlantic Hurricane was a destructive and powerful tropical cyclone that swept across a large portion of the east coast in the United States in September 1944. While this hurricane caused 46 deaths and $100 million in damage in the United States, the worst effects occurred at sea where it wreaked havoc on World War Ⅱ shipping. Five ships, including a U. S. Navy destroyer and a minesweeper, two U. S. Coast

Fig.10-10 Typhoon Lekima
Source: http://www.kaishugushi.com/428946.html.

Guard cutters, and a light vessel, sank due to the storm, causing 344 deaths. Hurricane Floyd 1999 (shown in Fig.10-11) was a very powerful Cape Verde hurricane which struck the Bahamas and the east coast of the United States. While wind gusts of 194km/h and storm surges of 9ft to 10ft were reported from the North Carolina Coast, Floyd will be most remembered in the United States for its rainfall. The combination of Floyd and a frontal system over the eastern United States produced widespread rainfalls in excess of 10 inches from North Carolina northeastward, with amounts as high as 19.06 inches in Wilmington, North Carolina and 13.70 inches at Brewster, New York. These rains, aided by rains from Tropical Storm Dennis two weeks earlier, caused widespread severe flooding that caused the majority of the $3 billion to $6 billion in damage caused by Floyd. These floods were also responsible for 50 of the 56 deaths caused by Floyd in the United States.

(3) Tornado Disaster

Tornadoes is defined as a violently rotating column of air extending from a thunderstorm to the ground, which are often formed when warm and cold air masses clash. They are capable of tremendous destruction, creating damage paths in excess of one mile wide and 50 miles long. Tornados' speed can vary from nearly stationary to up to 112km/h; however, the wind speed from these formations can exceed 400km/h. There are on average more than 1000 tornadoes reported all over the world each year. The United States is the country with the most frequently tornadoes in the world. The 1999 Oklahoma Tornado outbreak was a significant tornado outbreak that affected much of the central and parts of the eastern United States, with the highest record-breaking wind speed of 486±35km/h. 36 people died in this tornado, and over 8000 homes were badly damaged or destroyed. The tornado caused $1.5 billion in damage, making it the second-costliest tornado in the U.S. history. On the afternoon of June 23, 2016, a severe thunderstorm produced a large, violent tornado (shown in Fig.10-12) over Jiangsu province, China, striking areas along the outskirts of Yancheng at around 2:30 p.m. Wind speeds of up to 125km/h were measured at the edge of the city. Thousands of masonry-construction homes were damaged or destroyed, with many completely leveled. Manufacturing plants, businesses and rice mills suffered from similar destruction, and multiple large factory buildings were severely damaged at a Canadian solar plant. A large school

Fig.10-11 Hurricane Floyd
Source: https://en.wikipedia.org/wiki/Hurricane_Floyd.

Fig.10-12 Jiangsu tornado
Source: https://en.wikipedia.org/wiki/Jiangsu_tornado.

building sustained major structural damage as well. Many vehicles were tossed and destroyed; trees were completely denuded and debarked; and numerous metal power line pylons and truss towers were bent and crumpled to the ground. The tornado killed at least 99 people and 846 others were injured(152 critically), and damages were nearly at $760 million.

On November 7, 1940, the Tacoma Narrows Bridge (shown in Fig.10-13) collapsed due to high winds and this event stunned everyone, especially engineers. Since then wind effects on structures have attracted significant attention. Gradually, people have realized that wind effects should be considered during design, especially for slender and tall buildings and structures. Actually, from a structural design perspective, the most important aspects of building responses to wind are wind-induced loads and wind-induced building motions. The former are normally considered for the structural safety and the latter normally for the structural serviceability.

Since structural systems of high rise buildings and structures are higher and slenderer, they are sensitive to the effects of wind, which means structural dynamic responses are large in horizontal direction when lateral wind loadings exert on high rise buildings and structures. Winds in nature consist of normal wind and abnormal wind (such as tornado).

Fig.10-13 Tacoma Narrows Bridge collapsed due to high winds
Source: https://www.seattlepi.com/science/article/A-Tacoma-Narrows-Galloping-Gertie-bridge-6617030.php.

The effect of wind on structures has the following characteristics:

(1) The wind loading applied on structures is composed of mean wind and fluctuating wind, and structural vibration due to fluctuating wind loading must be considered in structural design.

(2) The effect of wind to structures is closely associated with the shape of structures, especially chimney and higher slender structures.

(3) The effect of wind to structures is highly related to the environment around. Sometimes the buildings in the building complex is in adverse condition due to vibration caused by the airflow of buildings around.

(4) The wind effect on buildings are non-uniformed distributed along its height, and there will be more wind in the corner area and the inward area of the facade.

(5) Compared to seismic action, wind loading applied on structures can last from couple of ten minutes to hours.

As a result, the effect of wind to structures could lead to following consequences:

(1) Stronger wind loadings make structures or components of structures unstable.

(2) Wind loadings could crack structures or develop large residual deformation.

(3) Wind loadings maybe result in large deflection or deformation and damage exterior walls as well as decoration materials.

(4) Structures or components of structures could suffer from fatigue failure due to repeated action of wind effect.

(5) Structural aeroelasticity is unstable, which aggravates aerodynamic forces on structures.

(6) Excessive vibration of structures could make residents or staff feel discomfortable and unsafe.

10.4.3 Wind Disaster Prevention and Mitigation Strategy

High wind events can cause destructive damage to structures especially to high-rise buildings.The following sections describe various design strategies for resisting or minimizing wind forces and response.

(1) Aerodynamic Strategies

Whether for a supertall building or a domestic house, wind loads are greatly influenced by the shape of the design—the architecture. Good aerodynamic choices will result in reduced wind loads. For example, hip roofs on a house generally experience lower uplift pressures during an extreme wind event than gable roofs. They are simply more aerodynamically friendly.

The application of aerodynamics to architectural shape becomes even more important as a building becomes taller. The wind loads typically control tall building design. Varying the shape of the building with height will decorrelate the vortex shedding and so the crosswind response of a tall building. Cladding pressures may also be ameliorated with strategic design features. Corner balconies, as noted above, will reduce the design cladding pressures. Roof pressures may be reduced by the use of porous parapets or perimeter spoilers. Even placing a new building within a complex cityscape of similar structures will be aerodynamically advantageous.

(2) Structural Strategies

While all structures in the wind must be designed to resist both lateral forces and uplift forces, of which the more important factor for a particular building is generally a function of the structure's aspect ratio, or relative height to width, as well as its shape. The design of a broad, low building with a relatively large roof area in proportion to its height, must give attention to resisting the uplift forces on the roof surface, with careful detailing of the connections to create a continuous load path to anchor the structure to the foundation. In contrast, a tall structure on a relatively narrow base must be designed with primary concern for lateral force resistance and overturning.

The use of shear walls is a time-honored, traditional way of resisting lateral wind loads on buildings. Solid masonry bearing walls provided resistance to lateral shear forces in times predating the development of the modern building materials steel and concrete. For shear walls to be most effective, they should be continuous throughout the height of the building and have as few openings as possible. Shear walls may be staggered within a vertical plane when required for interior spatial arrangements, but this is not ideal.

Reinforced concrete or braced or rigid steel frames may be used to form shear walls in multi-storey buildings. Braced steel frames have diagonal members in the vertical shear walls which act in tension or compression to resist the tendency for side-to-side lateral movement of one floor relative to another. Rigid steel frames use moment-resistant connections instead of diagonals to withstand lateral forces.

As tall buildings reach greater heights, they require structural systems with greater stiffness to increase their resistance to wind-induced lateral motion. In tube structures, the major wind-resisting structural system is located in or near the perimeter walls of a tall building, rather than around an internal core. As such, the wind-resisting structure becomes a factor in the architectural expression of the building.

Tall building structures must satisfy criteria for both strength (safety) and serviceability (human comfort). Wind-induced motion can cause discomfort to building occupants and, thus, in today's very tall buildings, it is usually the serviceability criteria that control the design. If it is determined during the design process that top-floor accelerations could exceed acceptable limits, either the stiffness or the damping of the structure must be increased.

It has become common practice in very tall buildings to reduce top-floor accelerations by providing the structure with energy-dissipating damping devices, of which there are three basic types: (1) viscoelastic dampers, (2) tuned mass dampers and (3) tuned liquid dampers.

10.5 Fire Disaster

10.5.1 Basic Concepts and Types of Fire Disaster

Fire is indispensable in people's production and life, and the development of mankind society cannot do without fire. However, if fire is out of control, it will cause loss of life and property, and even become a fire disaster. Fire disaster is defined as a disaster caused by burning out of control in time and space.

A fire occurs when a combustible item comes into contact with oxygen in the presence of heat. The fire, which usually starts with the burning of one item, gradually spreads to other nearby items and grows in size and intensity as a pre-flashover fire. In an open environment, a peak intensity is reached very rapidly, then the rate of combustion gradually decreases until the available fuel has been consumed. In a closed environment, on the other hand, the size and intensity of the fire can increase until all items in the enclosed space are fully engulfed in flame at the time of flashover, which is a transition to the burning period, during which the peak intensity is maintained while the rate of burning is controlled by the availability of oxygen through ventilation openings. After most of the fuel has been consumed, the fire decays gradually. Post-flashover fires are the main concern for the design of structures in fire conditions.

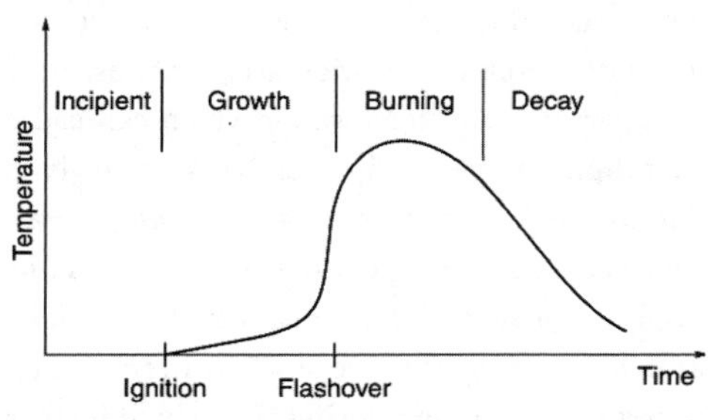

Fig.10-14 Time–temperature curve for full process of fire development [4]

According to Chinese national standard *Fire Classification* GB/T 4968-2008, fire disasters are grouped into four categories of A, B, C and D:

(1) Class A fire disaster involves solid material such as wood, cotton, wool, hemp, paper , etc.

(2) Class B fire disaster involves flammable liquids and meltable solid material such as gasoline, kerosene, crude oil, methanol, ethanol, etc.

(3) Class C fire disaster involves gases such as coal gas, natural gas, methane, ethane, propane, hydrogen, etc.

(4) Class D fire disaster involves metals such as potassium, sodium, magnesium, titanium, lithium, etc.

Fire safety objectives are usually met with a combination of active and passive fire protection systems. Depending on the design, active systems limit fire development and its effects by some actions taken by a person or an automatic device. Passive systems on the other hand control the fire or its effects by systems that are built into the structure or fabric of the building, not requiring specific operation at the time of a fire. Some building elements or materials cannot be easily classified as either active or passive systems, for example, intumescent coatings will react automatically in a fire, while fire doors may be shut automatically or by the occupants after a fire is detected. The typical development of a fire in a room is described in Fig.10-14 to emphasize the need for fire protection systems. Fig.10-14 shows a typical time–temperature curve for the complete process of fire development inside a typical room, assuming no fire suppression.

In the incipient period of fire development, heating of potential fuel is taking place. Ignition is the start of combustion, marking the transition to the growth period. In the growth period, most fires spread slowly at first on combustible surfaces, then more rapidly as the fire grows and there is radiant feedback from flames and hot gases to other fuel items. Hot gases rise by convection and spread across the ceiling, forming a hot upper layer which radiates heat to fuel items lower in the room. If upper layer temperatures reach about 600 °C, the burning rate increases rapidly, leading to flashover which is the transition to the burning period (often referred to as "full room involvement" or "fully developed fire"). The rate of burning in the growth period is generally controlled by the nature of the burning fuel surfaces, whereas in the burning period the temperatures and radiant heat flux within the room are so great that all exposed surfaces are burning and the rate of heat release is usually governed by the available ventilation. It is the burning period of the fire that predominantly impacts on structural elements and compartment boundaries. If the fire is left to burn, eventually the fuel burns out and temperatures drop in the decay period, where the rate of burning again becomes a function of the fuel itself rather than the ventilation.

10.5.2 Fire Effects

Fire is one of primary causes of loss of life and property throughout the world. During the past two decades, fires have damaged hundreds of thousands of buildings. Buildings constitute the majority of built infrastructure and play a pivotal role in socioeconomic development of a country. Fire disaster can cause partial or complete collapse of the building, and incapacitation of building operations. Such destruction or incapacitation in the event of a hazard can jeopardize the life safety of inhabitants and can cause significant direct and indirect monetary losses.

In the past two decades, a total of 86.4 million fire incidents have caused more than one million fire deaths, and total annual loss from global fire hazard accounts for about 1 percent of the world GDP. On an average, 3.8 million fires caused 44,300 fire deaths every year in both developed and developing countries across the globe. Between 2010~2014, maximum number of fires (600,000~1,500,000 per year) and the second highest number of fire deaths (1000~10000 per year) in the world occurred in a developed country such as the U.S. Whereas, developing countries such as India and Pakistan suffered highest number of fire causalities (10,000~25,000 per year) and second highest number of fires (100,000~600,000 per year). Therefore, to mitigate these adverse effects of fire hazard, it is important to provide necessary fire safety in buildings.

Buildings contain several direct and indirect sources that contribute to fire hazard; and in the event of a fire there is significant risk to life, structure, property and environment from the initial development stages of the fire itself.

(1) Impact on Life Safety

There is significant risk to life safety in both pre-flashover and post-flashover phases of building fires, and on an average about 44,300 fire deaths have occurred every year between 1993~2015. During pre-flashover phase of fire, combustion generates several toxic gases which are extremely deleterious to humans and inhalation (even in small quantities) can be fatal within minutes. Most common among these are carbon monoxide (generated from incomplete combustion), hydrogen cyanide (generated from burning plastics) and phosgene gas (generated from burning vinyl-based household materials). The smoke generated from combustion also contains small soot particles and toxic vapor which can cause irritation to eyes and digestive system. It is due to this high toxicity of smoke (toxic gases, soot particles and vapor) that more fire deaths occur from smoke than burning. Also, smoke and hot gases obscure and hinder escape routes from building during fire, which further increases risk to life safety from inhalation of toxic gases and burning.

Other threats to life safety are from reducing oxygen levels in room from combustion and inhaling hot air. Humans undergo impaired judgement and coordination when oxygen levels in room fall to 17 percent from normal 21 percent; headache, dizziness, nausea and fatigue will occur at 12 percent; unconsciousness will occur at 9 percent; and respiratory arrest, cardiac arrest and even death occurs when oxygen levels fall to 6 percent. Also, inhaling hot gases can burn respiratory tract, and one breath of hot air can even lead to death. During post-flashover phase, the concentration of toxic smoke is very high and fire temperatures are untenable for humans and can lead to certain death, thus, all life safety operations are usually targeted towards pre-flashover phase of fire. Apart from toxic smoke and burning, the biggest risk to life safety during post-flashover phase is partial or complete collapse of structure which can inhibit firefighting operations and kill trapped inhabitants under collapsed debris. Therefore, fire represents significant threat to life safety even when it is not fully developed, and every minute is critical in evacuating inhabitants during building fires.

(2) Impact on Structural Safety

During fully developed stage, fire temperatures can reach above 1000°C which can cause significant degradation in strength and stiffness properties of structural materials. This material degradation can incapacitate structural members to carry designed structural loads and lead to partial or complete collapse of building during or after fire. Also, material degradation has strong potential to cause permanent structural damage which can cause premature failure of building under other natural hazards for which it was originally designed for, thus, endangering structural safety.

(3) Impact on Property Safety

One of the biggest impacts of fire hazard is on property safety and it causes direct and indirect losses of billions of dollars in both developed and developing countries across the globe. Even if the building withstands fire without life losses, aftermath of almost every fire involves monetary losses, the magnitude of which depends on severity of fire. Direct losses from fire hazard include loss of property from burning, sprinkler operation, firefighting operations (damage to property from water of fire brigade, breaking of doors and windows, etc.), falling debris from partial or complete collapse of structure, and structural damage and cost of repair. Whereas, indirect losses include loss of use during time required for repairs, loss from temporary or permanent relocation, loss from demolishing structure, increase in insurance costs, environmental contamination, etc.

(4) Impact on Environmental Safety

Fire hazard generates several environmental pollutants from combustion, firefighting operations and spillage from containers of hazardous materials due to damage from fire. Most common fire pollutants include metals, particulates, polycyclic aromatic hydrocarbons, chlorinate dioxins , furans, brominated dioxins, polychlorinated biphenyls and polyfluorinated compounds. During fire, transmission of these pollutants occurs to the environment through fire plume (air contamination), firefighting water runoff (water contamination), and deposited air and water contaminants (land contamination); thus, causing environmental pollution. The magnitude of environmental pollution depends on the exposure duration, transmission medium, and susceptibility of receiving atmospheric, aquatic and terrestrial environments.

10.5.3 Fire Disaster Prevention and Mitigation Strategy

Most of the current fire protection measures are prescriptive and based on similar fire safety principles. Therefore, these provisions can be grouped under four generic categories :

general strategy for fire safety, building codes and standards, safety provisions within a building and firefighting operations.

(1) General Strategy for Fire Safety

The first line and foremost strategy to tackle fire hazards is the prevention of fire occurrence. Because it is not always possible to prevent fire, impact of fire should be managed by either managing fire itself or by managing exposed people and the property. The usual strategy for managing people is to evacuate exposed them from the building by making movement of people through a safe fire escape route. For people to evacuate safely, it is important that these requirements are met simultaneously: fire is detected in incipient or growth stage (the earlier the better), occupants are notified using a fire alarm and a safe fire escape route exists in the building. However, in case of high rise buildings, it is not possible to evacuate people through a safe fire escape passage in the time bound. Therefore, defend-in-place strategy is adopted by providing safe refuge on certain levels of a building, which are then evacuated by firefighting department. This allows firefighters to target evacuation operations to these specific refuge areas only and save precious time which can be a factor of life and death in fire situations. To manage fire and its impact, general strategy is to control the available fuel for combustion and use suppression by using various fire protection features installed in a building. Many building codes and standards specify a permissible limit of the available fuel load in a building (given as energy floor density in MJ/m^2), so that in case of ignition, fire growth is controlled by limited fuel supply. The fire severity corresponding to this limited fuel load is taken into consideration in the building design to withstand this certain level of fire severity. Therefore, the limit on the available combustible fuel load inside a building is dependent on the fire resistance requirement of the building and vice versa. The other effective method of controlling fire is through suppression using automated or manual fire protection provisions. In case of automatic fire suppression systems, it is essential that both fire detection equipment and fire suppression equipment work simultaneously. The automatic provisions for fire suppression include automated sprinklers, condensed aerosol fire suppression systems and gaseous fire suppression systems. On the other hand, manual fire suppression refers to manual fire extinguisher systems or standpipe systems. The suppression of fire depends upon early detection, functional reliability and performance reliability of fire protection measures. The last defense (for controlling fire and to manage its impact) is through compartmentation and structural stability. The structural stability is important as it helps in localizing fire, allows the firefighting operations to continue safely and prevent property losses arising from total collapse of the structure. To ensure structural stability, it is important to control the fire spread inside the building and to keep it to a localized zone only. This can be achieved by using fire compartmentation which contains the fire to a local area only and does not allow further movement of fire inside the building. Another possibility for controlling fire movement is by using fire venting which provides increased ventilation to the fire affected zone only and exhausts the available fuel.

(2) Building Codes and Standards

Detailed provisions in building codes are specified to avert the occurrence of fire, manage its impact, and ensure life and structural safety while keeping property and life losses to a minimum. Building codes and standards provide guidelines for both design and assessment of fire resistance for structural members and assemblies. In case of building fire design, codes specify the function of building elements under fire exposure, the permissible limit of fuel load density, the required fire ratings for building elements, the recommendations on type of materials, the minimum member dimensions to achieve required fire rating, and the guidelines for evacuation strategies. These recommendations vary with type of occupancy such as hospitals, commercial buildings, residential buildings, etc. Generally, for public buildings such as hospitals and nursing homes (where risk to life safety is higher and indirect monetary losses are very high), building codes and standards recommend much conservative solutions with high factor of safety.

To assess fire safety of a structural member or assembly, building codes and standards use three main fire safety criteria as per function of a building member. These include: stability criterion (R), which is the ability to withstand applied loads during fire exposure; integrity criterion (E), which is the ability to prevent fire propagation due to formation of cracks and fissures; and insulation criterion (I), which is the ability to insulate the unexposed faces during fire exposure. Considering these fire safety criteria, the fire resistance assessment can be carried out by prescriptive approach or advanced analysis. In prescriptive approach, fire resistance assessment is carried out by correlating member specifications (dimensions, clear cover, aggregate type) to fire safety criteria using data from standard fire tests. Whereas, in case of advanced analysis methods, building codes and standards provide parametric fire curves to be used in the fire resistance assessment, and recommend material properties at elevated temperatures to be used in the analysis while fire safety criteria remains the same.

(3) Safety Provisions within a Building

The safety provisions within a building are grouped under two main categories as active and passive fire protection systems. The active fire protection systems (sprinklers, smoke detectors, fire extinguishers, etc.) refer to the control of fire by taking some action using an automated device or by a person. On the other hand, passive fire protection systems refer to the fire protection measures which are built within the building itself, and do not require any operation by people or automated controls (for example, fire ratings of structural and non-structural members or assemblies). In the incipient stage of fire, fire extinguishers are used to contain the fire. If the fire goes into growth stage, the priority is to evacuate people out of the building as inhalation of toxic gases from fire can be fatal within minutes. In this stage, the fire management falls to automated or manual active fire protection systems. It should be noted that the timing for onset of all automated fire protection systems is crucial as any delay in fire alarm directly endangers life safety and reduces chances of containing fire once it grows in intensity. Therefore, ideally all evacuation process should be completed before fire gets out of control of active fire protection systems. After flashover, the fire temperature can reach as high as 1000°C and the resulting thermal expansion and degradation in material properties pose a serious threat to structural safety. During this stage of fire, the main target of passive fire protection systems is to contain the spread of fire while ensuring structural stability. These passive fire protection systems allow safe firefighting operations, safe evacuation operations and mitigate property losses.

(4) Firefighting Operations

If the fire is not extinguished through active fire protection systems, extinguishing or controlling fire as well as ensuring life safety comes down to the role of firefighting department. The time required by the firefighting department to reach the site and begin firefighting operations play a key role in firefighting and is known as response time. The firefighting department is equipped with specialized equipment to provide alternate entries into a building, and to perform rescue operations even in most inaccessible places. In some countries, firefighting department also has the legal powers to inspect and enforce building owners to comply with building fire safety provisions as specified in codes and standards. This allows for better enforcement of the fire safety provisions, and a continuous monitoring helps in improving fire safety.

10.6 New Trends of Disaster Prevention and Mitigation

In recent decades, research on natural hazards has moved into a new era driven by the rise of new technologies and techniques with potential use in risk assessment, management and mitigation. Nevertheless, in spite of these significant advances, it is recognized that the effects of natural hazards are rapidly increasing in frequency and extension. To tackle this global issue, there is a central need for developing basic and applied multi-disciplinary research that can lead to the development and implementation of more efficient risk mitigation strategies.

In many natural and man-made disasters, civil engineering works are not only the hazard bearing bodies, but their failure also constitutes further hazards. Especially in earthquake disasters, building collapse is the most significant cause of casualties and financial loss. This mechanism is yet to receive adequate emphasis in the investigation of disasters. Indeed, losses that have been attributed to many so-called natural disasters, such as earthquakes or winds, were actually caused by civil engineering factors rather than the actual natural phenomena. This is the key to developing proper methods for disaster prevention and mitigation.

A study of civil engineering disasters and their mitigation not only involves civil engineering disciplines, but also significantly draws on many emerging scientific fields. There are nevertheless two main goals of such a study, namely, to understand the scientific and technical mechanism of civil engineering disasters, and to mitigate the occurrences of the disasters in cities and rural areas. To fully understand the evolution mechanism of a civil engineering disaster, it is essential to quantify the effects of the hazards on the civil engineering work. It is equally important to determine how the civil engineering work is damaged or fails under the action of the hazard. However, these goals can only be achieved by proper numerical simulation of the failure of the civil engineering work. The ultimate goal of a study on civil engineering disasters and their mitigation is the protection of human life and property. This can only be achieved by comprehensive enhancement of the resistance capacity of civil engineering works to hazards. It must nevertheless be conceded that its full attainment is beyond the ability of natural science and technology. Socioeconomic development, education, regulations and local customs are all pertinent factors in this regard.

From a scientific and technological viewpoint, the following issues are essential to accomplish the ultimate goal of studies on disaster prevention and mitigation.

(1) Characteristics of Hazardous Actions

Taking earthquakes as an example, the hazardous action is the shaking of the ground. There has been gradual improvement in the understanding of earthquake actions more than the last 100 years. At the beginning of the 20th century, earthquake actions were considered as being equivalent to the application of horizontal static forces on buildings. This laid the foundation for the seismic design of buildings. It was later recognized that an earthquake action was dependent on the dynamic properties of the buildings. The current use of response spectrum analysis, time history analysis and nonlinear dynamic analysis has greatly extended the understanding of earthquake actions. Studies on earthquake actions have also progressed rapidly by exploiting the fast growing database on real earthquake ground motion records. Tens of parameters for quantifying earthquake actions have been proposed by researchers worldwide, who have demonstrated the complexity of the problem and contributed to the improved understanding in the earthquake engineering community.

(2) Zonation of Hazardous Actions and

Risk Analysis

The objective of hazard zonation is the provision of the time and spatial distributions of hazards for engineering design. Although much effort has been made along this line, the task is very challenging. In the case of earthquakes, the zonation of the ground motion parameters is the basis of seismic design and involves many scientific processes and factors related to seismic hazard analysis, ground motion attenuation modeling and local site effects. Large scatterings and uncertainties still exist in dealing with these issues. Along with the continuous effort to reduce the uncertainties, more efforts are needed to understand the inherent uncertainties of many hazardous actions and risks, and to develop robust civil engineering solutions to minimize the influence of such uncertainties.

(3) Response Characteristics and Damage Mechanism of Civil Engineering Works under Hazardous Actions

There has been an improvement in the understanding of the response characteristics and damage mechanisms of civil engineering works, the behaviors of which range from linear elastic to nonlinear. Many parameters based on force, displacement, energy, or combinations of these have been proposed to quantitatively describe the failure mechanism and dynamic behavior of civil engineering works under hazardous actions. In the process, many physical and numerical models have been developed, with the latter having obvious advantages and showing promise for application to the simulation of the damage of civil engineering works.

(4) Engineering Measures and Design Codes for Enhancing Hazard Resistance Capacity of Civil Engineering Works

New technologies and design codes are the most important basis for increasing the hazard resistance capacity of civil engineering works. Differing from scientific research, engineering practice gives consideration to safety, cost, simplicity, effectiveness and standardization. The provision of effective solutions to civil engineering problems at a reasonable cost is sometimes more important than the scientific quantification of the detailed parameters. In addition, research findings need to be implemented in civil engineering constructions for them to be of any benefit to the community. To this end, design codes and standards are developed for the use of scientific and technological innovations to protect civil engineering works under hazardous actions.

10.7 Conclusions

Disasters are inevitable. The fact lies in stating “we must all be prepared to try to survive the current and the forthcoming disasters”. Buildings and structures play a significant role in socioeconomic development of a country. During its long running services, buildings and structures are subjected to several natural (earthquake, wind, etc.) and manmade (fire, etc.) hazards which can cause partial or complete collapse of the building and structures, and incapacitation of building operations, which leads to severe losses of life and property. Hence, buildings and structures are designed to withstand actions from numerous anticipated hazards to ensure life and structural safety during their design life. As a result, it’s significantly important to pay much attention to the understanding and mitigation of disaster. At the same time, from a scientific and technological viewpoint, strategies are generally presented on disaster prevention and mitigation of earthquakes, winds and fires.

Exercises

10-1 How do you understand the term earthquakes? What are its causes? State different types of earthquakes.

10-2 Describe the various provisions to be made to make a building earthquake resistant.

10-3 What is the principle of base isolators and seismic dampers for earthquake resistance?

10-4 What kind of measures are to be taken to make buildings extremely wind resistant?

10-5 Write a short note on the effect of typhoon hurricane and tornado.

10-6 How a building can be made to be fire resistant? Describe your answers briefly.

10-7 What are building codes and why are they useful for structural design?

References

[1] MOHAMED S I. An overview on disasters [J]. Disaster Prevention and Management, 2007, 16 (5): 687-703.

[2] HAMADA M. Engineering for Earthquake Disaster Mitigation [M]. Tokyo: Springer Japan, 2014.

[3] XIE L L, QU Z. On civil engineering disasters and their mitigation [J]. Earthquake Engineering and Engineering Vibration, 2018, 17(1), 1-10.

[4] RICHTER C F. Elementary Seismology[M]. San Francisco:W.H. Freeman and Company, 1958.

[5]AMBROSE J, VERGUN D. Design for Earthquakes [M]. New York: John Wiley & Sons, 1999.

[6] BHAVIKATTI S S. Basic Civil Engineering [M]. Delhi: New Age International Publishers, 2010.

[7] ANDREASON K, ROSE J D. Northridge, California earthquake: structural performance of buildings in san Fernando Valley, California[R]. American Plywood Association, APA Report No. T94-5, Tacoma, Washington, D.C., USA,1994.

[8] XU X, WEN X, YU G, et al. Coseismic reverse and oblique-slip surface faulting generated by the 2008 M_w 7.9 Wenchuan earthquake, China. [J]. Geology, 2009, 37(6), 515-518.

[9] ELNASHAI AS, DI SARNO L. Fundamentals of Earthquake Engineering [M]. New York: Wiley, 2008.

[10] LYU H M, WANG G F, CHENG W C, et al. Tornado hazards on June 23 in Jiangsu Province, China: preliminary investigation and analysis [J]. Natural Hazards, 2017, 85(1), 597-604.

[11] KODUR V, KUMAR P, RAFI, M M. Fire hazard in buildings: review, assessment and strategies for improving fire safety [J]. PSU Research Review, 2019.

[12] BRUSHLINSKY, N N, et al. World fire statistics[R]. Center of Fire Statistics, 2016.

[13] BUCHANAN A H, ABU A K. Structural Design for Fire Safety (2nd Edition) [M]. New York: John Wiley & Sons, 2017.

[14] MARTIN D, TOMIDA M, MEACHAM B. Environmental impact of fire [J]. Fire Science Reviews, 2016, 5(1): 1-21.

[15] LATAILLE J. Fire Protection Engineering in Building Design [M]. New York: Elsevier, 2002.

[16] SIMIU E, ROBERT H S. Wind Effects on Structures: Fundamentals and Applications to Design (4th Edition) [M]. New York: John Wiley & Sons, 2019.

Chapter 11
Intelligent Construction and Management

11.1 Overview

In the Internet era, digitization has given birth to the transformation and innovation of various industries, and the construction industry is no exception.

Intelligent construction is one of the effective ways to solve the problem of low efficiency, high pollution and high energy consumption in the construction industry. It has been proposed and practiced in many projects. Therefore, it is necessary to summarize the characteristics of intelligent construction.

Intelligent construction covers the design, production and construction stages of constructing projects. With the help of Internet of things, big data, BIM and other advanced information technologies, data integration of the whole industry chain is realized to provide support for the whole life cycle management.

Intelligence is a special ability of advanced animals, which generally includes perception, recognition, transmission, analysis, decision-making, control, action, etc. If the system has the above capabilities, the system has intelligence. Intelligent construction aims to make full use of intelligent technology and related technologies in the construction process. The application of intelligent system can improve the intelligent level of construction process, reduce the dependence on people, achieve the purpose of safe construction and improve the cost performance and reliability of buildings.

This definition covers three aspects: (1) the purpose of intelligent construction is to improve the intelligent level of construction process; (2) the means of intelligent construction is to make full use of intelligent technology and related technologies; (3) the manifestation of intelligent construction is to apply intelligent system.

11.2 Introduction of Intelligent Construction

11.2.1 Purpose, Means and Form of Intelligent Construction

According to the role of each component, the building can be divided into four systems: structural system, envelope system, facility and pipeline system, and interior decoration system. An excellent building is the organic integration of these four systems, thus providing people with complete functions and excellent performance buildings.

In 2013, at the Hannover Industrial Exposition, the German Federal Ministry of Education and Research and the Federal Ministry of Economy proposed the concept of "Industry 4.0". It describes the future of the manufacturing industry: people will usher in the fourth industrial revolution based on the information physical fusion system and marked by highly digitized, networked and self-organized machines.

The concept of Industry 4.0 is introduced into the construction industry, which forms a new concept of intelligent construction.

Intelligent technology and related technologies can make machines have wisdom. The delivery robot can go on the road by itself, change lanes according to the road conditions, avoid obstacles, plan the route independently and deliver the goods to the designated address, which is no difference from the courier. It can achieve unmanned delivery, which is called smart express. Smart express is the product of comprehensive application of intelligent technology and related technologies. But, by contrast, intelligent building is much more complex.

Intelligent construction is the extension of smart city and intelligent building. "Wisdom" and "intelligence" extend to the construction process of engineering projects, and the concept of intelligent construction comes into being.

Intelligent construction means making full use of intelligent technology and related technologies in the construction process. The establishment and application of intelligent system can improve the intelligent level of construction process, reduce the dependence on people, realize safe construction and achieve better performance–price ratio and better quality buildings.

In other words, the purpose of intelligent construction is to improve the intelligent level of construction process, reduce the dependence on people and achieve better construction, which means that intelligent construction will make the construction process more economic, safer with less people and high quality. The form of intelligent construction is the application of intelligent system. "Less people" is mentioned here, which reflects the difference between engineering construction industry and manufacturing industry. Due to the complexity of the construction industry, it is difficult to achieve unmanned construction.

Intelligent construction is also based on the comprehensive application of intelligent technology and related technologies. Among them, it involves perception, including Internet of things, positioning and other technologies; involves transmission, including Internet, cloud computing and other technologies; involves analysis, including mobile terminal, touch terminal and other technologies; involves memory, including BIM, GIS and other technologies; involves analysis, including big data, artificial intelligence and other technologies; in addition, it also includes three-dimensional laser scanning, three-dimensional printing and robot technology. Through the application of these technologies, the intelligent system will have the following characteristics: sensitive perception, high-speed transmission, accurate identification, rapid analysis, optimized decision-making, automatic control and alternative operation.

11.2.2 Relationship between Intelligent Construction and Enterprise Informatization, Digitization and Intellectualization

In addition, we should also understand the relationship between intelligent construction and what we often call enterprise informatization, digitization and intellectualization, as well as the relationship between intelligent construction and highly intelligent construction. Enterprise informatization is mainly to realize the automation of work, and digitization makes it possible to share data between jobs, while intellectualization makes it possible to make intelligent decision using data. In the construction process, if it meets the basic requirements of intelligent construction, it can be called the realization of intelligent construction. The intelligent system used in the construction process is divided into two categories, namely management system and technical system. The examples of the former are ERP system, project integrated management system, etc. The examples of the latter are BIM platform software, BIM tool software, etc.

11.2.3 Construction Robot Technology

With the wide application of robot products, the construction field has begun to try to use robots to improve the construction efficiency. Therefore, professional construction robots are gradually moving from research and development to landing, playing an active role in the construction of houses, towers, bridges and subways.

Fig. 11-1 Robo Welder
Source: https://www.sdkrd.com/ShuKong-qzsglgjphjjqr01.htm.

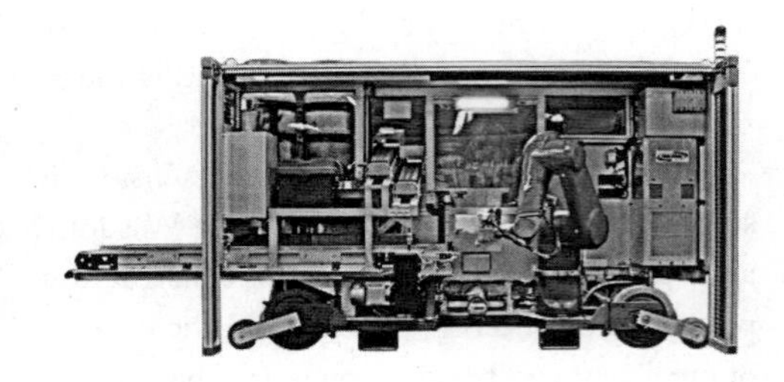

Fig. 11-2 Semi-Automated Mason Sam 100
Source: https://www.construction-robotics.com/sam100/.

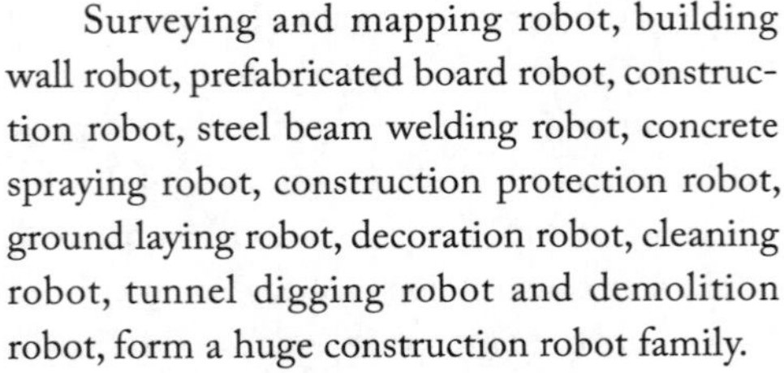

Surveying and mapping robot, building wall robot, prefabricated board robot, construction robot, steel beam welding robot, concrete spraying robot, construction protection robot, ground laying robot, decoration robot, cleaning robot, tunnel digging robot and demolition robot, form a huge construction robot family.

(1) Robo Welder (Fig. 11-1)

However, due to the diversity, delicacy and complexity of construction projects, this robot can only handle 1% of the total workload of construction projects. More studies are needed if the welding robot is to "realize the complete transformation of the construction industry". The robot is now deployed at construction sites, but only works the "night shift" for safety.

(2) Semi-Automated Mason Sam 100

The brick laying robot Sam 100, developed by an American construction robot company, can lay 300 to 400 bricks in an hour and 3000 bricks per day, which is five to six times faster than the average bricklayer, as shown in Fig. 11-2.

The Sam 100 system consists of a conveyor belt, a mortar pump, a nozzle and a mechanical arm. By the time the brick is delivered to the arm, the arm has been plastered with cement by the nozzle and only needs to lay the bricks.

The Sam 100 system is most suitable for building large flat walls, as well as suitable for some fine buildings. But the Sam 100 is a semi-automatic system that requires two more workers, one to put bricks on a conveyor belt and another to help remove excess cement from the wall.

(3) Fully Automatic Brick Laying Robot "Hadrian"

The prototype Hadrian X was developed in 2015 by the Australian engineer Mark Pivaac. It can work 24 hours a day, build 1000 bricks an hour and finish a house in two days, as shown in Fig. 11-3.

Hadrian X, which can scan the surrounding environment in 3D to accurately calculate the position of each brick, uses a 28m-long telescopic arm to pick up the brick, squeeze mortar or adhesive with pressure, apply it to the bricks to be glued, and place the bricks in order. It can also cut bricks.

The Hadrian X was mounted on the back of a truck and was able to move freely around the construction site. But it is still in the prototype stage and has yet to produce a commercial version.

(4) Japanese Construction Robot

The latest video from National Institute of Advanced Industrial Science and Technology (AIST) in Japan displays a prototype robot built to work on construction sites in scenarios where there is a scarcity of human workers. The machine is currently slow but surprisingly precise, revealing a future where human-like robots could swap a lot more human jobs.

The initial demonstration displays the robot, named HRP-5P, collecting a piece of plasterboard and screwing it into a partition, as shown in Fig. 11-4. This type of versatile humanoid robot is made to have the ability

Fig. 11-3 Fully automatic brick laying robot "Hadrian"
Source: https://www.constructionjunkie.com/blog/2020/9/9/brick-laying-robot-hadrian-x-completes-first-commercial-building.

Fig. 11-4 Japanese Construction Robot HRP-5P
Source: http://blog.sina.com.cn/s/blog_5509545f0102y4iy.html.

to imitate human motions in complex construction settings.

HRP-5P is not in any respect the most sophisticated robot we have ever witnessed. However, by instantly developing a robot that can perform heavy manual labor with similar motions to a human being, AIST is gesturing in the direction of a future where all the more granular work can be implemented over by robots.

11.3 Building Information Modeling (BIM)

Traditional building design largely relies upon two-dimensional drawings (plans, elevations, sections, etc.). BIM extends to 3D. It augments the primary three spatial dimensions (width, height and depth) by the fourth dimension of time and the fifth dimension of cost. BIM therefore covers more than just geometry. It also covers spatial relationships, light analysis, geographic information, and quantities and properties of building components. It has eight characteristics: information completeness, information relevance, information consistency, visualization, coordination, simulation, optimization and plotting. It is not a simple integration of digital information, but an application of digital information, and can be used for design, construction and management of digital methods.

CAD technology pushes architects and engineers from manual drawing to computer-aided drawing, and realizes the first information revolution in the field of engineering design. However, the supporting role of this information technology on the industrial chain is breakpoint, and there is no correlation between various fields and links. The comprehensive application of informatization is obviously insufficient in the whole industry.

BIM is a technology, a method and a process. It includes not only the information model of the whole life cycle of the building, but also the model of the management behavior of the construction project. It combines the two perfectly to realize the integrated management. The emergence of BIM will probably lead to the second revolution in the whole A/E/C (Architecture/Engineering/Construction) field.

11.3.1 BIM Technology Concept and Characteristics

BIM technology is a multi-dimensional model information integration technology, which enables all participants (including government departments in charge, owners,

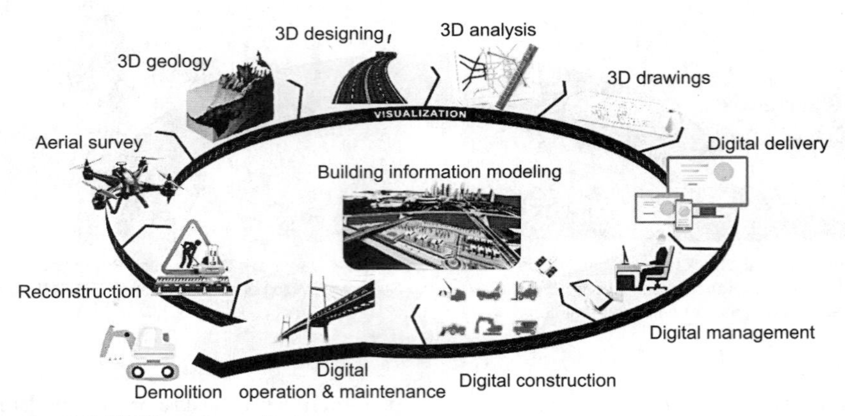

Fig. 11-5 BIM life-cycling application
Source: http://www.bimcn.org/uploads/allimg/181227/873_181227102305_1.jpg.

design, construction, supervision, cost, operation management, project users, etc.) in the construction project to operate the information with the model and operate the model with the information during the whole life cycle of buildings from concept generation to complete demolition, so as to fundamentally change the employees' working mode of project construction and operation management relying on symbolic text drawings, and achieve the goal of improving work efficiency and quality and reducing errors and risks in the whole life cycle of construction projects (Fig. 11-5).

The meanings of BIM are summarized as follows:

(1) BIM is an engineering data model based on three-dimensional digital technology, which integrates all kinds of related information of construction projects. It is a digital expression of the physical and functional characteristics of engineering facilities.

(2) BIM is a perfect information model, which can connect the data, process and resources in different stages of the construction project life cycle. It is a complete description of the engineering object. It provides real-time engineering data that can be automatically calculated, queried, combined and split. It can be widely used by all participants in construction projects.

(3) BIM has a single engineering data source, which can solve the problems of consistency and global sharing between distributed and heterogeneous engineering data, and support the dynamic engineering information creation, management and sharing in the life cycle of construction projects. It is a real-time data sharing platform for projects.

The benefits of BIM include the following aspects:

(1) Visualization: What you see is what you get. The three-dimensional physical graphics of the model can be seen, and the whole construction process such as project design, construction and operation can be seen. It can facilitate better communication, discussion and decision-making (Fig. 11-6).

(2) Coordination: The information of various specialties is often "incompatible", like the pipeline conflicts with the structure, uneven heating and cooling in each room, no reserved hole or the wrong size. BIM can effectively coordinate and integrate various disciplines and processes, and reduce unreasonable change scheme or problem change scheme (Fig. 11-7).

(3) Simulation: It includes 3D screen simulation, simulation of energy efficiency, emergency evacuation, sunshine, heat conduction, etc., 4D (development time) simulation, 5D (cost control) simulation, daily emergency treatment

Fig. 11-6 Visualization of BIM
Source: http://5b0988e595225.cdn.sohucs.com/images/20181127/fff98ca4683441e1b212a2f4259a2028.jpeg.

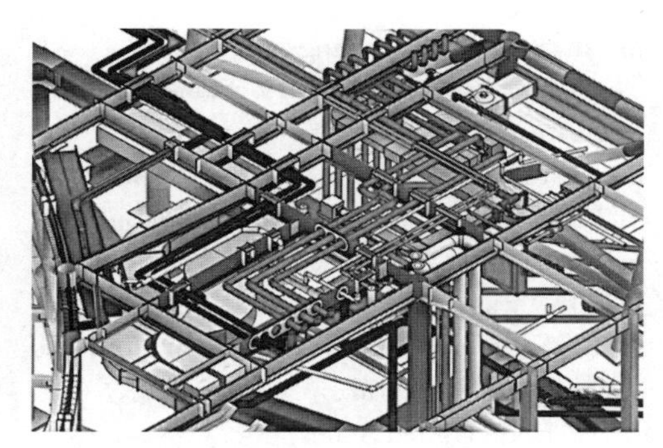

Fig. 11-7 Coordination of BIM
Source: https://www.uibim.com/24296.html.

methods such as earthquake personnel escape and fire personnel evacuation (Fig. 11-8).

(4) Optimization: BIM and its various optimization tools can optimize the project. We can use the information provided by the model to optimize the properties of the project, such as geometry, physics, rules, the information of various situations after the change of buildings.

(5) Printability: It can output construction drawings directly based on BIM model, such as comprehensive pipeline drawing, comprehensive structure hole drawing, collision detection report and improvement proposal, and other practical construction drawings.

11.3.2 Development History and Current Situation of BIM

BIM, as an important thought and reform on the workflow and working method of many industries including engineering construction industry, its rudiment can be traced back to the 1970s. Dr. Chuck Eastman put forward the concept of BIM in 1975. From the late 1970s to the early 1980s, the UK was also carrying out research and development work similar to BIM. At that time, Europe used to call it "product information model", while the United States usually called it "building product model". In 1986, Robert Aish used the word "building information modeling" for the first time in a paper. In this paper, he described the argument and implementation technology related to the BIM known to us today, and used building model system to analyze a case to express his concept.

Due to the limitation of computer hardware and software, BIM research before the 21st century could only be used as the object of academic research, and it was difficult to play a

Fig. 11-8 Construction simulation by BIM
Source: https://bbs.zhulong.com/106010_group_914/detail32888242/?f=bbsnew_BIM_1.

role in practical engineering application.

Since the 21st century, with the rapid development of computer software and hardware level and the in-depth understanding of building life cycle, BIM technology has been promoted. Since the method and concept of BIM was put forward and promoted in 2002, the reform wave of BIM technology has swept across the world.

(1) The United States

The United States is an early country to start the research on construction informatization. Since 2003, the General Services Administration of the United States (GSA) began to implement a project called National 3D-4D-BIM Program through the subordinate of Public Buildings Service (PBS). The objectives of the project are: 1) to realize technological transformation to provide more efficient, economical, safe and beautiful federal buildings; 2) to promote and support the application of open standards. According to the plan, GSA explores the application of BIM from the perspective of the whole life cycle of project, including space planning verification, 4D schedule control, laser scanning, energy analysis, pedestrian flow and safety verification, construction equipment analysis and decision support, etc. In order to ensure the smooth implementation of the plan, GSA has formulated a series of strategies for support and guidance, including: 1) develop a clear vision and value proposition; 2) use pilot projects to accumulate experience as a demonstration; 3) strengthen personnel training and establish organizational culture to encourage sharing and 4) select suitable software and hardware, applying open standard software and hardware systems to constitute the basic environment of BIM application.

(2) Singapore

In 1995, Singapore's Ministry of National Development launched an IT project called CoreNet (construction and real estate network). The main purpose is to improve the operation time, production efficiency and effect by Business Process Reengineering (BPR), and also focus on the use of advanced information technology to achieve efficient and seamless communication and information exchange among participants in the construction and real estate industry. CoreNet system consists of three parts: e-submission, E-plan check and e-info. In the whole system, the E-plan check subsystem occupies the core position, and it is also the most distinctive part of the whole system. The function of the subsystem is to use the automatic program to conduct digital inspection on the architectural design results, so as to find out the violation of the building code requirements. The whole plan involves eight related agencies of five government departments. The whole system adopts client – server architecture. With this system, designers can process and prepare the design results through BIM tools, and then submit them to the system for online automatic review.

In order to ensure the smooth implementation of CoreNet project (especially E-plan check system), the Singapore government has taken a series of policy measures and achieved good results. It mainly includes:

1) Extensive industry testing and trial to ensure the operation effect of the system;

2) Pay attention to communication with the industry through various forms, and strengthen personnel training;

3) Strengthen the cooperation with international organizations in the R & D (short for research and development) process of the system.

The Singapore government attaches great importance to cooperation with relevant international organizations, which can enable the system to receive all-round support from international organizations, and at the same time, it can be recognized in a wider range.

(3) The United Kingdom

Compared with most countries, the UK government requires the mandatory use of BIM. In May 2011, the British Cabinet Office released a document named *Government Construction Strategy*, which contains an entire chapter on building information model (BIM). In this chapter, it is clearly required that the government requires comprehensive and collaborative 3D BIM, and would manage all the documents with information technology by 2016.

(4) South Korea

South Korea is very advanced in the application of BIM technology. A number of government departments are committed to the development of BIM standards, such as the Public Procurement Service (PPS) and the Ministry of Land, Transportation and Maritime Affairs of the South Korea.

Major construction companies in South Korea have been actively adopting BIM technology, such as Hyundai Construction, Samsung Construction, Space Integrated Architecture Office, Daewoo Construction, GS Construction, Daelim Construction, etc. Among them, Daelim Construction company applies BIM technology to bridge construction management, and BMIS company uses digital project of BIM software to research and implement the integration of architectural design stage and construction stage.

(5) Northern Europe

Norway, Denmark, Sweden and Finland are home to some major IT software manufacturers in the construction industry, such as Tekla and Solibri.

The four Nordic governments have imposed but not required the full use of BIM. Due to the requirements of local climate and the promotion of advanced building information technology software, the development of BIM technology is mainly the conscious behavior of enterprises. For example, Senate Properties, a Finnish state-owned enterprise, is also the largest property asset management company in the Netherlands. In 2007, Senate Properties issued a BIM requirement for architectural design. Since October 1, 2007, the project of Senate Properties only requires the use of BIM in architectural design. Other design parts can decide to adopt BIM technology or not according to the project situation, but the goal will be to fully use BIM. The report also states that there will be mandatory BIM requirements in the design bidding, which will become part of the project contract and be legally binding. It is suggested that in project collaboration, modeling tasks need to create a common view and need accurate definition. The final BIM model needs to be submitted, and the collision between the building structure and the interior of the model needs to be filed. The modeling process is divided into four stages: spatial group BIM, spatial BIM, preliminary building element BIM and building element BIM.

11.3.3 Main Applications of BIM

The specific application contents of BIM are different for all roles involved in the project and different stages of the project.

(1) Main Contents of BIM Application in Design Enterprises

1) Scheme design. the BIM technology can not only analyze the shape, volume and space, but also analyze the energy consumption and construction cost at the same time, which makes the initial scheme decision more scientific.

2) Development design. BIM model is established for architecture, structure, mechanical and electrical specialty, which will be used to carry out the energy consumption, structure, acoustics, thermal engineering and sunshine analysis. Various interference inspection, specification inspection and engineering quantity statistics are conducted.

3) Construction drawings. All kinds of plane, elevation, section drawings and statistical reports are obtained from BIM model.

4) Design collaboration. There are more than ten or even dozens of specialties in design that need to be coordinated, including design plan, mutual submission of data, proofreading and auditing, version control, etc.

(2) Main Contents of BIM Application in Construction Enterprises

1) Collision check, reducing rework. The 3D technology of BIM is used to check the collision in the early stage, which can solve

the conflict of spatial relationship intuitively, optimize the engineering design, reduce the possible errors and rework in the construction stage, optimize the clearance and the pipeline layout scheme. Finally, the construction personnel can use the optimized scheme to carry out construction disclosure and construction simulation, so as to improve the construction quality and communication ability with the owner.

2) Simulation construction, effective coordination. The three-dimensional visualization function and time dimension can simulate the construction progress. Comparing the construction plan with the actual progress intuitively and quickly at any time and place can help the project participants know all of problems and situations of the project, so as to reduce construction quality problems, safety problems, reduce rework and rectification.

3) Three dimensional rendering, publicity and display. 3D rendering animation can let customers have a sense of substitution through virtual reality, give people a sense of reality and direct visual impact, and adjust the implementation scheme in the bidding demonstration and construction stage. The BIM model can be used as the model basis of secondary rendering development, greatly improving the accuracy and efficiency of 3D rendering effect, giving more intuitive publicity and introduction to the owner, and improving the probability of winning the bid in the bidding stage.

4) Knowledge management. The knowledge and skills that are not easy to be accumulated in the construction can be acquired by the simulation process of information preservation, which can be transformed into the knowledge base content accumulated by the construction unit for a long time.

(3) Main Contents of BIM Application in Operation and Maintenance Stage

1) Space management. Space management is mainly used in lighting, fire control, and other systems and equipment space positioning. We can obtain the space location information of each system and equipment, and change the original number or text representation into three-dimensional graphic location, which is intuitive and easy to find.

2) Facility management. It mainly includes the decoration, space planning and maintenance of facilities. The National Institute of Standards and Technology (NIST) in the United States conducted a study in 2004, which found that owners and operators spent almost two-thirds of the total cost of continuous facility operation and maintenance. The feature of BIM technology is that it can provide consistent and computable information about construction projects, so the information is worth sharing and reusing, and the owners and operators can reduce the cost loss caused by the lack of interoperability. In addition, it can also remote control important equipment.

3) Management of concealed works. In the architectural design stage, there will be some hidden pipeline information that the construction unit does not pay attention to, or these information may be in a corner that only a few people know. Especially with the increase of the service life of the building and the frequent replacement of personnel, these security risks become increasingly prominent, and sometimes directly lead to tragedy. The operation and maintenance based on BIM technology can manage complex underground pipe network, such as sewage pipe, drainage pipe, network cable, wire and related tub well, and can directly obtain the relative position relationship on the diagram. When the reconstruction or secondary decoration is undertaken, the existing pipe network can be avoided, which is convenient for pipe network maintenance, equipment replacement and positioning. Internal relevant personnel can share this electronic information, and can be adjusted at any time to ensure the integrity and accuracy of the information.

4) Emergency management. There will be no blind spots in the management based on BIM technology. In public buildings, large-scale buildings and high-rise buildings,

as the crowd gathering area, emergency response ability is very important. The traditional emergency management only focuses on response and rescue, but through the operation and maintenance management of BIM technology, the emergency management includes prevention, alarm and processing. Through the BIM system, we can quickly locate the location of facilities and equipment, avoid looking for information in the vast number of drawings, if not handled in time, it will lead to catastrophic accidents.

5) Energy saving and emission reduction management. The application of BIM combined with Internet of things technology makes daily energy management monitoring more convenient. Through the installation of electricity meter, water meter and gas meter with sensing function, the basic functions of real-time collection, transmission, preliminary analysis and fixed-point upload of building energy consumption data can be realized, and it has strong expansibility. The system can also realize the remote monitoring of indoor temperature and humidity, analyze the real-time temperature and humidity changes in the room, and cooperate with energy-saving operation management.

11.3.4 Application Prospect of BIM Technology

BIM technology in the future development must be combined with advanced communication technology and computer technology to greatly improve the efficiency of the construction industry. It is expected that BIM technology will have the following application prospects:

(1) Application of Mobile Terminal

With the popularity of the Internet and mobile intelligent terminals, people can now access information at any place and at any time. In the field of architectural design, many contractors will be able to provide their own staffs with these mobile devices, which can be designed on the job site.

(2) Wireless Sensor Networks

Now the monitor and sensor can be placed anywhere in the building to monitor the temperature, air quality and humidity in the building. Engineers can have a comprehensive and full understanding of the current situation of buildings by obtaining these information through wireless sensor networks, so as to provide effective decision-making basis for design and construction schemes.

(3) Cloud Computing Technology

Whether it is energy consumption or structure analysis, the processing and analysis of some information need to make use of the powerful computing power of cloud computing. Even, our rendering and analysis process can achieve real-time computation, helping designers compare different designs and solutions as soon as possible.

(4) Digital Reality Capture

This technology can quickly obtain early data by laser scanning bridges, roads, railways and buildings. In the future, designers can use this immersive and interactive way to work in a 3D space, and visually display product development.

(5) Collaborative Project Delivery

BIM is a workflow, and it is a technology based on changing the design method, and it has changed the whole project implementation method. It is a process of cooperation between designers, contractors and owners. Every party has their own valuable ideas and demands, so collaborative project delivery is a good way to meet the values and demands of all parties.

People are no longer satisfied with the completion of the design and construction tasks of the project, but pay more attention to whether the whole project from the design to the late implementation process meets the requirements of high efficiency and energy saving, and whether it can realize green and sustainable development. The integration of BIM technology and other technologies will create value from a more comprehensive field.

11.4 Artificial Intelligence

Artificial intelligence (AI) is a new technology science that researches and develops theories, methods, technologies and application systems for simulating and extending human intelligence. AI is a branch of computer science. It attempts to understand the essence of intelligence and produce a new kind of intelligent machine which can respond in a similar way to human intelligence. The research in this field includes robot, language recognition, image recognition, natural language processing, expert system, etc. Since the birth of artificial intelligence, the theory and technology are increasingly mature, and the application fields are also expanding. It can be imagined that the scientific and technological products brought by artificial intelligence in the future will be the "container" of human intelligence. Artificial intelligence can simulate the information process of people's consciousness and thinking.

Since it was put forward in the 1940s, it has been applied in many disciplines and formed a variety of algorithms. With the rapid development of deep learning (DL) since 2006, artificial intelligence has become a research and application hotspot in various fields. In 2012, Germany launched the "Industry 4.0" plan focusing on "smart factory". In 2015, China issued the "guiding opinions on actively promoting the 'Internet Plus' action", and explicitly identified "Internet plus AI" as one of the key actions. In December 2016, the UK released "artificial intelligence: opportunities and implications for the future of decision-making". France released the AI strategy in March 2017. Japan set 2017 as the first year of artificial intelligence to promote the "Super Intelligent Society 5.0". In 2018, the United States released the national strategy for artificial intelligence.

Artificial intelligence technology is bound to permeate all aspects of human social activities and production activities. A lot of physical labor and work in harsh environment will be replaced by machines or robots Compared with other industries, although steel (steel structure) and concrete (concrete engineering) are the products of industrialization, the degree of mechanization, automation, intelligence and informatization of infrastructure is still low. In the field of civil infrastructure, artificial intelligence technology deeply integrates the whole life cycle of planning, design, construction and maintenance of civil infrastructure, and profoundly changes the development of civil engineering.

(1) Intelligent Planning

The application of artificial intelligence in urban planning is a landmark change of urban planning discipline. Artificial intelligence will change the traditional urban planning methods. Through in-depth learning of the existing big data of urban environment, disaster, human and traffic behavior, combined with virtual reality situation reproduction technology, the intelligent urban planning can be realized. For example, machine learning and deep learning are used to plan urban generation and urban spatial laws, and "city tree" is generated to provide visual scientific support for urban development; or the planning and design problem is transformed into a network optimization problem, and urban planning and design is carried out through a planning and design paradigm based on artificial intelligence technology.

(2) Intelligent Design

The application of artificial intelligence technology in building structure, bridge structure and other engineering designs has become the main trend of intelligent civil engineering development. Artificial intelligence aided civil engineering design is in its infancy, but it has shown a good prospect.

The early application of artificial intelligence in structural design is mainly expert system, which makes judgments and decisions by simulating the reasoning thinking process of human experts. For example, artificial neural network can be used to design prestressed reinforced concrete slab, machine learning algorithm can be used for bridge selection, and deep learning and reinforcement learning can carry out structural topology optimization design. It can be foreseen that based on the big data of existing design, the design schemes for buildings, bridges, tunnels and other infrastructure will be generated automatically by computer through in-depth learning and intensive learning, combined with user demand parameters in the future.

(3) Intelligent Construction Management

In the field of construction site management, artificial intelligence has a wide range of applications. Researchers from the University of Michigan use augmented reality technology to help field personnel easily access project plans, charts, progress and budget information. In the field of construction intelligent monitoring, the in-depth learning target detection algorithm in computer vision can be used to automatically warn the construction site monitoring video and automatically identify whether the site personnel have worn safety helmets.

(4) Intelligent Monitoring and Detection

Computer vision technology is an important branch of artificial intelligence. The image or video is obtained by camera, and the computer processes the target image recognition, motion tracking, scene reconstruction, image restoration, image measurement and other tasks instead of human eyes, and the information and knowledge contained in the image can be obtained by understanding the image. The application of computer vision technology in civil engineering mainly includes crack identification, displacement measurement, modal parameter identification, vehicle load identification and so on.

Crack identification based on computer vision is a research hotspot in the field of damage identification of civil engineering structures. Displacement and vibration monitoring methods based on vision have been used in civil engineering for more than ten years. The typical visual vibration measurement methods include target tracking method, digital image correlation method, etc.

(5) Intelligent Disaster Prevention and Mitigation

Artificial intelligence technology can help people to develop integrated emergency positioning, multi-source emergency data fusion, disaster scene visualization, disaster model analysis services and other key technologies, build an integrated disaster reduction intelligent service system, realize the integrated perception, positioning, integration, analysis and service of emergency information, and provide technical support for comprehensive disaster reduction decision analysis. In addition, the risk management and control of urban large-scale infrastructure disaster is essentially a complex optimization decision-making problem. With the development of computer science, the use of deep reinforcement learning (DRL) technology has brought new ideas to solve this problem.

11.5 3D Printing Construction

As a rapid prototyping technology, 3D printing is a kind of additive manufacturing. It is based on the digital model file, using powder metal or plastic and other bondable materials to construct objects layer by layer.

11.5.1 Basic Principles of 3D Printing

Ordinary printers used in daily life can print plane objects designed by computers. The so-called 3D printers and ordinary printers work on the same principle, but the printing materials are somewhat different. The printing materials of ordinary printers are ink and paper, while 3D printers are filled with different "printing materials" such as metal, ceramics, plastics, sand, etc., which are real raw materials.

After the printer is connected with the computer, the "printing materials" can be superimposed layer by layer through computer control, and finally the blueprint on the computer can be turned into real objects. Generally speaking, 3D printer is a device that can "print" real 3D objects, such as printing a robot, printing toy cars, printing various models, even food, etc. The reason why it is commonly called "printer" refers to the technical principle of ordinary printer, because the process of layered processing is very similar to inkjet printing. This printing technology is called 3D printing technology.

There are many different technologies for 3D printing, such as fused deposition modeling (FDM), direct metal laser-sintering (DMLS), stereo litho-graphy appearance (SLA), etc. The difference between them is that they use different materials and use different methods to create objects. The commonly used materials for 3D printing include nylon glass fiber, durable nylon material, gypsum material, aluminum material, titanium alloy, stainless steel, silver plating, gold plating and rubber materials.

The design process of 3D printing is modeling by computer modeling software, and then dividing the 3D model into sections layer by layer, i.e. slice, so as to guide the printer to print layer by layer. The standard file format for collaboration between design software and printers is STL file format. A STL file uses triangular surfaces to approximate the surface of an object. The smaller the triangular surface is, the higher the resolution of the generated surface will be.

11.5.2 3D Printing Concrete Building

3D printing concrete technology is a new application technology which combines 3D printing technology with commercial concrete technology. The main principle is to use the computer for 3D modeling and segmentation of concrete components to produce three-dimensional information, and then the prepared concrete mixture is extruded by the nozzle for printing according to the set program through the extrusion device, and finally the concrete component is obtained. 3D printing concrete technology is a new type of concrete formless forming technology, which has the advantages of self compacting concrete without vibration, and the advantages of shotcrete in manufacturing complex components.

In 2014, NASA (National Aeronautics and Space Administration) cooperated with the University of Southern California to develop the "contour process" 3D printing technology, which can print about 232m^2 two-storey building in 24 hours (Fig. 11-9). It greatly saves construction time and construction cost, and opens a door for green manufacturing.

In the "contour process", there is a super printing robot shaped like a bridge crane hovering over a building with rails on both

Fig. 11-9 3D printing construction (design sketch)
Source: http://www.chinadaily.com.cn/hqzx/images/attachement/jpg/site1/20140124/001ec92bc767144be89703.jpg.

Fig. 11-10 3D printing construction on the moon
Source: https://www.xianjichina.com/data/editer/20190305/image/28605f9a76f63fd0bbd3cad39fc2f2db.jpg.

sides and a "print head" in the middle beam which can move up and down, back and forth to print out the house layer by layer.

It is worth mentioning that the "contour process" is not only limited to the earth, but also can be used in outer space. If human beings want to build a habitat on the moon in the future, 90% of the building materials are expected to come from the lunar soil, while the rest may need to be transported from the earth to the moon by spacecraft. As the "contour process" can be more quickly and environmentally friendly to mass build buildings suitable for human habitation, we can imagine that with this cutting-edge technology becoming more sophisticated, space migrants are expected to live a more comfortable life (Fig. 11-10).

In some European countries, there are many companies in the research and practice of 3D printing architecture. In January 2013, Dutch architect Janjaap Ruijssenaars and artist Rinus Roelofs designed the world's first 3D printing building. The design inspiration comes from the Mobius Ring. It is named landscape house because of its shape similar to Mobius Ring and its pleasant characteristics like scenery. The building uses Italian "D-shape" printer to make 6m×9m blocks, and finally splicing is completed. In 2018, Russia's Amt-Specavia announced the completion of the first habitable 3D printing house in Yaroslavl. The house covers an area of nearly 300 square meters and is a large single building. In 2020, a Belgian company has built a two-storey 3D printing model building using what is known as Europe's largest 3D printer (Fig. 11-11). The unnamed project, covering an area of about 90 square meters, was built using a 10m×10m COBOD BOD2 printer. In addition, there are some companies in China developing 3D printing building technology.

Fig. 11-11 The two-storey 3D printing building in Belgium.
Source: http://s11.sinaimg.cn/orignal/002ZLegtzy7EZf61Ax44a.

11.5.3 Concrete Materials for 3D Printing

Ordinary cement concrete has been difficult to meet its technical requirements for 3D printing concrete building, so it puts forward higher requirements for concrete performance to meet the needs of 3D printing construction technology.

In order to meet the needs of 3D printing buildings, concrete mixture must meet specific requirements. The following analysis is made from the composition of concrete.

First of all, ordinary Portland cement in strength, setting time and other aspects may not meet the requirements of 3D printing, which needs to do further research on this basis, such as changing the mineral composition of cement composition, fineness of clinker, etc. Furthermore, we can use sulphoaluminate cement or aluminate modified Portland cement to obtain faster setting time and better early strength.

Secondly, 3D printing is realized by nozzle. The size of the nozzle determines the particle size in the preparation of concrete mixture, and the most appropriate size of aggregate must be found. If the particle size of aggregate is too large, the nozzle will be blocked; if the particle size is too small, the specific surface area of slurry required for wrapping the aggregate is large, the hydration rate is fast, and the hydration heat per unit time is high, which will lead to the deterioration of various properties of concrete.

Thirdly, the prepared concrete mixture should have appropriate mix proportion. As the raw material for 3D printing, the new concrete has been different from the traditional concrete, and its various properties have changed greatly, which can not be determined by the traditional water binder ratio and sand ratio, and its basic performance has changed greatly. At present, theories related to concrete, such as strength, durability and hydration, can not meet the requirements of 3D printing concrete. In order to make printing concrete obtain ideal state, such as high strength, good durability, good mixing performance, appropriate setting time, good workability, pumpability and constructability, it is necessary to perfect the theory from a new angle.

Finally, the admixture is one of the essential components of modern concrete and an important method and technology of concrete modification. 3D printing concrete must have better rheological properties to facilitate extrusion and rapid condensation in the air, so as to prevent damage to the structure of printing concrete due to its own gravity. Moreover, the maximum particle size of aggregate will become smaller and its morphology will be more round, which will lead to more complex gradation. Finally, the problem of condensation between layers needs to be solved, which requires new admixtures to solve. From the point of view of material rheology, 3D printing concrete should have higher plastic viscosity and lower ultimate shear stress, so that it has good plasticity without fluidity, at the same time, it should have faster setting time and higher early strength.

11.5.4 Problems in 3D Printing Concrete Buildings

Compared with traditional buildings, 3D printing architecture has several outstanding advantages, such as high strength, free building form, rapid construction time, environmental protection, energy saving and so on. However, as a new technology which is in the stage of R&D and trial, there are inevitably the following problems.

(1) The Raw Materials

Compared with the traditional concrete construction technology, 3D printing concrete puts forward higher requirements on the rheology and plasticity of raw materials. Ordinary cement may not be able to meet the requirements of building performance and printing technology at the same time. The aggregate may adopt new crushing technology to produce aggregate with smaller particle size and more round particle morphology. The admixture should not only retain the existing performance in concrete, but also solve the problem of how to perfectly combine each layer.

(2) Printing Accuracy

Whether there will be deviation in 3D printing technology and how to prevent it are the problems that should be paid attention to in the application of 3D printing technology in architectural design model. Because the development of 3D printing concrete technology is not perfect, the precision and surface quality of rapid prototyping parts can not meet the requirements of the project, and can not be used as functional components, only used as prototype.

(3) Printing Device

With the development of technology, 3D printing equipment is developing rapidly. The price of a 3D printing device has changed from several hundred thousand dollars at the beginning to several thousand dollars now, and then to the price of more than 5000 RMB in China. 3D printing equipment is constantly moving towards the public and into various fields. However, the current 3D printing concrete equipment can not fully meet the special requirements of its application environment. For example, the current printing equipment can only meet the plane expansion stage, and can be used for the construction of low-rise and large-area buildings, but for the widely used high-rise buildings, it can not print.

In addition, compared with the traditional concrete, 3D printing concrete has changed a lot, and a set of construction technology and safety regulations need to be reestablished.

11.6 Construction Site of Intelligentization

Intelligent construction site is a new concept of life cycle management of engineering, which is a new concept of intelligent earth concept in engineering field.

The intelligent construction site is a construction project information ecosystem with interconnection and coordination, intelligent production and scientific management based on the construction process management through the accurate design and construction simulation of the project using the three-dimensional design platform. The engineering information collected by Internet of things in virtual reality environment is used in data-mining and analyzing to provide process trend prediction and expert plan. Finally, the visual intelligent management of engineering construction will be realized, and the informatization level of project management will be improved, so as to gradually realize green construction and ecological construction (Fig.11-12).

Intelligent construction site integrates more high-tech technologies such as artificial intelligence, sensing technology and virtual reality into various objects such as buildings, machinery, personnel wearing facilities, entrance and exit of the site (Fig. 11-13), and is widely interconnected to form the "Internet of things", which is then integrated with the "Internet" to realize the integration of project management stakeholders and engineering construction site.

The core of intelligent construction site is to improve the interaction mode of various stakeholders and post personnel with a "smarter" method, so as to improve the clarity, efficiency, flexibility and response speed of interaction.

Fig. 11-12 Management system for construction site of intelligentization
Source: http://wwwcdn.glodon.com/case_images/2018-08-06/5b67a766oeo4e.png.

Fig. 11-13 Intelligent access control with face brush system
Source: http://5b0988e595225.cdn.sohucs.com/images/20181201/d1ab591d378c46cdb99dbae5327a54d8.jpeg.

11.6.1 Key Elements of Intelligent Construction Site

The specific application content of intelligent construction site can be considered from the following aspects.

(1) Labor Management

Attendance is the most basic condition of human management. Through the intelligent equipment for personnel examination, we can accurately and real-time understand the number of people on the project site, and what area each person works in, as well as the work content and progress.

(2) Machinery and Equipment Management

Through the information management system and on-site monitoring equipment, various machinery and equipment on the construction site can be visually and accurately managed. It can effectively improve the utilization rate of machinery and ensure the overall management of special operation personnel. And we can use intelligent processing machinery instead of manual, which can greatly improve the construction efficiency and management efficiency.

(3) Material Management

First of all, material management is the property information of material itself, such as name, type, model, quantity, etc. Then the material inquiry is carried out. The basis of guiding price can only be the average price generated by big data. Finally, the quantity control of materials in and out of the project site, on the one hand, controls the consumption, on the other hand, controls the purchase quantity.

(4) Scheme and Construction Method Management

The project is becoming more and more complex. Through the modern technical means, the visual construction scheme and risk plan are formulated, and the construction method management is carried out through artificial intelligence technology, which requires all the process methods to be carried out strictly in accordance with the process to achieve quality control.

(5) Environmental Management

Using intelligent sensors and environmental monitoring system can monitor the environmental assessment indicators of the construction site in real time and quickly, such as PM2.5, noise, dust, etc., and automatically take response measures through the Internet of things and intelligent equipment to reduce the environmental pollution problems caused by construction projects.

11.6.2 Framework of Intelligent Construction Site

The overall framework of intelligent construction site can be divided into three layers.

(1) The First Layer is the Terminal Layer

The using of the Internet of things technology and mobile applications can improve on-site management and control capabilities. Through radio frequency identification (RFID), sensors, cameras, mobile phones and other terminal equipment, real-time monitoring, intelligent perception, data collection and efficient collaboration of the project construction process are realized, and the management ability of the operation site is improved.

(2) The Second Layer is the Platform Layer

How to improve the processing efficiency of complex business, big model and big data generated in each system? This has a huge demand for the server to provide high-performance computing power and low-cost mass data storage capacity. Through the cloud platform for efficient computing, storage and services, it can make the project participants more convenient to access data and work together, making the construction process more intensive, flexible and efficient.

(3) The Third Layer is the Application Layer

The core content of application layer should always focus on the key business of improving project management (PM). Therefore, PM system is one of the key systems of site management. The visualization, parameterization and data characteristics of BIM make the management and delivery of construction projects more efficient and lean, and it is an effective means to achieve lean management of project site. BIM and PM system provides a lot of data information for deep processing and reuse for the production and management of the project. How to effectively manage and use this massive information and big data needs the support of data management (DM) data management system to give full play to the value of data. Therefore, the application layer is closely combined with PM, BIM and DM to support each other to realize the intelligent management of the construction site.

11.6.3 Technical Supports for Intelligent Construction Site

Intelligent construction site is the comprehensive application of modern advanced technology in construction site. In addition to the BIM technology introduced above, the necessary supporting technology also needs a lot of advanced technology support, which is listed as follows.

(1) Data Exchange Standard

In order to realize intelligent construction site, it is necessary to exchange information and data between different project members and different software products,

Because this kind of information exchange involves many kinds of project members, complex project stages, long project life cycle, and a large number of application software products, only by establishing an open information exchange standard, can all software products exchange information with each other through this open standard, and realize the information exchange between different project members and different application software information flows. The standard format of object-based open information exchange includes defining the format of information exchange, defining the information to be exchanged, determining the information to be exchanged and the information needed, which are three standards of the same thing.

(2) Visualization Technology

Visualization technology can change scientific data (including measured values, field acquired images or digital information generated in calculation) into intuitive, graphic and image information, physical phenomena or physical quantities and present to managers. This technology is the premise of intelligent construction site to realize 3D display.

(3) 3S Technology

This is a general term for remote sensing (RS), geographic information systems (GIS) and global positioning systems (GPS). It is a modern information technology combining space technology, sensor technology, satellite positioning and navigation technology, computer technology and communication technology, and highly integrated multi-disciplinary to collect, process, manage, analyze, express, disseminate and apply spatial information. It is a centralized display platform for the achievements of intelligent construction sites.

(4) Virtual Reality Technology

Virtual reality (VR) is a technology that uses computer to generate a simulation environment, and makes users "immerse" into the environment through a variety of sensor devices, so as to realize the natural interaction between users and the environment. It can make the designers who apply BIM feel in their own environment, interact with the computer-generated environment in a natural way, and experience more rich feelings than the real world.

(5) Digital Construction System

Digital construction system is based on the establishment of digital geographic basic platform, geographic information system, remote sensing technology, site data acquisition system, site machinery guidance and control system, global positioning system and other basic platforms to integrate site information resources, break through the limitations of time and space, and establish an open information environment, so as to make the participants of engineering construction project more effective for real-time information exchange and using BIM model results for digital construction management.

(6) Internet of things (IOT)

"The Internet of things" is an important part of the new generation of information technology. As the name implies, the Internet of things is the Internet connected with things. This has two meanings: firstly, the core and foundation of the Internet of things is still the Internet, which is an extension and expansion network on the basis of the Internet; secondly, its user end extends to any item to carry out information exchange and communication. The Internet of things is widely used in the integration of network through intelligent perception, identification technology and pervasive computing, which is also known as the third wave of world information industry development after computer and Internet.

(7) Cloud Computing Technology

Cloud computing is the product of the integration of computer technology and network technology, such as grid computing, distributed computing, parallel computing, utility computing, network storage, virtualization and load balancing. It aims to integrate several relatively low-cost computing entities into a perfect system with strong computing power through the network, and distribute these powerful computing capabilities to end users. It is the best technical means to solve BIM big data transmission and processing.

(8) Information Management Platform Technology

The main purpose of information management platform technology is to integrate the existing management information system and make full use of the data in BIM model for management interaction, so that all parties involved in the project construction can work together on a unified platform.

(9) Database Technology

The application of BIM technology will rely on the database technology that can support big data processing, including the comprehensive application of massively parallel processor (MPP) database, data mining power grid, distributed file system, distributed database, cloud computing platform, Internet and scalable storage system.

(10) Network Communication Technology

Network communication technology is the communication bridge of BIM technology application and the channel of BIM data circulation, which constitutes the basic network of the whole BIM application system. According to the actual project construction situation, mobile phone network, wireless WiFi network, radio communication and other programs can be used to realize the communication needs of the project construction.

11.7 Smart City

Smart city originated in the field of media. It uses various information technologies or innovative concepts to connect and integrate urban systems and services, so as to improve the efficiency of resource utilization, optimize urban management and services, and improve the quality of life of citizens.

Smart city is an advanced form of urban informatization in which the new generation of information technology is fully applied to all walks of life in the city based on the next generation innovation (Innovation 2.0) of the knowledge society. The deep integration of informatization, industrialization and urbanization is helpful to alleviate the "big city disease", improve the quality of urbanization, realize fine and dynamic management, and improve the effectiveness of urban management and the people's livelihood of the city quality life.

11.7.1 Introduction to Smart City

Smart city often intersects with regional development concepts such as digital city, perceptual city, wireless city, ecological city, low-carbon city, and even mixes with e-government, intelligent transportation, smart grid and other industrial informatization concepts. The interpretation of the concept of smart city often has different emphasis. Some hold that the key lies in the application of technology, some believe that the key lies in the construction of network, some believe that the key lies in the participation of people, and some hold that the key lies in the effect of wisdom. Some leading cities in the informatization construction of cities emphasize people-oriented and sustainable innovation. In a word, smart city is not only the intelligent application of information technology, but also the connotation of human intelligent participation, people-oriented and sustainable development.

From the perspective of technology development, smart city uses a new generation of information technology such as Internet of things infrastructure, cloud computing infrastructure, geospatial infrastructure, and tools and methods such as Wiki, social network, Fab Lab, living lab, comprehensive integration method, and mobile all media convergence communication termina.

From the perspective of social development, smart city is to realize the sustainable innovation characterized by comprehensive and thorough perception, ubiquitous broadband interconnection, intelligent integration application and user innovation, open innovation, mass innovation and collaborative innovation. With the rise of network empire, the integration and development of mobile technology and the democratization process of innovation, smart city under the environment of knowledge society is the advanced form of information city after digital city.

11.7.2 International Practice of Smart City

In 2006, the European Union launched a European Living Lab, which uses new tools and methods, advanced information and communication technologies to mobilize "collective wisdom and creativity" in all aspects and provide opportunities for solving social problems. It also launched the European smart city network. Living lab is completely user-centered, with the help of the creation of an open innovation space, to help residents improve their quality of life by using information technology and mobile application services, so that people's needs can be respected and met to the greatest extent.

In November 2008, at the Council on Foreign Relations held in New York, IBM put forward the concept of "smart earth", which led to the upsurge of smart city construction.

In 2009, Dubuque cooperated with IBM to establish the first smart city in the United States. Using the Internet of things technology in a community of 60,000 residents connects various kinds of urban public resources (water, electricity, oil, gas, transportation, public services, etc.) to monitor, analyze and integrate various data to make intelligent response and better serve the citizens. The first step is to install numerical control water and electricity meters to all households and shops, including low flow sensor technology to prevent waste caused by water and electricity leakage. At the same time, a comprehensive monitoring platform is built to analyze, integrate and display the data in a timely manner, so that the use of resources in the whole city is clear at a glance. What's more, the city publishes this information to individuals and enterprises, so that they have a clearer understanding of their energy consumption and a greater sense of responsibility for sustainable development.

Based on the network, South Korea has built an ecological and intelligent city with green, digital and seamless mobile connection. Through the integration of public communication platform and ubiquitous network access, consumers can easily carry out remote education, medical treatment, tax, and realize intelligent monitoring of household building energy consumption.

Smart cities in Europe pay more attention to the role of information and communication technology in urban ecological environment, transportation, medical care, intelligent building and other fields of people's livelihood. They hope to achieve emission reduction goals with the help of knowledge sharing and low-carbon strategy, promote low-carbon, green and sustainable development of cities, invest in the construction of smart cities, develop low-carbon housing, intelligent transportation and smart grid, and improve energy efficiency to cope with climate change and build green smart city.

Copenhagen, Denmark's smart city, aims to be the first carbon neutral city by 2025. To achieve this goal, the city's climate action plan, which launched 50 initiatives, aimed to achieve its medium-term goal of reducing carbon by 20% by 2015. While striving for sustainable development of cities, the challenge for many cities is to maintain the balance between environmental protection and economy. With sustainable urban solutions, Copenhagen is approaching its goals. A study shows that revenue from green industries in the capital region has increased by 55% in five years.

11.7.3 Main Applications of Smart City

The application of smart city involves all aspects of urban management and residents' life, and also develops with the continuous update of science and technology. So smart city is widely used. Here are just a few common applications.

(1) Smart Public Service

Building smart public service and urban management system, strengthening the construction of professional application systems such as employment, medical treatment, culture and comfortable housing, and improving the standardization, precision and intelligence level of urban construction and management, can effectively promote the sharing of urban public resources in the whole city, actively promote the coordinated and efficient operation of urban people flow, logistics, information flow and capital flow, and improve the urban operation efficiency and public service level to promote the transformation and upgrading of urban development.

(2) Smart City Complex

The intelligent visual Internet of things is constructed by using top technologies such as visual acquisition and identification, various sensors, wireless positioning system, RFID, barcode identification, visual label, etc., to carry out intelligent perception and automatic data collection for the elements of urban complex, covering business, office, residence, hotel, exhibition, catering, conference, entertainment and transportation, lighting, information communication and display, etc. The collected data are visualized and standardized, so that managers can carry out visual urban complex management.

(3) Intelligent Education and Cultural Services

Focus on the construction of educational comprehensive information network, network school, digital courseware, teaching resource library, virtual library, teaching integrated management system, distance education system and other resource sharing database and application platform system, so as to promote the development of intelligent education. Promote the reeducation project, provide multi-channel education, training and employment services, and build a learning society. Promote the development of advanced network culture, speed up the pace of informatization in the industries of press and publication, radio, film and television, and electronic entertainment, strengthen the integration of information resources, and improve the public cultural information service system. Build a tourism public information service platform, provide more convenient tourism services, and enhance the brand of tourism culture.

(4) Smart Living Service

In order to make residents' life "intelligent development", developing smart application systems such as community government affairs, smart home systems, smart building management, smart community services, community remote monitoring, security management, and smart business office is an effective means. It needs to fully consider the different needs of public areas, business areas and residential areas, and integrate the application of Internet of things, Internet, mobile communication and other information technologies.

(5) Smart Logistics

With the information construction of the city, promote the application of RFID, multi-dimensional barcode, satellite positioning, goods tracking, e-commerce and other information technologies in the logistics industry, accelerate the construction of logistics information platform based on Internet of things and fourth party logistics information platform, integrate logistics resources, realize the integration of logistics government service and logistics business service, and promote the development of information, standardized and intelligent logistics enterprises and logistics industry.

(6) Smart Transportation

The project of "digital transportation" will be constructed. Through monitoring, traffic flow distribution optimization and other technologies, the monitoring system and information network system of public security, urban management and highway will be improved. A unified intelligent urban traffic comprehensive management and service system focusing on traffic guidance, emergency command, intelligent travel, taxi and bus management system will be established. The full sharing of traffic information, real-time monitoring and dynamic management of highway traffic conditions can comprehensively improve the monitoring strength and intelligent management level, thus ensuring the safety and smooth transportation.

(7) Smart Health Security System

It is necessary to establish a "digital health" system. The establishment of health service network, urban community health service system and information platform with the core of the regional health information management, can promote the communication and interaction between the information systems of medical and health units. The electronic health records for residents in the whole city is established focusing on hospital management and electronic medical records. The construction of intelligent medical systems, such as remote registration, electronic charging, digital telemedicine services, graphic physical examination and diagnosis system, can improve the level of medical and health services.

11.8 Conclusions

Human society has experienced three industrial revolutions, each of which has greatly promoted the progress of human society. Now the world is entering the fourth industrial revolution. Human society will once again usher in a period of explosive growth. In recent years, the rapid development of emerging technologies represented by big data, artificial intelligence, mobile Internet, Internet of things, cloud computing and other technologies has brought great impact and change to all fields of human society. The field of civil engineering will also be deeply affected.

To achieve better construction with the purpose of greening, industrialization as the way and informatization as the guarantee will be the only way and development direction for the construction industry to transform and upgrade itself under the background of the fourth industrial revolution. It will greatly promote the development of the construction industry to improve the intelligent level of the construction process, reduce the dependence on people, and effectively improve and manage all the processes in the whole life cycle of engineering projects through technological innovation and management innovation.

The deep integration of civil engineering professional technology and modern scientific and technological means is not only the external cause of intelligent construction, but also the external power to realize intelligent construction. The connotation of intelligent construction and management is constantly enriched and clarified, and the extension of intelligent construction and management is also constantly expanding. The development of the industry requires more countries and personnel to work together.

Exercises

11-1 What is the purpose of intelligent construction? How to realize intelligent construction?

11-2 What is BIM? What are the characteristics of BIM?

11-3 What changes can artificial intelligence bring to the construction industry?

11-4 What do you think is the difficulty of 3D printing concrete building?

11-5 What are the supporting technologies of construction site of intelligentization?

11-6 Besides the application contents introduced in this paper, what are the possible applications of smart city?

References

[1] MA Z L. What is the intelligent construction?[EB/OL]. [2020-09-21]. https://baijiahao.baidu.com/s?id=1634109494258756222&wfr=spider&for=pc.

[2] MAO Z B, LI Y G, GUO H S, et al. Intelligent construction technology system [EB/OL]. [2020-09-21]. https://www.sohu.com/a/319488896_787199.

[3] Building information modeling[EB/OL]. [2020-09-21]. https://new.siemens.com/global/en/products/buildings/digitalization/bim.html.

[4] Designing and building better with BIM[EB/OL]. [2020-09-21]. https://www.autodesk.com/solutions/bim.

[5] CATHERINE J. 3-D printing and the future of stuff[EB/OL]. [2020-09-21]. https://www.wipo.int/wipo_magazine/en/2013/02/article_0004.html.

[6] Construction site of intelligentization [EB/OL]. [2020-09-21]. https://baike.baidu.com/item/%E6%99%BA%E6%85%A7%E5%B7%A5%E5%9C%B0/18615677?fr=aladdin.

[7] Smart city[EB/OL]. [2020-09-21]. https://baike.baidu.com/item/%E6%99%BA%E6%85%A7%E5%9F%8E%E5%B8%82/9334841?fr=aladdin.

[8] IEEE Smart Cities. White papers[EB/OL]. [2020-09-21]. https://smartcities.ieee.org/resources/white-papers.